# Riemannian Submersions, Riemannian Maps in Hermitian Geometry, and their Applications

# Riemannian Submersions, Riemannian Maps in Hermitian Geometry, and their Applications

**Bayram Şahin**

Department of Mathematics,
Ege University, İzmir, Turkey

ACADEMIC PRESS

An imprint of Elsevier
elsevier.com

Academic Press is an imprint of Elsevier
32 Jamestown Road, London NW1 7BY, UK
525 B Street, Suite 1800, San Diego, CA 92101-4495, USA
50 Hampshire Street, 5th Floor, Cambridge, MA 02139, USA
The Boulevard, Langford Lane, Kidlington, Oxford OX5 1GB, UK

**Notices**

Knowledge and best practice in this field are constantly changing. As new research and experience broaden our understanding, changes in research methods, professional practices, or medical treatment may become necessary.

Practitioners and researchers must always rely on their own experience and knowledge in evaluating and using any information, methods, compounds, or experiments described herein. In using such information or methods they should be mindful of their own safety and the safety of others, including parties for whom they have a professional responsibility.

To the fullest extent of the law, neither the Publisher nor the authors, contributors, or editors, assume any liability for any injury and/or damage to persons or property as a matter of products liability, negligence or otherwise, or from any use or operation of any methods, products, instructions, or ideas contained in the material herein.

**British Library Cataloguing-in-Publication Data**
A catalogue record for this book is available from the British Library

**Library of Congress Cataloging-in-Publication Data**
A catalog record for this book is available from the Library of Congress

ISBN: 978-0-12-804391-2

For information on all Academic Press Publications
visit our website at https://www.elsevier.com/

Working together
to grow libraries in
developing countries

www.elsevier.com • www.bookaid.org

*Publisher:* Nikki Levy
*Acquisition Editor:* Graham Nisbet
*Editorial Project Manager:* Susan Ikeda
*Production Project Manager:* Poulouse Joseph
*Designer:* Christian J. Bilbow

Typeset by SPi Global, India

*To Fulya, Cemre, and İdil*

# CONTENTS

# ACKNOWLEDGMENTS

First of all, I would like to thank my precious wife Fulya for her love, understanding, and support throughout.

I would like to express my deep gratitude to my sister Sebahat and my elder brother Emin for their patience and loyalty.

I gratefully acknowledge my appreciation and gratitude to various colleagues who have collaborated with me on the research relevant to this book. I also gratefully acknowledge the research funding I have received over the years from The Scientific and Technological Research Council of Turkey (TÜBİTAK).

I am thankful to all authors of books and articles whose work I have used in preparing this book.

Last, but not least, I am grateful to the publishers for their effective cooperation and their excellent care in publishing this volume. I thank Susan Ikeda and Graham Nisbet at Elsevier for guidance throughout the publishing process. Any further comments and suggestions by the readers will be gratefully received.

Bayram Şahin
August 2016

# PREFACE

Smooth maps between Riemannian manifolds are useful for comparing geometric structures between two manifolds. As is indicated in [122], a major flaw in Riemannian geometry (compared to other subjects) is a shortage of suitable types of maps between Riemannian manifolds that will compare their geometric properties. In Riemannian geometry, there are two basic maps; isometric immersions and Riemannian submersions. Isometric immersions (Riemannian submanifolds) are basic such maps between Riemannian manifolds and they are characterized by their Riemannian metrics and Jacobian matrices. More precisely, a smooth map $F : (M, g_M) \longrightarrow (N, g_N)$ between Riemannian manifolds $(M, g_M)$ and $(N, g_N)$ is called an isometric immersion (submanifold) if $F_*$ is injective and

$$g_N(F_*X, F_*Y) = g_M(X, Y)$$

for vector fields $X, Y$ tangent to $M$; here $F_*$ denotes the derivative map. A smooth map $F : (M_1, g_1) \longrightarrow (M_2, g_2)$ is called a Riemannian submersion if $F_*$ is onto and it satisfies the above equation for vector fields tangent to the horizontal space $(ker F_*)^\perp$. Riemannian submersions between Riemannian manifolds were first studied by O'Neill [205] and Gray [125]; see also [31], [80], and [111]. The differential geometry of isometric immersions is well known and available in many textbooks, (see, for example [64], [69], [70], [74], [87] and [310]). However, the theory of isometric immersions (or submanifold theory) is still a very active research field. On the other hand, compared with isometric immersions, the geometry of Riemannian submersions has not been studied extensively. This theory is yet available in a few textbooks very recently (see [80] and [111]).

One of the most active and important branches of differential geometry of submanifold theory is the theory of submanifolds of Kähler manifolds. In the early 1930s, Riemannian geometry and the theory of complex variables were synthesized by Kähler [154], which developed (during 1950) into the complex manifold theory [104, 190]. A Riemann surface, $C^n$ and its projective space $CP^{n-1}$ are simple examples of the complex manifolds. This interrelation between the above two main branches of mathematics developed into what is now known as Kählerian geometry. Almost complex manifolds [306] and their complex, totally real, CR, generic, slant, semi-slant, and hemi-slant submanifolds [26, 53, 68, 70, 83, 202, 214, 238, 308] are some of the most interesting topics of Riemannian geometry. It is known that the complex techniques in relativity have been also very effective tools for understanding spacetime geometry [174]. Indeed, complex manifolds have two interesting classes of Kähler manifolds.

One is Calabi-Yau manifolds, which have their applications in superstring theory [51]. The other one is Teichmüller spaces applicable to relativity [288]. It is also important to note that CR-structures have been extensively used in spacetime geometry of relativity [223]. For complex methods in general relativity, see [108].

A submanifold of a Kähler manifold is a complex (invariant) submanifold if the tangent space of the submanifold at each point is invariant with respect to the almost complex structure of the Kähler manifold. In addition to complex submanifolds of a Kähler manifold, there is another important class of submanifolds called totally real submanifolds. A totally real submanifold of a Kähler manifold $\bar{M}$ is a submanifold of $\bar{M}$ such that the almost complex structure $J$ of $\bar{M}$ carries the tangent space of the submanifold at each point into its normal space and the main properties of such submanifolds established in [83], [202], and [308]. On the other hand, CR-submanifolds were defined by Bejancu [26] as a generalization of complex and totally real submanifolds. By a CR submanifold we mean a real submanifold $M$ of an almost Hermitian manifold $(\bar{M}, J, \bar{g})$, carrying a $J$-invariant distribution $D$ (i.e., $JD = D$) and whose $\bar{g}$-orthogonal complement is $J$-anti-invariant (i.e., $JD^{\perp} \subseteq T(M)^{\perp}$), where $T(M)^{\perp} \to M$ is the normal bundle of $M$ in $\bar{M}$. The CR submanifolds were introduced as an umbrella of a variety (such as invariant, anti-invariant and anti-holomorphic) of submanifolds. A CR-submanifold is called proper if it is neither a complex nor a totally real submanifold. The geometry of CR-submanifolds has been studied in several papers since 1978. Later generic submanifolds were introduced by Chen by releasing an anti-invariant condition on the distribution $D^{\perp}$. As another generalization of holomorphic submanifolds and totally real submanifolds, slant submanifolds were introduced and studied by Chen in [70]. Recently, semi-slant submanifolds ([214]) and hemi-slant submanifolds ([53], [238]) have been also introduced. Details on these may be seen in [26, 69, 74, 310, 311].

Riemannian submersions as a dual notion of isometric immersions have been studied in complex settings in the early 1970s. First note that a smooth map $\phi : M \longrightarrow N$ between almost complex manifolds $(M, J)$ and $(N, \bar{J})$ is called an almost complex (or holomorphic) map if $\phi_*(JX) = \bar{J}\phi_*(X)$ for $X \in \Gamma(TM)$, where $J$ and $\bar{J}$ are complex structures of $M$ and $N$, respectively. As an analogue of holomorphic submanifolds and almost complex maps, Watson [300] defined almost Hermitian submersions between almost Hermitian manifolds and he showed that the base manifold and each fiber have the same kind of structure as the total space, in most cases. We note that almost Hermitian submersions have been extended to the almost contact manifolds [84], locally conformal Kähler manifolds [183], quaternion Kähler manifolds [146], paraquaternionic manifolds [50], statistical manifolds [293], almost product manifolds [131], and paracontact manifolds [137]; see [111] for holomorphic submersions on complex manifolds and their extensions to other manifolds. We also note that Riemannian submersions have their applications in spacetime of unified theory and robotics. In the theory of the Klauza-Klein type, a general solution of the non-linear sigma model is given by Riemannian submersions from the extra-dimensional space to the space in which the scalar fields of the nonlinear sigma model take values; for details, see [111]. In robotic theory, the forward kinematic map can be given by Riemannian submersions; see [12] for details. Moreover, modern Kaluza-Klein theories are given in terms of principal fiber bundle and Riemannian submersion [303].

Although holomorphic submersions, as a corresponding version of holomorphic submanifolds, have been studied by many authors, the other versions of submanifolds have not been studied in the submersion theory. Therefore we introduced and studied anti-invariant Riemannian submersions in [239]. It is shown that holomorphic submersions are useful to determine the geometric structure of the base space; however, anti-invariant submersions are useful to study the geometry of the total space. As a generalization of holomorphic submersions and anti-invariant Riemannian submersions, we introduced semi-invariant submersions from almost Hermitian manifolds onto Riemannian manifolds in [249], then we studied the geometry of such maps. We recall that a Riemannian submersion $F$ from an almost Hermitian manifold $(M, J_M, g_M)$ with an almost complex structure $J_M$ to a Riemannian manifold $(N, g_N)$ is called a semi-invariant submersion if the fibers have differentiable distributions $D$ and $D^\perp$ such that $D$ is invariant with respect to $J_M$ and its orthogonal complement $D^\perp$ is a totally real distribution, i.e., $J_M(D_p^\perp) \subseteq (kerF_{*p})^\perp$. Obviously, invariant Riemannian submersions and anti-invariant Riemannian submersions are semi-invariant submersions with $D^\perp = \{0\}$ and $D = \{0\}$, respectively. As another generalization of invariant submersions and anti-invariant Riemannian submersions, slant submersions were introduced in [245]. Moreover, new submersions and their extensions in this new directions have been studied by various authors; see [3], [5], [6], [7], [8], [30], [105], [106], [131], [132], [133], [134], [135], [136], [169], [171], [172], [215], [216], [219], [221], [248], [265], [283], [284] and [286].

Both these branches, Riemannian submanifolds and Riemannian submersions, are extremely important in differential geometry and as well as in applications in mechanics and general theory of relativity. One of the interesting approach is to study the most general case, which includes both the geometry of submanifolds as well as Riemannian submersions, and such a study was initiated by Fischer [117], where he defines Riemannian maps between Riemannian manifolds as a generalization of the notions of isometric immersions and Riemannian submersions as follows. Let $F : (M, g_M) \longrightarrow (N, g_N)$ be a smooth map between Riemannian manifolds. Then we denote the kernel space of $F_*$ at $p \in M$ by $kerF_{*p}$ and consider the orthogonal complementary space $\mathcal{H}_p = (kerF_{*p})^\perp$ to $kerF_{*p}$. Then the tangent space of $M$ at $p \in M$ has the following decomposition:

$$T_pM = kerF_{*p} \oplus \mathcal{H}_p.$$

We denote the range of $F_*$ at $p \in M$ by $range F_{*p}$ and consider the orthogonal complementary space $(range F_{*p})^\perp$ to $range F_{*p}$ in the tangent space $T_{F(p)}N$ of $N$. Thus the tangent space $T_{F(p)}N$ of $N$ for $p \in M$ has the following decomposition:

$$T_{F(p)}N = (range F_{*p}) \oplus (range F_{*p})^\perp.$$

Now, a smooth map $F : (M^m, g_M) \longrightarrow (N^n, g_N)$ is called a Riemannian map at $p_1 \in M$ if the horizontal restriction $F^h_{*p_1} : (ker F_{*p_1})^\perp \longrightarrow (range F_{*p_1})$ is a linear isometry between the inner product spaces

$$((ker F_{*p_1})^\perp, g_M(p_1)\,|_{(ker F_{*p_1})^\perp})$$

and

$$(range F_{*p_1}, g_N(p_2)\,|_{(range F_{*p_1})}),$$

$p_2 = F(p_1)$. Therefore, Fischer stated in [117] that a Riemannian map is a map that is as isometric as it can be. In other words, $F_*$ satisfies the equation

$$g_N(F_*X, F_*Y) = g_M(X, Y)$$

for $X, Y$ vector fields tangent to $\mathcal{H}$. It follows that isometric immersions and Riemannian submersions are particular Riemannian maps with $ker F_* = \{0\}$ and $(range F_*)^\perp = \{0\}$. It is known that a Riemannian map is a subimmersion, which implies that the rank of the linear map $F_{*p} : T_p M \longrightarrow T_{F(p)}N$ is constant for $p$ in each connected component of $M$ [1] and [117]. A remarkable property of Riemannian maps is that a Riemannian map satisfies the generalized eikonal equation $\| F_* \|^2 = rank F$ which is a bridge between geometric optics and physical optics. Since the left-hand side of this equation is continuous on the Riemannian manifold $M$ and since $rank F$ is an integer-valued function, this equality implies that $rank F$ is locally constant and globally constant on connected components. Thus if $M$ is connected, the energy density $e(F) = \frac{1}{2} \| F_* \|^2$ is quantized to integer and half-integer values. The eikonal equation of geometrical optics is solved by using Cauchy's method of characteristics, whereby, for real valued functions $F$, solutions to the partial differential equation $\| dF \|^2 = 1$ are obtained by solving the system of ordinary differential equations $x' = grad f(x)$. Since harmonic maps generalize geodesics, harmonic maps could be used to solve the generalized eikonal equation [117].

In [117], Fischer also proposed an approach to build a quantum model and he pointed out that the success of such a program of building a quantum model of nature using Riemannian maps would provide an interesting relationship between Riemannian maps, harmonic maps, and Lagrangian field theory on the mathematical side, and Maxwell's equation, Schrödinger's equation, and their proposed generalization on the physical side.

Riemannian maps between semi-Riemannian manifolds have been defined in [122] by putting in some regularity conditions. On the other hand, affine Riemannian maps were also investigated and some interesting decomposition theorems were obtained by using the existence of Riemannian maps. Moreover curvature relations for Riemannian maps were obtained in [123]. For Riemannian maps and their applications in spacetime geometry, see [122].

Since Riemannian maps include isometric immersions and Riemannian submersions as subclasses, starting from [241], we introduce and study new Riemannian maps as generalizations of holomorphic submanifolds, totally real submanifolds, CR-submanifolds, slant submanifolds, holomorphic submersions, anti-invariant submersions, semi-invariant submersions, and slant submersions. Since then a number of papers published on the Riemannian submersions and Riemannian maps ([218], [222], [243], [244], [247], [250], [251], [254] and [260]) have demanded the publication of this volume as an update on submersion theory and Riemannian maps.

As we have seen from previous paragraphs, although submanifolds of complex manifolds have been an active field of study for many years, there are few books ([111] and [311]) on the geometry of Riemannian submersions defined on complex manifolds. Moreover, these books only cover holomorphic submersions. The objective of this book is to focus on all new geometric results on Riemannian submersion theory and Riemannian maps with proofs and their applications, and to bring the reader to the frontiers of active research on related topics.

The book consists of six chapters. Chapter 1 covers preliminaries followed by up-to-date mathematical results in Chapters 4 on Riemannian maps. Chapter 2 is focused on applications of Riemannian submersions in active ongoing research area in Robotic theory and their brief applications in Kaluza-Klein theory. First, we deal with the geometry of certain matrix Lie groups and relate them with a significant work of Altafini [12] on forward kinematic maps.

In Chapter 4, we study the geometry of Riemannian maps and introduce some new notions (umbilical Riemannian map, pseudo-umbilical Riemannian map) for Riemannian maps. We also give a new Bochner identity and, by using this identity we obtain divergence theorem for a map. We also show that it is possible to obtain new totally geodesic conditions by using this divergence theorem. Moreover, some well-known results of isometric immersions have been carried into the geometry of Riemannian maps, like the Chen inequality, circle theorems, and Clairaut conditions.

The motivation of the rest of the chapters comes from the development of the general theory of Cauchy-Riemann (CR) submanifolds [26] and slant submanifolds [70], and it mainly contains a series of results obtained by the author and collaborators in recent years.

In Chapter 3, we study invariant Riemannian submersions, anti-invariant Riemannian submersions, semi-invariant Riemannian submersions, slant submersions, semi-slant submersions, generic submersions, hemi-slant submersions, and pointwise slant submersions from almost Hermitian manifolds onto arbitrary Riemannian manifolds.

We show that these new submersions produce new conditions for submersions to be harmonic and totally geodesic. We give several examples and show that such submersions have rich geometric properties.

In Chapter 5, we study new Riemannian maps from almost Hermitian manifolds to Riemannian manifolds as generalizations of invariant Riemannian submersions, anti-invariant Riemannian submersions, semi-invariant submersions, generic submersions, slant submersions, semi-slant submersions, and hemi-slant submersions, and we give many examples and study the effect of such maps on the geometry of the domain manifold and the target manifold.

In Chapter 6, we study Riemannian maps from Riemannian manifolds to almost Hermitian manifolds as generalizations of holomorphic submanifolds, totally real submanifolds, CR-submanifolds, generic submanifolds, slant submanifolds, semi-slant submanifolds, and hemi-slant submanifolds. We obtain many new properties of the domain manifolds and the target manifolds, and investigate the harmonicity and totally geodesic conditions for such maps. We also relate these maps with pseudo-harmonic maps.

The results included in this book should stimulate future research on Riemannian submersions and Riemannian maps. To the best of our knowledge, there does not exist any other book covering the material in this volume. Our approach, in this book, has the following special features:

- The first chapter of the book has been also designed to incorporate the latest developments in the Riemann geometry.
- Extensive list of cited references on the Riemannian geometry is provided for the readers to understand easily the main focus on the Riemannian submersions and Riemannian maps and their applications.
- Each chapter begins with a brief introduction to the background and the motivation of the topics under consideration. This will help readers grasp the main ideas more easily.
- There is an extensive subject index.
- The sequence of chapters is arranged so that the understanding of a chapter stimulates interest in reading the next one and so on.
- Applications are discussed separately (see Chapter 2) from the mathematical theory.
- Overall, the presentation is self-contained, fairly accessible, and, in some special cases supported by references.

This book is intended for graduate students and researchers who have good knowledge of Riemannian geometry and its submanifolds, and, interest in Riemannian submersions and Riemannian maps.

Bayram Şahin
Ege University
August 2016

# ABOUT THE AUTHOR

**Bayram Şahin** is Professor of Mathematics in the Department of Mathematics at Ege University of Izmir at Turkey. Prior to coming to Ege University, he was a full professor in the Department of Mathematics at Inonu University of Malatya at Turkey. After he obtained his Ph.D. (Differential Geometry) from the University of Inonu, he spent 12 months at the University of Windsor, Windsor, Ontario, Canada as a Postdoctorate and a visiting research scholar. He is editorial board member for the *Turkish Journal of Mathematics* and *Mediterranean Journal of Mathematics*. He is the co-author of the book *Differential Geometry of Lightlike Submanifolds* published in 2010 by Springer-Verlag and he has written more than 80 articles on manifolds, submanifolds and maps between them in various peer-reviewed publications, including (among others) *Geometriae Dedicata, Acta Applicandae Mathematica, International Journal of Geometric Methods in Modern Physics, Indagationes Mathematica, Central Europen Journal of Mathematics, Turkish Journal of Mathematics, Journal of the Korean Mathematical Society, Mediterranean Journal of Mathematics, Chaos, Solitons and Fractals*.

# CHAPTER 1

# Basic Geometric Structures on Manifolds

## Contents

## Abstract

In this chapter, we give brief information about geometric structures which will be used in the following chapters. In addition to the basic concepts, some geometric notions such as symmetry conditions, parallelity conditions, new product structures on manifolds, generalized Einstein manifolds, and biharmonic maps introduced very recently will be presented in this chapter. The chapter consists of six sections. In the first section, the main notions and theorems of Riemannian geometry are reviewed, such as the Riemannian manifold, Riemannian metric, Riemannian connection, curvatures (Riemannian, sectional Ricci, Scalar), derivatives ( exterior, covariant, inner), operators (Hessian, divergence, Laplacian), Einstein manifolds and their generalizations, symmetry conditions, and very brief notions from integration. In the second section, we look at vector bundles and related notions. In the third section, we recall basic notions from submanifold theory and distributions on manifolds. In the fourth section, we mention Riemannian submersions, O'Neill's tensor fields, and curvature relations of Riemannian submersions. In this section, we also recall the notion of horizontally weakly conformal maps, which will be a crucial tool for a characterization of harmonic maps. In the fifth section, we study various product structures including warped product, twisted product, oblique warped product, and convolution product on Riemannian manifolds. We also provide connection relations and curvature expressions of each case. In the last section, we give a general setting for a map between manifolds such as a vector field along a map, a connection along a map, a curvature tensor field along a map, a second fundamental form of a map, a tension field of a map. We also recall harmonic maps and biharmonic maps in detail.

**Keywords:** Riemannian manifold, Riemannin connection, Riemannian curvature tensor, locally symmetric manifold, semisymmetric Riemannian manifold, pseudosymmetric manifold, Chaki-type pseudosymmetric manifold, Ricci tensor, Scalar curvature, Hesian, Ricci soliton, gradient Ricci soliton, gradient Ricci almost soliton, quasi-Einstein manifold, generalized quasi-Einstein manifold, quasi-Einstein Riemannian manifold, generalized $m-$ quasi-Einstein

Reimannian Submersions, Reimannian Maps in Hermitian Geometry, and their Applications
http://dx.doi.org/10.1016/B978-0-12-804391-2.50001-4

manifold, divergence, Laplacian, vector bundle, pullback bundle, pullback connection, Riemannian submanifold, Gauss formula, Weingarten formula, parallel submanifold, semi-parallel submanifold, pseudo-parallel submanifold, Chen inequality, distribution, Riemannian submersion, O'Neill's tensors, weakly conformal maps, horizontally weakly conformal maps, usual product, warped product, multiply warped product, doubly warped product, twisted product, multiply twisted product, doubly twisted product, D-homothetic warping, oblique warped product, convolution product, second fundamental form, harmonic map, biharmonic map, tension field, bi-tension field.

*Nothing in this world can take the place of persistence. Talent will not; nothing is more common than unsuccessful men with talent. Genius will not; unrewarded genius is almost a proverb. Education will not; the world is full of educated derelicts. Persistence and determination alone are omnipotent.*

*Calvin Coolidge*

## 1.  Riemannian manifolds and related topics

In this section, we set out the basic materials needed in this book. We begin by reviewing some results of Riemannian geometry.

Let $M$ be a real $n$-dimensional smooth manifold. First of all, we emphasize that throughout this book, all manifolds and structures on (between) them are supposed to be differentiable of class $C^\infty$. We also note that $C^\infty(M, \mathbb{R})$ and $\chi(M)$ are, respectively, the algebra of differentiable functions on $M$ and the module of differentiable vector fields of $M$. A *linear connection* on $M$ is a map

$$\nabla : \chi(M) \times \chi(M) \to \chi(M),$$

such that

$$\nabla_{fX+hY}Z = f(\nabla_X Z) + h(\nabla_Y Z), \qquad \nabla_X f = Xf,$$
$$\nabla_X(fY + hZ) = f\nabla_X Y + h\nabla_X Z + (Xf)Y + (Xh)Z,$$

for arbitrary vector fields $X$, $Y$, $Z$ and smooth functions $f$, $h$ on $M$. $\nabla_X$ is called the *covariant derivative operator* and $\nabla_X Y$ is called the *covariant derivative* of $Y$ with respect to $X$. Define a tensor field $\nabla Y$, of type $(1, 1)$, and given by $(\nabla Y)(X) = \nabla_X Y$, for any $Y$. Also, $\nabla_X f = Xf$ is the covariant derivative of $f$ along $X$. The covariant derivative of a 1-form $\omega$ is given by

$$(\nabla_X \omega)(Y) = X(\omega(Y)) - \omega(\nabla_X Y). \tag{1.1}$$

The covariant derivative of a tensor $T$ of type $(r, s)$ along a vector field $X$ is a tensor field $\nabla_X T$ , of type $(r, s)$, given by

$$
\begin{aligned}
(\nabla_X T)(\omega^1, \ldots, \omega^r, Y_1, \ldots, Y_s) &= X(T(\omega^1, \ldots, \omega^r, Y_1, \ldots, Y_r)) \\
&\quad - \sum_{\alpha=1}^{r} T(\omega^1, \ldots, \nabla_X \omega^\alpha, \ldots, \omega^r, Y_1, \ldots, Y_s) \\
&\quad - \sum_{t=1}^{s} T(\omega^1, \ldots, Y_1, \ldots, \nabla_X Y_t, \ldots, Y_s)
\end{aligned}
$$

for any vector field $X$, $r$ covariant vectors $\omega^1, \ldots, \omega^r$ and $s$ contravariant vectors $Y_1, \ldots, Y_s$. Note that $\nabla T$ of $T$ is a tensor of type $(r, s+1)$. A vector field $Y$ on $M$ is said to be parallel with respect to a linear connection $\nabla$ if for any vector field $X$ on $M$ it is covariant constant, i.e., $\nabla_X Y = 0$. In general, a tensor field $T$ on $M$ is parallel with respect to $\nabla$ if it is covariant constant with respect to any vector field $X$ on $M$. Let $\alpha$ be a smooth curve on $M$ given by the following equations:

$$
x^i = x^i(t), \qquad t \in I \subset R, \qquad i = 1, \ldots, n.
$$

Then a tangent vector field $V$ to $\alpha$ is given by

$$
V = \frac{dx^i}{dt} \partial_i,
$$

where $(x^i)$ is a local coordinate system on $M$ and $\partial_i = \frac{\partial}{\partial x_i}$. Thus, a vector field $Y$ is said to be parallel along $\alpha$ if $\nabla_V Y = 0$. The curve $\alpha$ is called a *geodesic* if $V$ is parallel along $\alpha$, i.e., if $\nabla_V V = fV$ for some smooth function $f$ along $\alpha$. It is possible to find a new parameter $s$ along $\alpha$ such that $f$ is zero along $\alpha$. The parameter $s$ is called an *affine parameter*. Two affine parameters $s_1$ and $s_2$ are related by $s_2 = a s_1 + b$, where $a$ and $b$ are constants. For a smooth or $C^r$ $\nabla$, the theory of differential equations certifies that, given a point $p$ of $M$ and a tangent vector $X_p$, there is a *maximal geodesic* $\alpha(s)$ such that $\alpha(0) = p$ and $\frac{dx^i}{ds}|_{s=0} = X_p^i$. If $\alpha$ is defined for all values of $s$, then it is said to be *complete*, otherwise it is incomplete. Let $V$ be a vector field on a real $n$-dimensional smooth manifold. The *integral curves (orbits)* of $V$ are given by the following system of ordinary differential equations:

$$
\frac{dx^i}{dt} = V^i(x(t)), \qquad i \in \{1, \ldots, n\}, \tag{1.2}
$$

where $(x^i)$ is a local coordinate system on $M$ and $t \in I \subset \mathbb{R}$. It follows from the well-known theorem on the existence and uniqueness of the solution of (1.2) that for any given point, with a local coordinate system, there is a unique integral curve defined over a part of the real line.

Consider a mapping $\phi$ from $[-\delta, \delta] \times \mathcal{U}$ ( $\delta > 0$ and $\mathcal{U}$ an open set of $M$) into $M$

defined by $\phi : (t, x) \rightarrow \phi(t, x) = \phi_t(x) \in M$, satisfying:

**(1)** $\phi_t : x \in \mathcal{U} \rightarrow \phi_t(x) \in M$ is a diffeomorphism of $\mathcal{U}$ onto the open set $\phi_t(\mathcal{U})$ of $M$, for every $t \in [-\delta, \delta]$; and

**(2)** $\phi_{t+s}(x) = \phi_t(\phi_s(x))$, $\forall t, s, t + s \in [-\delta, \delta]$ and $\phi_s(x) \in \mathcal{U}$.

In the above case, the family $\phi_t$ is a 1-*parameter group of local transformations* on $M$. The mapping $\phi$ is then called a *local flow* on $M$. Using (1.2) it has been proved (see Kobayashi-Nomizu [161, page 13]) that the vector field $V$ generates a local flow on $M$. If each integral curve of $V$ is defined on the entire real line, we say that $V$ is a *complete vector field* and it generates a *global flow* on $M$. A set of local (resp. complete) integral curves is called a *local congruence* ( *resp. congruence) of curves* of $V$. We note that $\chi(M)$ is a vector space under natural addition and scalar multiplication. If $X$ and $Y$ are in $\chi(M)$, then the Lie bracket $[X, Y]$ is defined as a mapping from the ring of functions on $M$ into itself by

$$[X, Y]f = X(Yf) - YX(f).$$

If $f$ is a function and $X$ is a vector field on $M$, then $fX$ is a vector field on $M$ defined by $(fX)_p = f(p)X_p$, for $p \in M$. It is easy to see that $[,]$ is skew-symmetric. We also list some properties of the Lie bracket as follows:

$$[fX, gY] = fg[X, Y] + f(Xg)Y - g(Yf)X,$$

$$[[X, Y], Z] + [[Y, Z], X] + [[Z, X], Y] = 0 \quad \text{(Jacobi's identity)}$$

for $X, Y, Z \in \chi(M)$ and $f, g \in C^\infty(M, \mathbb{R})$.

A Riemannian metric $g$ on $M$ is a tensor field $g$ of type $(0, 2)$ which satisfies

$$g(X, Y) = g(Y, X), \text{ for vector fields } X, Y \in \chi(M)$$

and for each $p \in M$, $g_p$ is positive definite bilinear form on $T_pM \times T_pM$, i.e., $g(X_p, X_p) \geq 0$ ($= 0$, if and only if $X_p = 0$) for all $X_p \in T_pM$. Then $(M, g)$ is called a *Riemannian manifold*. For an example, Euclidean space is a Riemannian manifold with its usual inner product. The sphere $S^n = \{(x_1, ..., x_{n+1}) \in \mathbb{R}^{n+1} \mid \sum_{i=1}^{n+1} x_i^2 = r^2\}$ with radius $r$ is an example of a Riemannian manifold with induced inner product from $\mathbb{R}^{n+1}$. This Riemannian manifold is defined as a hypersurface of a Euclidean space $\mathbb{R}^{n+1}$.

Let $M$ be a real smooth manifold and $\nabla$ a linear connection on $M$. The *Riemannian curvature tensor*, denoted by $R$, of $M$ is a $(1, 3)$ tensor field defined by

$$R(X, Y)Z \quad = \quad \nabla_X \nabla_Y Z - \nabla_Y \nabla_X Z - \nabla_{[X, Y]}Z \qquad (1.3)$$

for any $X, Y, Z \in \chi(M)$. The *torsion tensor*, denoted by $T$, of $\nabla$ is a $(1, 2)$ tensor defined by

$$T(X, Y) = \nabla_X Y - \nabla_Y X - [X, Y].$$

$R$ is skew-symmetric in the first two slots. In case $T$ vanishes on $M$, we say that $\nabla$ is *torsion-free* or *symmetric connection* on $M$. A linear connection $\nabla$ on $(M, g)$ is called a *compatible connection* if $g$ is parallel with respect to $\nabla$, i.e.,

$$(\nabla_X g)(Y, Z) = X(g(Y, Z)) - g(\nabla_X Y, Z) - g(Y, \nabla_X Z) = 0, \tag{1.4}$$

for any $X, Y, Z \in \chi(M)$. In terms of local coordinates system, we have

$$g_{ij;k} = \partial_k g_{ij} - g_{ih}\Gamma^h_{jk} - g_{jh}\Gamma^h_{ik} = 0,$$

where

$$\Gamma^h_{ij} = \frac{1}{2}g^{hk}\left\{\partial_j g_{ki} + \partial_i g_{kj} - \partial_k g_{ij}\right\}, \quad \Gamma^h_{ij} = \Gamma^h_{ji}.$$

Furthermore, if we set $\Gamma_{k|ij} = g_{kh}\Gamma^h_{ij}$, then, the above equation becomes

$$g_{ij;k} = \partial_k g_{ij} - \Gamma_{i|jk} - \Gamma_{j|ik} = 0.$$

The connection coefficients $\Gamma_{k|ij}$ and $\Gamma^h_{ij}$ are called the *Christoffel symbols of first* and *second type*, respectively. We are ready to state the fundamental result of Riemannian geometry.

**Theorem 1.** *Given a Riemannian manifold M, there exists a unique linear connection $\nabla$ on M satisfying the following conditions:*
**(a)** $\nabla$ *is symmetric.*
**(b)** $\nabla$ *is compatible with the Riemannian metric.*

The connection given by the above theorem is called *Levi-Civita* (metric or Riemannian) connection on $M$. It is known that a metric connection $\nabla$ satisfies the following identity, the so-called *Koszul formula*

$$\begin{aligned}2g(\nabla_X Y, Z) = {} & X(g(Y, Z)) + Y(g(X, Z)) - Z(g(X, Y)) \\ & + g([X, Y], Z) + g([Z, X], Y) - g([Y, Z], X)\end{aligned} \tag{1.5}$$

for any $X, Y, Z \in \chi(M)$. The two *Bianchi's identities* are

$$R(X, Y)Z + R(Y, Z)X + R(Z, X)Y = 0, \tag{1.6}$$
$$(\nabla_X R)(Y, Z, W) + (\nabla_Y R)(Z, X, W) + (\nabla_Z R)(X, Y, W) = 0.$$

The Riemannian curvature tensor of type $(0, 4)$ is defined by

$$R(X, Y, Z, U) = g(R(X, Y)Z, U), \quad \forall X, Y, Z, U \text{ on } M.$$

Then, by direct calculations, we get

$$
\begin{aligned}
R(X, Y, Z, U) + R(Y, X, Z, U) &= 0, \\
R(X, Y, Z, U) + R(X, Y, U, Z) &= 0, \\
R(X, Y, Z, U) - R(Z, U, X, Y) &= 0.
\end{aligned}
$$

Let $\{E_1, \ldots, E_n\}_x$ be a local orthonormal basis of $T_xM$. Then,

$$
g(E_i, E_j) = \delta_{ij} \text{ (no summation in } i\text{)}, \quad X = \sum_{i=1}^{n} g(X, E_i)\, E_i.
$$

Thus, we obtain

$$
g(X, Y) = \sum_{i=1}^{n} g(X, E_i)\, g(Y, E_i).
$$

The *gradient* of a smooth function $f$ is defined as a vector field, denoted by *grad f*, and given by

$$
g(grad\, f, X) = X(f), \quad \text{i.e.,} \quad grad\, f = \sum_{i,j=1}^{n} g^{ij} \partial_i f \partial_j. \tag{1.7}
$$

In particular, for $f_1, f_2 \in C^\infty(M, \mathbb{R})$, we have

$$
\begin{aligned}
(grad(f_1))(f_2) &= g(grad\, f_1, grad\, f_2) \\
grad(f_1 + f_2) &= grad\, f_1 + grad\, f_2 \\
grad(f_1 f_2) &= f_2(grad(f_1)) + f_1(grad(f_2)). \tag{1.8}
\end{aligned}
$$

We also use the symbol $\nabla f$ for the gradient operator in this book. The *divergence*, denoted by *div X*, is given by

$$
divX = \sum_{i=1}^{n} g(\nabla_{X_i} X, X_i), \tag{1.9}
$$

where $\{X_1, ..., X_n\}$ is a local orthonormal frame of $\chi(M)$. For any $f \in C^\infty(M, \mathbb{R})$ and $X, Y \in \chi(M)$, we have

$$
\begin{aligned}
div(X + Y) &= divX + divY \\
div(fX) &= g(\nabla f, X) + f\, divX. \tag{1.10}
\end{aligned}
$$

Let $f : (M, g) \longrightarrow \mathbb{R}$ be a map. Then the *Hessian tensor* $h_f : \chi(M) \longrightarrow \chi(M)$ of $f$ is defined by

$$
h_f(X) = \nabla_X \nabla f, \tag{1.11}
$$

for $X \in \chi(M)$. The *Hessian form*, denoted by $H_f = \nabla(\nabla f)$, of a function $f$, is its second covariant differential given by

$$H^f(X, Y) = XYf - (\nabla_X Y)f = g(\nabla_X(grad f), Y) == g(h_f(X), Y), \tag{1.12}$$

for all $X, Y \in \chi(M)$. We also note that the Hessian form $H^f$ is symmetric, $H^f(X, Y) = H^f(Y, X)$. The *Laplacian* of $f$, denoted by $\triangle f$, is given by

$$
\begin{aligned}
\triangle f &= div(grad\, f) \\
&= \sum_{i,j=1}^{n} g^{ij} \{ \frac{\partial^2 f}{\partial x^i \partial x^j} - \Gamma_{ij}^{k} \frac{\partial f}{\partial x^k} \}.
\end{aligned}
\tag{1.13}
$$

(1.13) shows that the Laplacian $\triangle f$ is the trace of $h_f$. For gradient, divergence and Laplacian, the following relations are satisfied.

$$
\begin{aligned}
h_{fk}(X) &= fh_k(X) + kh_f(X) + g(\nabla f, X)\nabla k \\
&\quad + g(\nabla k, X)\nabla f
\end{aligned}
$$

$$
\begin{aligned}
H_{fk}(X, Y) &= fH_k(X, Y) + kH_f(X, Y) + g(\nabla f, X)g(\nabla k, Y) \\
&\quad + g(\nabla k, X)g(\nabla f, Y)
\end{aligned}
$$

$$\Delta(fk) = f\Delta k + k\Delta f - 2g(\nabla f, \nabla k),$$

where $f, k$ are differentiable functions on manifold $M$; for more properties on the above operators, see [85].

A two-dimensional subspace $\pi$ of the tangent space $T_pM$ is called a *tangent plane* to $M$ at $p \in M$. For tangent vectors $u, v$, define

$$G_(u, v) = g(u, u)g(v, v) - g(u, v)^2.$$

The absolute value of $| G(u, v) |$ is the square of the area of the parallelogram with sides $u$ and $v$. Define a real number

$$K(\pi) = K_p(u, v) = \frac{R(u,\ v,\ v,\ u)}{det(G_p)},$$

where $R(u, v, v, u)$ is the 4-linear mapping on $T_pM$ by the curvature tensor. The smooth function $K$ which assigns to each tangent plane $\pi$ the real number $K(\pi)$ is called the *sectional curvature* of $M$, which is independent of the basis $B = \{u, v\}$. If $K$ is a constant $c$ at every point of $M$, then $M$ is of constant sectional curvature $c$, denoted by $M(c)$, whose curvature tensor field $R$ is given by [311]:

$$R(X,\ Y)Z = c\{g(Y, Z)X - g(X, Z)Y\}. \tag{1.14}$$

In particular, if $K = 0$, then $M$ is called a *flat manifold* for which $R = 0$.

Euclidean space $\mathbb{R}^n$ is a flat manifold. A sphere $\mathbf{S}^n(r)$ of radius $r$ is defined as a hypersurface in a Euclidean space $\mathbf{R}^{n+1}$ given by

$$\mathbf{S}^n(r) = \{v \in \mathbf{R}^{n+1} | \langle v, v \rangle = \sum_i (v^i)^2 = r^2\}$$

whose sectional curvature $K_\sigma = \frac{1}{r^2}$ at every point $p$ and every plane $\pi$.

We now recall some important symmetry conditions for Riemannian manifolds. Riemannian manifolds satisfying certain symmetry conditions can be considered as a generalization of Riemannian manifolds of constant curvature. We first recall the following operators. Let $(M, g)$ be a connected $n$-dimensional, $n \geq 3$, Riemannian manifold of class $C^\infty$. For a $(0, k)$-tensor field $T$ on $M$, $k \geq 1$, we define the $(0, k + 2)$-tensors $R \circ T$ and $Q(g, T)$ by

$$\begin{aligned}
(R \circ T)(X_1, ..., X_k; X, Y) &= (\tilde{R}(X, Y) \circ T)(X_1, ..., X_k) \\
&= -T(\tilde{R}(X, Y)X_1, X_2, ..., X_k) \\
&\quad - ... - T(X_1, ..., X_{k-1}, \tilde{R}(X, Y)X_k),
\end{aligned}$$

and

$$\begin{aligned}
Q(g, T)(X_1, ..., X_k; X, Y) &= ((X \wedge Y) \circ T)(X_1, ..., X_k) \\
&= -T((X \wedge Y)X_1, X_2, ..., X_k) \\
&\quad - ... - T(X_1, ..., X_{k-1}, (X \wedge Y)X_k)
\end{aligned}$$

respectively, for $X_1, ..., X_k, X, Y \in \Gamma(TM)$, where $\tilde{R}$ is the curvature tensor field of $M$ and $R$ is the Riemannian curvature tensor of type $(0, 4)$, $\wedge$ is defined by

$$(X \wedge Y)Z = g(Y, Z)X - g(X, Z)Y.$$

A Riemannian manifold $M$ is called a *locally symmetric manifold* if its Riemannian curvature tensor field is parallel, that is, $\nabla R = 0$. Locally symmetric Riemannian manifolds are a generalization of manifolds of constant curvature. As a generalization of locally symmetric Riemannian manifolds, *semisymmetric Riemannian manifolds* were defined by the condition

$$R \circ R = 0,$$

where the first $R$ acts as a derivation on the second $R$, which stands for the Riemannian curvature tensor field. It is known that locally symmetric manifolds are semisymmetric manifolds but the converse is not true [231]. Such manifolds have been investigated by E. Cartan and they have been locally classified by Szabo [276], [277] (see also [101]). Moreover, pseudosymmetric manifolds were introduced by Deszcz [230]. The Riemannian manifold $(M, g)$ is called a *pseudosymmetric manifold* if at every point of

$M$ the following condition is satisfied: the tensor $R \circ R$ and $Q(g, R)$ are linearly dependent. The manifold $(M, g)$ is pseudosymmetric if only if $R \circ R = LQ(g, R)$ on the set $U = \{x \in M \mid Q(g, R) \neq 0 \ at \ x\}$, where $L$ is some function on $U$. Pseudosymmetric manifolds have been discovered during the study of totally umbilical submanifolds of semisymmetric manifolds [2]. It is clear that every semisymmetric Riemannian manifold is pseudosymmetric manifold but the converse is not true. On the other hand, Chaki-type pseudosymmetric manifolds were introduced by Chaki [59] in 1987 as follows: A Riemannian manifold $(M, g)$, $n \geq 3$, is called a *Chaki-type pseudosymmetric manifold* if it satisfies the following condition:

$$(\nabla_W R)(X, Y)Z = 2\alpha(W)R(X, Y)Z + \alpha(X)R(W, Y)Z + \alpha(Y)R(X, W)Z$$
$$+\alpha(Z)R(X, Y)W + g(R(X, Y, Z), W)U, \tag{1.15}$$

where $\alpha$ is a non-zero 1-form, called the associated 1-form, and

$$g(X, U) = \alpha(X),$$

for any vector field $X$. It is known that pseudosymmetric manifolds in the sense of Deszcz may not be Chaki-type pseudosymmetric manifolds, vice versa [268]. As a generalization of pseudosymmetric manifolds in the sense of Chaki, weakly symmetric Riemannian manifolds were defined by Tamassy and Binh [282]. A non-flat Riemannian manifold $(M, g)$ of dimension $n > 2$ is called a *weakly symmetric Riemannian manifold* if there exists 1−forms $A, B, D, E$ and a vector field $F$ such that

$$(\nabla_W R)(X, Y)Z = A(W)R(X, Y)Z + B(X)R(W, Y)Z + D(Y)R(X, W)Z$$
$$+E(Z)R(X, Y)W + g(R(X, Y, Z), W)F, \forall X, Y, Z, W \in \chi(M).$$

It follows that if $B = D = E = \frac{1}{2}A$, then the weakly symmetric manifold reduces to a pseudosymmetric manifold in the sense of Chaki. It was proved by Pranović in [227] that these 1−forms and the vector field must be related as follows: $B(X) = D(X) = E(X)$ and $g(X, F) = E(X)$ for any $X$. Thus the weakly symmetric manifold is characterized by the condition

$$(\nabla_W R)(X, Y)Z = A(W)R(X, Y)Z + B(X)R(W, Y)Z + B(Y)R(X, W)Z$$
$$+B(Z)R(X, Y)W + g(R(X, Y, Z), W)F, \forall X, Y, Z, W \in \chi(M)$$

For more on weakly symmetric Riemannian manifolds and their generalizatios, see survey [89].

Geometrically, the locally symmetric spaces imply that the sectional curvature of every plane is preserved after parallel transport of the plane along any curve, up to the first order [175].

Recently, the geometric interpretations of semisymmetry conditions and pseudosymmetry conditions were given in [139] in terms of the notion of double sectional curvature. We first give the definition of this notion.

**Definition 1.** [139] Let $(M, g)$ be an n-dimensional Riemannian manifold $(n > 3)$ which is not of constant curvature, and denote by $\mathcal{U}$ the set of points where the Tachibana tensor $Q(g, R)$ is not identically zero, i.e., $\mathcal{U} = \{x \in M | Q(g, R)_x \neq 0\}$. Then, at a point $p \in U$, a plane $\pi = u \wedge v \subset T_pM$ is said to be *curvature-dependent* with respect to a plane $\pi' = x \wedge y \in T_pM$ if $Q(g, R)(u, v, u, v; x, y) \neq 0$. This definition is independent of the choice of bases for $\pi$ and $\pi'$.

**Definition 2.** [139] At a point $p \in \mathcal{U}$, let the tangent plane $\pi = u \wedge v$ be curvature-dependent with respect to $\pi' = x \wedge y$. Then, we define the *double sectional curvature* $L(p, \pi, \pi')$ of the plane $\pi$ with respect to $\pi'$ at $p$ as the scalar

$$L(p, \pi, \pi') = \frac{(R \cdot R)(u, v, v, u; x, y)}{Q(g, R)(u, v, v, u; x, y)}.$$

This definition is again independent of the choice of bases for the tangent planes $\pi$ and $\pi'$

According to [139] (see also [96]), geometrically semisymmetry condition implies that sectional curvatures of the tangent planes $\pi$ and $\pi'$ are invariant under parallel transport around infinitesimal coordinate parallelograms. For in short, the semisymmetric spaces $M$ are characterized by the vanishing of the double sectional curvatures $L(p, \pi, \pi')$ [139, Corollary 2]. In [139], the interpretation of the pseudosymmetry in the sense of Deszcz is given as follows. A Riemannian manifold $M$ $(n > 3)$ is pseudosymmetric in the sense of Deszcz if and only if at all of its points $p \in \mathcal{U}$ all the double sectional curvatures $L(p, \pi, \pi')$ are the same, i.e., for all curvature-dependent planes $\pi$ and $\pi'$ at $p$, $L(p, \pi, \pi') = \phi(p)$ for some function $\phi : M \to \mathbb{R}$ [139, Theorem 3].

The *exterior derivation* is a differential operator, denoted by $d$, which assigns to each $p$-form $\omega$, a $(p + 1)$-form $d\omega$ defined by

$$(d\omega)(X_1, \ldots, X_{p+1}) = \frac{1}{p + 1} \{ \sum_{i=1}^{p+1} (-1)^{i+1} X_i \omega(X_1, \ldots, \widehat{X}_i, \ldots, X_{p+1})$$

$$+ \sum_{1 \leq i < j \leq 1+p} (-1)^{i+j} \omega([X_i, X_j], X_1, \ldots, \widehat{X}_i, \ldots, \widehat{X}_j, \ldots, X_{p+1}) \},$$

where the hood means the term in that particular slot is omitted. In particular, for a

1-form $\omega$ and a 2-form $\Omega$, one can find the following formulas:

$$(d\omega)(X, Y) = \frac{1}{2}\{X(\omega(Y)) - Y(\omega(X) - \omega([X, Y])\},$$

$$(d\Omega)(X, Y, Z) = \frac{1}{3}\{X(\Omega(Y, Z)) - Y(\Omega(X, Z)) + Z(\Omega(X, Y))$$
$$- \Omega([X, Y], Z) + \Omega([X, Z], Y) - \Omega([Y, Z], X)\}.$$

The exterior derivation $d$ has the following properties:

**(1)** For a smooth function $f$ on $M$, $df$ is a 1-form (also called the gradient-form of $f$) such that $(df)X = Xf$, for any vector field $X$.

**(2)** For a $p$-form $\omega$ and a $q$-form $\theta$

$$d(\omega \wedge \theta) = (d\omega) \wedge \theta + (-1)^p \omega \wedge d\theta.$$

**(3)** $d(d\omega) = 0$ for any $p$-form $\omega$, (Poincare Lemma).

**(4)** $d$ is linear with respect to the addition of any two $p$-forms.

Let $A(M) = \sum_{r=0}^{n} A^p(M)$ be the exterior algebra of smooth differential forms on $M$, where $A^p(M)$ denotes the $\mathbb{R}$ linear space of smooth differential forms on $M$. The *inner derivation* is the operator $i_X$ of $A(M)$, which is defined on smooth functions by,

$$i_X(f) = 0, f \in C^{\infty}(M, \mathbb{R}).$$

For 1$-$forms by

$$i_X\omega = \omega(X).$$

For $p-$forms by

$$(i_X\Omega)(Y_2, ..., Y_p) = \Omega(X, Y_2, ..., Y_p),$$

where $Y_2, ..., Y_p$ are vector fields on $M$. It also satisfies the following properties

**(1)** $i_X \circ i_X = 0$.

**(2)** For $\Omega_1 \in A^p(M)$ and $\Omega_2 \in A^q(M)$, $i_X(\Omega_1 \wedge \Omega_2) = (i_X\Omega_1) \wedge \Omega_2 + (-1)^p\Omega_1 \wedge (i_X\Omega_2)$.

On the other hand, the Lie derivation is an $\mathbb{R}$ linear map:

$$\pounds_X : A^p(M) \longrightarrow A^p(M).$$

For a $f \in C^{\infty}(M, \mathbb{R})$, it is defined by

$$\pounds_X f = X(f).$$

Let $Y$ be a vector field on $M$. Then,

$$\pounds_X Y = [X, Y].$$

Let $\omega = \omega_i dx^i$ be a 1-form on $M$. Then,

$$
\begin{aligned}
(\pounds_V \omega)(X) &= V(\omega(X)) - \omega[V, X] \\
(\pounds_V \omega)_i &= V^j \partial_j(\omega_i) + \omega_j \partial_i(V^j).
\end{aligned}
$$

Using the theory of tensor analysis, we obtain the following general formulae for the tensor field's Lie derivative with respect to a vector field $V$,

$$
\begin{aligned}
(\pounds_V T)(\omega^1, \ldots, \omega^r, X_1, \ldots, X_s) &= V(T(\omega^1, \ldots, \omega^r, X_1, \ldots, X_s)) \\
&\quad - \sum_{a=1}^{r} T(\omega^1, \ldots, \pounds_V \omega^a, \ldots, \omega^r, X_1, \ldots, X_s) \\
&\quad - \sum_{A=1}^{s} T(\omega^1, \ldots, \omega^r, X_1, \ldots, [V, X_A], \ldots, X_s),
\end{aligned}
$$

where $\omega^1, \ldots, \omega^r$ and $X_1, \ldots, X_s$ are $r$ 1-forms and $s$ vector fields, respectively. In particular, if $T$ is a tensor of type $(1, 1)$ then

$$
(\pounds_V T)(X) = V(T(X)) - T([V, X]),
$$

for an arbitrary vector field $X$ on $M$. The Lie derivative of the metric tensor $g$ is

$$
\begin{aligned}
(\pounds_V g)(X, Y) &= V(g(X, Y)) - g([V, X], Y) - g(X, [V, Y]) \\
&= g(\nabla_X V, Y) + g(\nabla_Y V, X)
\end{aligned}
\tag{1.16}
$$

for arbitrary vector fields $X$ and $Y$ on $M$ and Levi-Civita (metric) connection $\nabla$ of $g$.

The following are basic properties of Lie derivatives with other operators defined in this section.

$$
\begin{aligned}
\pounds_V &= d \circ i_X + i_X \circ d \\
\pounds_V \circ d &= d \circ \pounds_V \\
\pounds_V \circ i_X &= i_X \circ \pounds_V \\
\pounds_V(\Omega_1 \wedge \Omega_2) &= \pounds_V(\Omega_1) \wedge \Omega_2 + \Omega_1 \wedge \pounds_V(\Omega_2)
\end{aligned}
$$

for $V \in \chi(M)$ and $\Omega_1 \in A^p(M)$ and $\Omega_2 \in A^q(M)$.

Let $(M, g)$ be an $m + 2-$ dimensional Riemannian manifold. The *Ricci tensor*, denoted by $Ric$, is defined by

$$
Ric(X, Y) = trace\{Z \to R(X, Z)Y\},
\tag{1.17}
$$

for any $X, Y \in \chi(M)$. Locally, $Ric$ and its *Ricci operator* $Q$ are given by

$$
Ric(X, Y) = \sum_{i=1}^{m+2} g(R(E_i, X)Y, E_i), \quad g(\bar{Q}X, Y) = Ric(X, Y).
\tag{1.18}
$$

$M$ is *Ricci flat* if its Ricci tensor vanishes on $M$. If $\dim(M) > 2$ and

$$Ric = kg, \quad k \text{ is a constant,} \tag{1.19}$$

then $M$ is an *Einstein manifold*. For $\dim(M) = 2$, any $M$ is Einstein but $k$ in (1.19) is not necessarily constant. The *scalar curvature* $r$ is defined by

$$r = \sum_{i=1}^{m+2} Ric(E_i, E_i). \tag{1.20}$$

(1.19) in (1.20) implies that $M$ is Einstein if and only if $r$ is constant and

$$Ric = \frac{r}{m+2} g.$$

For more on Einstein manifolds, see [31]. Recently, by modifying condition (1.19), new kinds of Einstein manifolds have been defined. The first natural generalization of Einstein manifolds is Ricci solitons. A Riemannian manifold $(M, g)$ is called to admit a *Ricci soliton* structure, denoted by $(M, g, \zeta, \lambda)$, if there exists a vector field $\zeta \in \mathfrak{X}(M)$ and a scalar $\lambda$ satisfying

$$\frac{1}{2}\pounds_\zeta g + Ric = \lambda g, \tag{1.21}$$

where $Ric$ denotes the Ricci tensor of $M$ and $\pounds_\zeta$ denotes the Lie derivative in the direction of $\zeta$. A Ricci soliton is said to be shrinking, steady, or expanding if the scalar $\lambda$ is positive, zero or negative respectively If $\zeta = grad f$, for a smooth function $f$, the Ricci soliton $(M, g, \zeta, \lambda) = (M, g, f, \lambda)$ is called a *gradient Ricci soliton* and the function $f$ is called the *potential function*. Gradient Ricci solitons are natural generalizations of Einstein manifolds. The study of Ricci solitons was first introduced by Hamilton [141] as fixed or stationary points of the Ricci flow in the space of the metrics on $M$ modulo diffeomorphisms and scaling; see [52]. Recently, a generalization of (1.21) has been considered in [224], allowing $\lambda$ to be a smooth function on $M$. Thus $(M, g, f)$ is said to be a *gradient Ricci almost soliton* if (1.21) is satisfied for some $\lambda \in C^\infty(M)$. Since gradient Ricci almost solitons contain gradient Ricci solitons as a particular case, if the function $\lambda$ is non-constant, the gradient Ricci almost soliton is called *proper*.

A non-flat Riemannian manifold $(M, g)$ is said to be a *quasi-Einstein manifold* [62] if its Ricci tensor $Ric^M$ satisfies the condition

$$Ric^M(X, Y) = ag(X, Y) + bA(X)A(Y)$$

for every $X, Y \in \Gamma(TM)$, where $a, b$ are real scalars and $A$ is a non-zero 1-form on $M$ such that $A(X) = g(X, U)$ for all vector fields $X \in \Gamma(TM)$, $U$ being a unit vector field which is called the generator of the manifold. If $b = 0$, then the manifold reduces to an Einstein space. A non-flat Riemannian manifold $(M, g)$ is said to be a *general-*

*ized quasi-Einstein manifold* [60] if its Ricci tensor $Ric^M$ is non-zero and satisfies the condition

$$Ric^M(X, Y) = ag(X, Y) + bA(X)A(Y) + cB(X)B(Y)$$

for every $X, Y \in \Gamma(TM)$, where $a, b, c$ are real scalars and $A, B$ are two non-zero 1-form on $M$. The unit vector fields $U_1$ and $U_2$ corresponding to the 1-forms $A$ and $B$, respectively, are defined by $A(X) = g(X, U_1)$, $B(X) = g(X, U_2)$, and are orthogonal, i.e., $g(U_1, U_2) = 0$. If $c = 0$, then the manifold reduces to a quasi-Einstein manifold. In literature, quasi-Einstein manifolds [54] and generalized quasi-Einstein manifolds [58] were also defined in a different way as follows. Let $(M^n, g)$, for $n \geq 3$, be a complete Riemannian manifold. $(M^n, g)$ is called a *quasi-Einstein Riemannian manifold*, if there exists a smooth function $f : M^n \to \mathbb{R}$ and two constants $\mu, \lambda \in \mathbb{R}$, such that

$$Ric + H^f - \mu \, df \otimes df = \lambda g.$$

When $\mu = 0$, quasi-Einstein manifolds correspond to gradient Ricci solitons and when $f$ is constant, then it gives the Einstein equation and we call the quasi-Einstein metric trivial. We also notice that, for $\mu = \frac{1}{2-n}$, the metric $\widetilde{g} = e^{-\frac{2}{n-2}f}g$ is Einstein [54]. A complete Riemannian manifold $(M^n, g)$, $n \geq 3$, is called a *generalized quasi-Einstein manifold* (GQE) [58], if there exist three smooth functions $f, \mu, \lambda$ on $M$, such that

$$Ric + H^f - \mu \, df \otimes df = \lambda g. \tag{1.22}$$

Natural examples of GQE manifolds are given by Einstein manifolds (when $f$ and $\lambda$ are two constants), gradient Ricci solitons (when $\lambda$ is constant and $\mu = 0$), gradient Ricci almost solitons (when $\mu = 0$), and quasi-Einstein manifolds (when $\mu$ and $\lambda$ are two constants) in the sense of [54]. If the function $f$ is constant, then it will clearly imply that $g$ is an Einstein metric. In particular, a Riemannian manifold $(M^n, g)$, $n \geq 3$, is called a *a generalized m– quasi-Einstein manifold* [198], if there exists a constant $m$ with $0 < m \leq \infty$ as well as two smooth functions $f, \lambda$ on $M$, such that

$$Ric + H^f - \frac{1}{m}df \otimes df = \lambda g.$$

If $m = \infty$ and $\lambda$ is constant, the equation becomes the fundamental equation of a gradient Ricci soliton. Further, if $m = \infty$ and $\beta$ is a smooth function, the above equation reduces to the Ricci almost soliton equation.

**Remark 1.** In the literature, there are some more notions of generalized Einstein manifolds. A Riemannian manifold $(M^n, g)$ $(n > 2)$ is called *nearly quasi-Einstein* [90] if its Ricci tensor $Ric^M$ is not identically zero and satisfies the condition

$$Ric^M(X, Y) = Kg(X, Y) + LD(X, Y),$$

where $K, L$ are non-zero scalars and $D$ is a symmetric non-zero $(0, 2)$ tensor. The scalars $K, L$ are known as associated scalars and $D$ is called the associated tensor of the manifold. In addition a non-flat Riemannian manifold $(M^n, g), n > 2$ is called a *nearly Einstein manifold* [236] if its Ricci tensor $Ric^M$ of type $(0, 2)$ is not identically zero and satisfies the condition

$$Ric^M(LX, Y) = \lambda g(X, Y)$$

for all vector fields $X, Y$, where $\lambda$ is a non-zero scalar called the associated scalar and $L$ is the symmetric endomorphism of the tangent space at each point corresponding to the Ricci tensor $Ric^M$ of type $(0, 2)$ defined by $g(LX, Y) = Ric^M(X, Y)$ for all vector fields $X, Y$. See also [33] for the notion of a mixed quasi-Einstein manifold, [61] for a super quasi-Einstein manifold, and [267] for a pseudo-quasi Einstein manifold.

Finally we now give brief information for integration theory of Riemannian manifolds. Our aim is to state divergence or Green's theorem. Let $(M, g)$ be a compact Riemannian manifold with atlas $\mathcal{A} = \{\psi_\alpha, U_\alpha)_{\alpha \in I}\}$. A *partition of unity* to $\mathcal{A}$ is a collection for smooth functions $(\phi_\alpha : M \longrightarrow \mathbb{R}_{\alpha \in I})$ satisfying the following properties:

**(i)**   $0 \leq \phi_\alpha(x) \leq 1, x \in M, \alpha \in I$,

**(ii)**   the support $supp \, \phi_\alpha : \overline{\{x \in M : \phi_\alpha(x) \neq 0\}}$ is contained in $U_\alpha$ for all $\alpha \in I$, where overline denotes the topological closure, and

**(iii)**   for any point $x \in M$, $\sum_{\alpha \in I} \phi_\alpha(x) = 1$.

For a continuous function $f$ on $U_\alpha$, we define

$$\int_{U_\alpha} f \mathcal{V}_g = \int_{\psi(U_\alpha)} f_\alpha \sqrt{g} \, dx_1^\psi ... dx_n^\psi, f_\alpha = f \circ \psi^{-1}, \qquad (1.23)$$

where $g = det(g(\frac{\partial}{\partial x_i}, \frac{\partial}{\partial x_i}))$ and the right side is the Lebesque integration of a continuous function $f_\alpha \sqrt{g}$ on an open set $\psi(U_\alpha)$ of $\mathbb{R}^n$. In general, for a continuous function $f$, using the partition unity $(\phi_\alpha, \alpha \in I)$, $\int_M f \mathcal{V}_g$ is defined by

$$\int_M f \mathcal{V}_g = \int_M \sum_{\alpha \in I} \phi_\alpha f \mathcal{V}_g = \sum_{\alpha \in I} \int_M \phi_\alpha f \mathcal{V}_g \qquad (1.24)$$

This definition is well defined. If $f \equiv 1$, then $Vol(M) = \int_M \mathcal{V}_g$ is called *volume* of $(M, g)$. If $M$ is not compact, the volume is not necessarily finite.

**Theorem 2.** *(Divergence theorem) Let $(M, g)$ be an oriented Riemannian manifold with boundary $\partial M$ (possibly $\partial M = 0$) and Riemannian volume form $dV_g$. If $Z$ is a vector field on $M$ with compact support, then*

$$\int_M div(Z) \mathcal{V}_g = \int_{\partial M} g(Z, N) \mathcal{V}_{\tilde{g}},$$

*where N is the unit outward normal vector field to $\partial M$ and $\mathcal{V}_{\tilde{g}} = i_N \mathcal{V}_g \mid_{\partial M}$.*

**Proposition 1.** *Let $(M, g)$ be a compact Riemannian manifold. Then, for $X \in \chi(M)$ and $f, f_1, f_2 \in C^\infty(M, \mathbb{R})$, the following hold:*

$$\int_M f \, div(X) \mathcal{V}_g = - \int_M g(grad f, X) \mathcal{V}_g, \tag{1.25}$$

$$\int_M (\triangle f_1) f_2 \mathcal{V}_g = \int_M g(grad f_1, grad f_2) \mathcal{V}_g = \int_M f_1(\triangle f_2) \mathcal{V}_g, \tag{1.26}$$

*and*

$$\int_M div(X) \mathcal{V}_g = 0 \quad \text{(divergence formula).} \tag{1.27}$$

## 2. Vector bundles

The notion of vector bundles is going to be very important for other chapters of this book. Therefore this section is devoted to the notion of vector bundles. Roughly speaking, a vector bundle $E$ can be thought of as a manifold $M$ with a vector space $E_p$ attached to each point $p \in M$.

**Definition 3.** Let $E$ and $B$ differentiable manifolds and $\pi : E \to B$ a smooth map. $(E, B, \pi)$ is called a *vector bundle* of rank $k$ if the following holds

**(i)** there is a $k-$ dimensional vector space $V$, called fiber of $(E, \pi)$, such that for any point $p \in M$ the fiber $E_p = \pi^{-1}(p)$ of $\pi$ over $p$ is a vector space isomorphic to $V$,

**(ii)** any point $p \in M$ has a neighborhood $U$ such that there is a diffeomorphism $\phi_U$ and

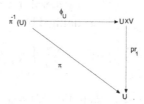

Figure 1.1 Condition (ii) means that every fiber $E_p$ is mapped to $p \times V$

the diagram in Figure 1.1 is commutative.

**(iii)** $\phi_U|_{Ep} : Ep \to V$ is an isomorphism of vector spaces.

The spaces $E$, $B$, and $\pi$ are called the total and the base space and the projection map, respectively. $\pi^{-1}(p)$ is a fiber over the point $p$, and $(U, \phi)$ is a coordinate chart of

*E*. *U* can be the whole base manifold *B*. In this case, $(E, \pi)$ is called a trivial bundle. A vector bundle may be considered as a family of vector spaces (all isomorphic to a fixed model $\mathbb{R}^n$) parametrized (in a locally trivial manner) by a manifold.

**Definition 4.** Let $(E, \pi, M)$ be a vector bundle. A map $s : M \longrightarrow E$ such that $\pi \circ s = id$ is called a (global) section of *E*. If *s* is defined on an (open) subset *U* of *M*, then $s : U \longrightarrow E$ is called a local section of *E* over *U*.

From now on, we denote by $\Gamma(E)$ the module of differentiable sections of a vector bundle *E*.

**Example 1.** If *M* is a smooth manifold and k is a nonnegative integer, then $F : M \times \mathbb{R}^k \longrightarrow M$ is a real vector bundle of rank *k* over *M*. It is called the trivial rank-k real vector bundle over *M*.

**Example 2.** [304] Let *M* be a differentiable manifold of dimension *n*. Then the tangent bundle *TM* of *M* is the disjoint union of the tangent spaces $T_p M$, $p \in M$, i.e.

$$TM = \bigcup_{p \in M} T_p M.$$

First let $\pi : TM \longrightarrow M$ with $\pi(W) = p$ for $W \in T_p M$ be the projection. Then for each open set *U* of *M*, $\pi^{-1}(U) = TU$. For $t_1, t_2 \in \mathbb{R}$, $V_1, V_2 \in T_p M$, and $p \in M$, define

$$t_1(p, V_1) + t_2(p, V_2) = (p, t_1 V_1 + t_2 V_2).$$

With this operation, $\pi^{-1}(p)$ becomes a vector space. On the other hand, for $p \in M$ there is a coordinate map $\phi$ and coordinate neighborhood *U* such that $\phi(U) = \{(x_1, ..., x_n)\} \subset \mathbb{R}^n$. At $p \in M$, define a tangent vector $\frac{\partial}{\partial x_i}|_p$ and a map

$$h: \quad U \times \mathbb{R}^n \longrightarrow \pi^{-1}(U)$$
$$(p, V) \longrightarrow h(p, V) = \sum_{i=1}^n V_i \frac{\partial}{\partial x_i}|_p .$$

Then for $e = \sum_{i=1}^n e_i \frac{\partial}{\partial x_i}|_p$, we have

$$h^{-1}(e) = (\pi(e), (e_1, ..., e_n)).$$

This shows that *h* is a diffeomorphism. As a result, *TM* is a vector bundle on *M*. The sections of *TM* are the vector fields on *M*.

**Example 3.** [304] Let $E_1 = (M_1, M_2, \pi_E)$ be a vector bundle and $N_2$ another manifold. Let $F : N_2 \to M_2$ be a smooth map. Then the induced vector bundle $E_1 = F^{-1}(E)$ over $N_2$ ( or the pullback of $E_1$ along *F*) is constructed as follows. (a) The total manifold

$N_1$ consists of the elements $(p, e) \in N_2 \times M_1$ satisfying

$$F(p) = \pi_E(e).$$

(b) The projection $\pi_{E_1} : N_1 \rightarrow N_2$ is defined by $\pi_{E_1}(p, e) = p$. If we define $\hat{F}(p, e) = e$, then the following diagram is commutative. (c) Choose two points $(p, e_1)$, $(p, e_2)$

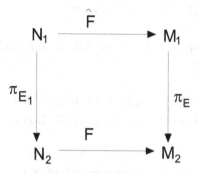

Figure 1.2  The projection of the induced vector bundle

on the same fiber. Since $\pi_E(e_1) = F(p) = \pi_E(e_2)$, we can define linear operation for elements of a fiber $N_1$ as

$$t_1(p, e_1) + t_2(p, e_2) = (p, t_1 e_1 + t_2 e_2).$$

From $\pi_E(t_1 e_1 + t_2 e_2) = f(p)$, it follows that

$$
\begin{aligned}
\hat{F}(t_1(p, e_1) + t_2(p, e_2)) &= \hat{F}(p, t_1 e_1 + t_2 e_2) \\
&= t_1 e_1 + t_2 e_2 \\
&= t_1 \hat{F}(p, e_1) + t_2 \hat{F}(p, e_2).
\end{aligned}
$$

Thus $\hat{F}$ is a linear map between $F_p(F^{-1}E)$ and $F_{F(p)}(E)$. Moreover it is a surjective map. Now suppose that for any $e \in M_1$, its inverse images are $(p, e)$ and $(q, e)$ under $\hat{F}$. Since we consider $\pi_{E_1}^{-1}(p)$, it is necessary that $p = q$, which means $\hat{F}$ is injective. As a result, $F_p(F^{-1}E)$ and $F_{F(p)}(E)$ are isomorphic. (d) For any $p \in M_2$ and $F(p) \in N_2$, let $(U, h)$ be a coordinate chart of $E$ near $F(p)$. Denote $U_1 = F^{-1}(U)$. Suppose that $U_1$ is a coordinate neighborhood of $p$ in $M$. Define $h_1$ as

$$
\begin{aligned}
h_1 : \quad & U_1 \times \mathbb{R}^n \longrightarrow \pi_{E_1}^{-1}(U_1) \\
& (p, V) \longrightarrow h_1(p, V)) = (p, h(F(p), V)) \in N_2 \times \pi_E^{-1}(U) \subset N_2 \times M_1.
\end{aligned}
$$

Since

$$F(p) = \pi_E \circ h(F(p), V), \ h_1(p, V) \in \pi^{-1}(U_1) \subset N_1,$$

$h_1$ is a diffeomorphism. In fact, it has an inverse differentiable map. Choose $e_1 \in \pi^{-1}(U_1)$, then $e_1 = (p, V) \in N_2 \times M_2$ such that $F(p) = \pi_E(e)$. If $h^{-1}(e) = (F(p), V)$, $V \in \mathbb{R}^n$, then we define $h_1^{-1}(e_1) = (p, V)$. Hence we obtain

$$
\begin{aligned}
h_1 \circ h_1^{-1}(e_1) &= h_1(p, V) \\
&= (p, h(F(p), V)) \\
&= (p, h \circ h^{-1}(e)) \\
&= (p, e) \\
&= e_1
\end{aligned}
$$

and

$$
\begin{aligned}
h_1^{-1} \circ h_1(p, V) &= h_1^{-1}(p, h(F(p), V)) \\
&= (p, V).
\end{aligned}
$$

Thus $h_1 : U_1 \times \mathbb{R}^n \longrightarrow \pi_{E_1}^{-1}(U_1)$ is a diffeomorphism and $(U_1, h_1)$ is a local coordinate neighborhood of $N_1$.

**Definition 5.** A *subbundle* of $(E_2, \pi_2, B_2, F_2)$ is a vector bundle $(E_1, \pi_1, B_1, F_1)$ with the same base such that each of its fibers $F_{1p}$ is a linear subspace of $F_{2p}$, and for which the induced inclusion map $i : E_1 \to E_2$ of total spaces is smooth.

**Definition 6.** Let $(E_1, \pi_1, M)$ and $(E_2, \pi_2, M)$ be vector bundles over $M$. Let the differentiable map $F : E_1 \longrightarrow E_2$ be fiber preserving, i.e.,

$$
\pi_2 \circ F = \pi_1
$$

and let the fiber maps $F_p : E_{1,p} \longrightarrow E_{2,p}$ be linear, i.e. vector space homomorphisms. Then $F$ is called a *vector bundle homomorphism*. A vector bundle isomorphism is a vector bundle homomorphism which is invertible and whose inverse is also a vector bundle homomorphism.

We note that if $E_1$ and $E_2$ are vector bundles and $\phi : E_1 \to E_2$ is a vector bundle homomorphism, then $ker\phi$ and $im\phi$ are vector subbundles of $E_1$ and $E_2$, respectively. For more details on bundle theory, see [126] and [145].

## 3. Riemannian submanifolds and distributions

In this section we are going to present basic information for Riemannian submanifolds of Riemannian manifolds. Let $(\bar{M}, g)$ be an $n$-dimensional Riemannian manifold and $M$ an $m$-dimensional manifold isometrically immersed in $\bar{M}$. If $i$ is an immersion from $M$ to $\bar{M}$, we will identify vector fields in $M$ and their images under $i$, i.e., if $X$

is a vector field on $M$, we identify $X$ and $i_*(X)$. Then it is easy to see that immersed submanifold $M$ is also a Riemannian manifold with the Riemannian metric

$$g'(X, Y) = g(i_*(X), i_*(Y)) = g(X, Y) \tag{1.28}$$

for vector fields $X, Y$ on $M$. Throughout this section, the induced metric $g'$ will be also denoted by the same $g$ as that of ambient manifold $\bar{M}$. If a vector $N_p$ at point $p \in M$ satisfies $g(X_p, N_p) = 0$ for any vector $X_p$ at a point $p$ of $M$, then $N$ is called a normal vector of $M$ at a point $p$. We denote the vector bundle of all normal vectors of $M$ in $\bar{M}$ by $TM^\perp$. Then the tangent bundle of $\bar{M}$ restricted to $M$ is the direct sum of the tangent bundle $TM$ and the normal bundle $TM^\perp$ in $\bar{M}$, i.e.,

$$T\bar{M} = TM \oplus TM^\perp. \tag{1.29}$$

Let $\bar{\nabla}$ and $\nabla$ be the covariant differentiations on $\bar{M}$ and $M$, respectively. Then the *Gauss* and *Weingarten* formulas are given respectively by

$$\bar{\nabla}_X Y = \nabla_X Y + h(X, Y) \tag{1.30}$$

and

$$\bar{\nabla}_X V = -A_V X + \nabla^\perp_X V, \tag{1.31}$$

where $X$ and $Y$ are vector fields tangent to $M$ and $V$ normal to $M$. Moreover, $h$ is the second fundamental form, $\nabla^\perp$ the linear connection induced in the normal bundle $TM$, called the *normal connection*, and $A_V$ the *second fundamental tensor* or *Weingarten map* at $V$. It is easy to see that $\nabla$ is a metric connection in the tangent bundle $TM$ of $M$ in $\bar{M}$ with respect to the induced metric $g$ on $M$. $\nabla$ is called the induced connection on $M$. $\nabla^\perp$ is also a metric connection in the normal bundle $TM^\perp$ of $M$ in $\bar{M}$ with respect to the induced metric on $TM^\perp$. Furthermore, $h$ is symmetric. From Equations (1.30) and (1.31), we have

$$g(A_V X, Y) = g(h(X, Y), V) \tag{1.32}$$

for $X, Y \in \Gamma(TM)$ and $V \in \Gamma(TM^\perp)$. This relation implies that $A_V$ is a self-adjoint and linear operator on $M$ due to $h$ is bilinear and symmetric. $A_V$ is called the *Weingarten fundamental tensor*.

For an $m$–dimensional submanifold $M$ in $\bar{M}$, the *mean curvature vector field $H$* is given by

$$H = \frac{1}{m} traceh. \tag{1.33}$$

Let $e_1, ... e_m$ be an orthonormal frame field in $M$, then $traceh = \sum_{i=1}^n h(e_i, e_i)$, which is independent of the choice of frames. If we denote $A_{e_\alpha}$ by $A_\alpha$, then by using (1.32) we obtain

$$H = \frac{1}{n} \sum_{\alpha} [trace(A_{\alpha})]e_{\alpha}$$

where $e_{n+1}, ..., e_m$ is an orthonormal frame of $TM^{\perp}$.

A submanifold $M$ is said to be *minimal* (respectively, *totally geodesic*) if $H = 0$ (respectively, $h = 0$). If we have

$$h(X, Y) = g(X, Y)H \tag{1.34}$$

for $X, Y$ in $\Gamma(TM)$, $M$ is said to be *totally umbilical*. A submanifold $M$ is called a *pseudo-umbilical submanifold* if the following relation is satisfied:

$$g(h(X, Y), H) = g(X, Y)g(H, H) \tag{1.35}$$

for $X, Y \in \Gamma(TM)$.

Let $\bar{R}$ and $R$ be the Riemannian curvature tensor fields of $\bar{M}$ and $M$, respectively. Then the Gauss and Weingarten formulas imply

$$\begin{aligned} \bar{R}(X, Y)Z &= R(X, Y)Z - A_{h(Y,Z)}X + A_{h(X,Z)}Y \\ &\quad + (\nabla_X h)(Y, Z) - (\nabla_Y h)(X, Z), \end{aligned} \tag{1.36}$$

where $(\nabla_X h)(Y, Z)$ is defined by

$$(\nabla_X h)(Y, Z) = \nabla_X^{\perp} h(Y, Z) - h(\nabla_X Y, Z) - h(Y, \nabla_X Z). \tag{1.37}$$

For $T \in \chi(M)$, (1.36) gives the *equation of Gauss*

$$\begin{aligned} g(\bar{R}(X, Y)Z, T) &= g(R(X, Y)Z, T) - g(h(Y, Z), h(X, T)) \\ &\quad + g(h(X, Z), h(Y, T)). \end{aligned} \tag{1.38}$$

Taking the normal part of (1.36), we get the *equation of Codazzi*

$$(\bar{R}(X, Y)Z)^{\perp} = (\nabla_X h)(Y, Z) - (\nabla_Y h)(X, Z) \tag{1.39}$$

For $X, Y \in \chi(M)$ and $V \in \chi(M)^{\perp}$, we define the curvature tensor field $R^{\perp}$ of the normal bundle of $M$ by

$$R^{\perp}(X, Y)V = \nabla_X^{\perp} \nabla_Y^{\perp} V - \nabla_Y^{\perp} \nabla_X^{\perp} V - \nabla_{[X,Y]}^{\perp} V.$$

Then, using the Gauss and Weingarten formulas, in a similar way, we get

$$\begin{aligned} \bar{R}(X, Y)V &= R^{\perp}(X, Y)V - h(X, A_V Y) + h(Y, A_V X) \\ &\quad - (\nabla_X A)_V Y + (\nabla_Y A)_V X. \end{aligned} \tag{1.40}$$

Then, for $W \in \chi(M)^{\perp}$, we obtain

$$g(\bar{R}(X,Y)V,U) = g(R^{\perp}(X,Y)V,U) + g([A_U, A_V]X, Y), \qquad (1.41)$$

where $[A_U, A_V] = A_U A_V - A_V A_U$. (1.41) is called the *equation of Ricci*. If $R^{\perp} = 0$, then $\nabla^{\perp}$ is called a *flat normal connection*. An immersion is called *parallel* if $\nabla h = 0$, where $\nabla$ is is an affine connection and $h$ is the second fundamental form of the immersion. A general classification of parallel submanifolds of Euclidean space was obtained in [116]. It is clear that every parallel submanifold of a space form is locally symmetric [179]. In a space of constant curvature, they are the submanifolds which are invariant with respect to the reflections in the normal space and they are the simplest surfaces in Euclidean 3−space. Indeed, the only surfaces in $E^3$ that are parallel are the open parts of planes, spheres, and round cylinders [116]. In [93], Deprez defined and studied semi-parallel immersions. We recall that a submanifold $M$ of a Riemannian manifold $\bar{M}$ is said to be *semi-parallel* if the following condition is satisfied for every point $p \in M$ and every vector fields $X, Y \in \Gamma(TM)$, $\bar{R}(X,Y)h = 0$, where $h$ is the second fundamental form. This condition is equivalent to the condition

$$(\bar{R}(X,Y).h)(Z,W) = R^{\perp}(X,Y)h(Z,W) - h(R(X,Y)Z,W) - h(Z, R(X,Y)W)$$

where $R$ is the curvature tensor field of $M$. It is clear that every parallel immersion is semi-parallel, but the converse is not true. We note that every semi-parallel submanifold $M^m$ of a space form $N^n(c)$ is intrinsically a semisymmetric Riemannian manifold, but the converse does not hold [179]. As a generalization of semi-parallel submanifolds, the notion of pseudo-parallel submanifolds was introduced in [18] as follows. An isometric immersion $f$ from a Riemannian manifold $M^n$ into a space form $N(c)$ of constant sectional curvature $c$ is called *pseudo-parallel* if $R^N(X \wedge Y)h = \phi(X \wedge Y)h$ holds for all $X, Y \in \Gamma(TM)$, where $R^N$ denotes the Riemannian curvature tensor of $N(c)$ and $h$ the second fundamental form of $f$, and $\phi$ is a smooth real-valued function on $M^n$. More precisely, a submanifold $M$ of a Riemannian manifold $\bar{M}$ is called a pseudo-parallel submanifold if its second fundamental form $h$ satisfies the following condition for some function $\phi$ on $M$:

$$R^{\perp}(X,Y)h(Z,W) - h(R(X,Y)Z,W) - h(Z, R(X,Y)W) = \phi(h((X \wedge Y)U, V)$$
$$-h(U, (X \wedge Y)V)),$$

where $\wedge$ is defined by

$$(X \wedge Y)Z = g(Y,Z)X - g(X,Z)Y,$$

and $R^{\perp}$ is the normal curvature tensor and $R$ is the Riemann curvature tensor of $M^n$. We note that every pseudo-parallel submanifold of a space form is pseudosymmetric, but the converse is not true. Thus there are relations among symmetry conditions and parallelity conditions. Therefore we list table 1.1.

| Symmetry condition | Parallelity condition |
|---|---|
| Locally symmetric | Parallel submanifolds |
| Semi-symmetric | Semi-parallel submanifolds |
| Pseudo symmetric | Pseudo-parallel submanifolds |
| Chaki-symmetric | ? |

Table 1.1 Comparison for the symmetry conditions and parallelity conditions

In [96], the authors gave the geometric interpretation of parallelity conditions in terms of parallel transport. We only quote the following interpretation of parallel immersions; we refer to [96] for other parallelity conditions. *A submanifold M in $\bar{M}$ is parallel if and only if the parallel transport of the second fundamental form with respect to $\nabla^{\perp}$ along any curve in M is equal to the second fundamental form acting on the parallel transport of two tangent vectors to M along the same curve* [96, Proposition 3.2].

In [72], B. Y. Chen established the sharp inequality for a submanifold in a real space form involving intrinsic invariants of the submanifolds and squared mean curvature, the main extrinsic invariant, and this inequality is called the Chen inequality. Before we give the inequality, let us recall some notions. Let $M$ be an $n-$dimensional Riemannian manifold and $p \in M^n$. Let $\{e_1, ..., e_n\}$ be an orthonormal basis of the tangent space $T_p M$ and $\{e_{n+1}, ..., e_m\}$ be the orthonormal basis of $T^{\perp}M$. We denote by $H$, the mean curvature vector at $p$. Also, we set

$$h_{ij}^r = g(h(e_i, e_j), e_r), \qquad i, j \in \{1, ..., n\}, \quad r \in \{n+1, ..., m\}$$

and

$$\|h\|^2 = \sum_{i,j=1}^{n} (h(e_i, e_j), h(e_i, e_j)).$$

We denote by $K(\pi)$ the sectional curvature of plane section $\pi \subset T_p M$ and we put $(\inf K)(\pi) = \{K(\pi) \mid \pi \subset T_p M, dim\pi = 2\}$. In [72], Chen obtained the following result.

**Theorem 3.** *[72]Let M be an n−dimensional ($n \geq 2$) submanifold of a Riemannian manifold $\bar{M}(c)$ of constant sectional curvature c. Then*

$$inf K \geq \frac{1}{2}\{r - \frac{n^2(n-2)}{(n-1)}\|H\|^2 - (n+1)(n-2)c\}. \tag{1.42}$$

*Equality holds if and only if there exists an orthonormal basis $\{e_1, e_2, ..., e_n\}$ of $T_p M$ and orthonormal basis $\{e_{n+1}, e_{n+2}, ..., e_m\}$ of $T^{\perp}M$ such that the shape operator takes* *the following forms:*

$$A_{n+1} = \begin{pmatrix} a & 0 & 0 & \cdots & 0 \\ 0 & b & 0 & \cdots & 0 \\ 0 & 0 & \mu & \cdots & 0 \\ \vdots & \vdots & \vdots & \ddots & \vdots \\ 0 & 0 & 0 & \cdots & \mu \end{pmatrix}, a+b=\mu,$$

*and*

$$A_r = \begin{pmatrix} h_{11}^r & h_{12}^r & 0 & \cdots & 0 \\ h_{12}^r & -h_{11}^r & 0 & \cdots & 0 \\ 0 & 0 & 0 & \cdots & 0 \\ \vdots & \vdots & \vdots & \ddots & \vdots \\ 0 & 0 & 0 & \cdots & 0 \end{pmatrix}, r=n+2,...,m.$$

After that work, many research articles were published by various authors for different submanifolds in different ambient spaces, and all the results of those research papers have been included in the monograph [80]. But this subject is still a very active research area; see [19], [110], [130], [138], [157], [168], [187], [211], [212], [296], [314], [315] and [316], for recent research papers.

We now state the well-known Frobenius theorem on integrable distributions. Let $(M, g)$ be a real $n$–dimensional smooth manifold with a symmetric tensor field $g$ of type $(0, 2)$ on $M$. We identify each $T_x M$ as a vector space at $x$. A *distribution* of rank $r$ on $M$ is a mapping $D$ defined on $M$ which assigns to each point $x$ of $M$ an $r$-dimensional linear subspace $D_x$ of $T_x M$. Let $f : M' \rightarrow M$ be an immersion of $M'$ in $M$. This means that the tangent mapping $(f_*)_x : T_x M' \rightarrow T_{f(x)} M$, is an injective mapping for any $x \in M'$. Suppose $D$ to be a distribution on $M$. Then $M'$ is called an *integral manifold* of $D$ if for any $x \in M'$ we have $(f_*)_x (T_x M') = D_{f(x)}$. If $M'$ is a connected integral manifold of $D$ and there exists no connected integral manifold $\bar{M}'$, with immersion $\bar{f} : \bar{M}' \rightarrow M$, such that $f(M') \subset \bar{f}(\bar{M}')$, we say that $M'$ is a *maximal integral manifold* or a *leaf* of $D$. The distribution $D$ is said to be *integrable* if for any point $x \in M$ there exists an integral manifold of $D$ containing $x$. Recall that the distribution $D$ is involutive if for two vector fields $X$ and $Y$ belonging to $D$, the Lie-bracket $[X, Y]$ also belongs to $D$. We quote the following well-known theorem:

**Theorem 4.** *A distribution $D$ on $M$ is integrable if and only if, it is involutive. Moreover, through every point $x \in M$ there passes a unique maximal integral manifold of $D$ and every other integral manifold containing $x$ is an open submanifold of the maximal one.*

For the proof of this theorem, as well as the other facts on distributions, we refer to [42], and for the submanifold theory we refer to [74].

## 4. Riemannian submersions

In this section, we give brief information for Riemannian submersions between Riemannian manifolds and also recall the notion of horizontally weakly conformal maps very briefly. Riemannian submersions between Riemannian manifolds were first studied by O'Neill [205] and Gray [125]. Later, such submersions were considered between manifolds with differentiable structures. We recall the main notions and theorems from [205] and [111]. Let $M$ be an $m + n$–dimensional Riemannian manifold and $N$ an $n$–dimensional Riemannian manifold. Then a differentiable map $F : M \to N$ is called a submersion if $F$ is surjective and for all $p \in M$, the differential of $F$ at $p$, $F_*$ is surjective. Let $(M, g)$ and $(N, g')$ be Riemannian manifolds with dimensions $m$ and $n$, respectively, and $\pi : M \longrightarrow N$ a submersion. Then $rank\pi = dimN < dimM$. For any point $x \in N$, the leaf $F_x = \pi^{-1}(x)$ is a submanifold of $(M, g)$ with $r = (m - n)$–dimensional. Submanifolds $\pi^{-1}(x)$ are called the leaves of of the submersion $F$. The integrable distribution of $F$ is defined by

$$\mathcal{V}_p = ker\pi_{*p}$$

and $\mathcal{V}_p$ is called the *vertical distribution* of submersion $F$. We note that for any point $p \in M$, $\mathcal{V}_p$ and the tangent space of submanifold $\pi^{-1}()$ coincides. The distribution

$$\mathcal{H}_p = (\mathcal{V}_p)^{\perp}$$

which is complementary and orthogonal distribution to $\mathcal{V}$ is called *horizontal distribution*. Thus for every $p \in M$, $M$ has the following decomposition:

$$T_pM = \mathcal{V}_p \oplus \mathcal{H}_p = \mathcal{V}_p \oplus \mathcal{V}_p^{\perp}.$$

**Definition 7.** Let $(M^m, g_M)$ and $(N^n, g_N)$ be Riemannian manifolds, where $dim(M) = m$, $dim(N) = n$ and $m > n$. A Riemannian submersion $F : M \longrightarrow N$ is a surjective map of $M$ onto $N$ satisfying the following axioms:
(S1)   $F$ has maximal rank.
(S2)   The differential $F_*$ preserves the lengths of horizontal vectors.

For each $q \in N$, $F^{-1}(q)$ is an $(m - n)$–dimensional submanifold of $M$. The submanifolds $F^{-1}(q)$, $q \in N$, are called *fibers*. A vector field on $M$ is called *vertical* if it is always tangent to fibers. A vector field on $M$ is called *horizontal* if it is always orthogonal to fibers. A vector field $X$ on $M$ is called *basic* if $X$ is horizontal and $F$–related to a vector field $X_*$ on $N$, i.e., $F_*X_p = X_{*F(p)}$ for all $p \in M$. Note that we denote the projection morphisms on the distributions $kerF_*$ and $(kerfF_*)^{\perp}$ by $\mathcal{V}$ and $\mathcal{H}$, respectively.

**Lemma 1.** *[205] Let $F : M \longrightarrow N$ be a Riemannian submersion between Riemannian manifolds and $X, Y$ be basic vector fields of $M$. Then:*

(a) $g_M(X, Y) = g_N(X_*, Y_*) \circ F$,

(b) *the horizontal part $[X, Y]^{\mathcal{H}}$ of $[X, Y]$ is a basic vector field and corresponds to $[X_*, Y_*]$, i.e., $F_*([X, Y]^{\mathcal{H}}) = [X_*, Y_*]$,*

(c) *$[V, X]$ is vertical for any vector field $V$ of $ker F_*$, and*

(d) *$(\nabla_X^M Y)^{\mathcal{H}}$ is the basic vector field corresponding to $\nabla_{X_*}^N Y_*$.*

The geometry of Riemannian submersions is characterized by O'Neill's tensors $\mathcal{T}$ and $\mathcal{A}$ defined for vector fields $E, F$ on $M$ by

$$\mathcal{A}_E F = \mathcal{H}\nabla_{\mathcal{H}E}\mathcal{V}F + \mathcal{V}\nabla_{\mathcal{H}E}\mathcal{H}F \tag{1.43}$$

$$\mathcal{T}_E F = \mathcal{H}\nabla_{\mathcal{V}E}\mathcal{V}F + \mathcal{V}\nabla_{\mathcal{V}E}\mathcal{H}F, \tag{1.44}$$

where $\nabla$ is the Levi-Civita connection of $g_M$.

It is easy to see that a Riemannian submersion $F : M \longrightarrow N$ has totally geodesic fibers if and only if $\mathcal{T}$ vanishes identically. For any $E \in \Gamma(TM)$, $\mathcal{T}_E$ and $\mathcal{A}_E$ are skew-symmetric operators on $(\Gamma(TM), g)$ reversing the horizontal and the vertical distributions. It is also easy to see that $\mathcal{T}$ is vertical, $\mathcal{T}_E = \mathcal{T}_{\mathcal{V}E}$, and $\mathcal{A}$ is horizontal, $\mathcal{A} = \mathcal{A}_{\mathcal{H}E}$. We note that the tensor fields $\mathcal{T}$ and $\mathcal{A}$ satisfy

$$\mathcal{T}_U W = \mathcal{T}_W U, \forall U, W \in \Gamma(ker F_*) \tag{1.45}$$

$$\mathcal{A}_X Y = -\mathcal{A}_Y X = \frac{1}{2}\mathcal{V}[X, Y], \tag{1.46}$$

for $X, Y \in \Gamma((ker F_*)^{\perp})$. On the other hand, from Equations (1.43) and (1.44) we have

$$\nabla_V W = \mathcal{T}_V W + \hat{\nabla}_V W \tag{1.47}$$

$$\nabla_V X = \mathcal{H}\nabla_V X + \mathcal{T}_V X \tag{1.48}$$

$$\nabla_X V = \mathcal{A}_X V + \mathcal{V}\nabla_X V \tag{1.49}$$

$$\nabla_X Y = \mathcal{H}\nabla_X Y + \mathcal{A}_X Y \tag{1.50}$$

for $X, Y \in \Gamma((ker F_*)^{\perp})$ and $V, W \in \Gamma(ker F_*)$, where $\hat{\nabla}_V W = \mathcal{V}\nabla_V W$. If $X$ is basic, then $\mathcal{H}\nabla_V X = \mathcal{A}_X V$.

For $E, F \in \chi(M)$, the Schouten connection is defined as

$$\bar{\nabla}_E F = v(\nabla_E v F) + h(\nabla_E h F). \tag{1.51}$$

It is easy to see that the Schouten connection satisfies the following relation:

$$\nabla_E F = \bar{\nabla}_E F + \mathcal{T}_E F + \mathcal{A}_E F. \tag{1.52}$$

We also have the following theorem, which characterizes the geodesic on the total space.

**Theorem 5.** *[111] Let $(M, g)$ and $(N, g')$ be Riemann manifolds and $F : (M, g) \to (N, g')$ a Riemannian submersion. If $\alpha : I \longrightarrow M$ is a regular curve and $E(t)$, $W(t)$ denote the horizontal and vertical parts of its tangent vector field, then $\alpha$ is a geodesic on M if and only if*

$$(\bar{\nabla}_{\dot\alpha} W + \mathcal{T}_W E)(t) = 0 \tag{1.53}$$

*and*

$$(\bar{\nabla}_E E + 2\mathcal{A}_E W + \mathcal{T}_W W)(t) = 0. \tag{1.54}$$

We now state the covariant derivatives of fundamental tensor fields $\mathcal{A}$ and $\mathcal{T}$ which will be useful when we deal with curvatures.

**Lemma 2.** *[205] Let $(M, g)$ and $(N, g')$ be Riemannian manifolds and $F : (M, g) \to (N, g')$ a Riemannian submersion. Then for $X, Y \in \chi^h(M)$ and $V, W \in \chi^v(M)$, we have*

$$(\nabla_V \mathcal{A})_W = -\mathcal{A}_{\mathcal{T}_V W}, \tag{1.55}$$

$$(\nabla_X \mathcal{T})_Y = -\mathcal{T}_{\mathcal{A}_X Y}, \tag{1.56}$$

$$(\nabla_X \mathcal{A})_W = -\mathcal{A}_{\mathcal{A}_X W}, \tag{1.57}$$

*and*

$$(\nabla_V \mathcal{T})_Y = -\mathcal{T}_{\mathcal{T}_V Y}. \tag{1.58}$$

We note that if $\mathcal{A}$ (resp. $\mathcal{T}$) is parallel, then $\mathcal{A}$ (resp. $\mathcal{T}$) vanishes. We now state the following curvature relations between the base manifold and the total manifold [205].

$$g(R(U, V)W, F) = g(\hat{R}(U, V)W, F) - g(\mathcal{T}_U F, \mathcal{T}_V W) + g(\mathcal{T}_V F, \mathcal{T}_U W) \tag{1.59}$$

$$g(R(U, V)W, X) = g((\nabla_U \mathcal{T})_V W, X) - g((\nabla_V \mathcal{T})_U W, X) \tag{1.60}$$

$$g(R(X, Y)Z, H) = g(R^*(X, Y)Z, H) + 2g(\mathcal{A}_Z H, \mathcal{A}_X Y)$$
$$+ g(\mathcal{A}_Y H, \mathcal{A}_X Z) - g(\mathcal{A}_X H, \mathcal{A}_Y Z) \tag{1.61}$$

$$g(R(X,Y)Z, V) = -g((\nabla_Z \mathcal{A})_X Y, V) - g(\mathcal{T}_V Z, \mathcal{A}_X Y)$$
$$-g(\mathcal{A}_X Z, \mathcal{T}_V Y) + g(\mathcal{A}_Y Z, \mathcal{T}_V X) \qquad (1.62)$$

$$g(R(X,Y)V, W) = -g((\nabla_V \mathcal{A})_X Y, W) + g((\nabla_W \mathcal{A})_X Y, V)$$
$$-g(\mathcal{A}_X V, \mathcal{A}_Y W) + g(\mathcal{A}_X W, \mathcal{A}_Y V)$$
$$+g(\mathcal{T}_V X, \mathcal{T}_W Y) - g(\mathcal{T}_W X, \mathcal{T}_V Y) \qquad (1.63)$$

$$g(R(X,V)Y, W) = -g((\nabla_X \mathcal{T})_V W, Y) - g((\nabla_V \mathcal{A})_X Y, W)$$
$$+g(\mathcal{T}_V X, \mathcal{T}_W Y) - g(\mathcal{A}_X V, \mathcal{A}_Y W) \qquad (1.64)$$

for $X, Y, Z, H \in \in (M)$ and $U, V, W \in \chi^v(M)$. As a result of the above equations, we have the following relations for sectional curvatures $K$, $K'$, $\hat{K}$ of total space, base space and fibers, respectively.

(i)  $K(U,V) = \hat{K}(U,V) + \|\mathcal{T}_U V\|^2 - g(\mathcal{T}_U U, \mathcal{T}_V V)$,

(ii)  $K(X,Y) = K'(X', Y') \circ \pi - 3\|\mathcal{A}_X Y\|^2$, and

(iii)  $K(X,V) = g((\nabla_X \mathcal{T})_V V, X) + \|\mathcal{T}_V X\|^2 - \|\mathcal{A}_X V\|^2$.

We now recall that a Riemannian submersion $F$ from a Riemannian manifold $(M_1, g_1)$ onto a Riemannian manifold $(M_2, g_2)$ is called a Riemannian submersion with totally umbilical fibers if

$$\mathcal{T}_V W = g(V,W)H, \qquad (1.65)$$

for $V$, $W \in \Gamma(ker F_*)$, where $H$ is the mean curvature vector field of the fibers .

## Weakly Conformal Maps

In the rest of this section, we present the notions of weakly conformal maps and horizontally weakly conformal maps between Riemannian manifolds from [20]. This notions will be important when we deal with harmonic morphisms.

**Definition 8.** [20] Let $(M, g_M)$ and $(N, g_N)$ be Riemannian manifolds. A smooth map $\varphi : (M^m, g_M) \longrightarrow (N^n, g_N)$ is said to be *weakly conformal* at $p$ if there is a number $\Lambda(p) = \lambda^2(p)$ such that

$$g_N(\varphi_* X, \varphi_* Y) = \lambda^2(p) g_M(X, Y) \qquad (1.66)$$

for $X, Y \in T_p M$.

It is known that a smooth map $\varphi : (M^m, g_M) \longrightarrow (N^n, g_N)$ is weakly conformal at $p$ if and only if precisely one of the following holds: (a)$\varphi_{*p} = 0$, (b)$\varphi_{*p}$ is a conformal injection from $T_p M$ into $T_{\varphi(p)} N$. A point $p$ of type (a) in the above characterization is called a *branch point* of $\varphi$ and a point of type (b) is called a *regular point*. If the weakly conformal map $\varphi$ has no branch points, then it is an immersion on its whole domain. Such a map is called a *conformal immersion*. We note that $dim(M) < dim(N)$ for a nonconstant weakly conformal map [20, Proposition 2.3.4]. It is clear that an isometric immersion is a weakly conformal map with $\sqrt{\Lambda} = 1$.

**Definition 9.** [20] Let $(M^m, g_M)$ and $(N^n, g_N)$ be Riemannian manifolds. Suppose that $\varphi : (M^m, g_M) \longrightarrow (N^n, g_N)$ is a smooth map between Riemannian manifolds and $p \in M$. Then $\varphi$ is called a *horizontally weakly conformal map* at $p$ if either (i) $\varphi_{*p} = 0$ or (ii) $\varphi_{*p}$ maps the horizontal space $\mathcal{H}_p = (ker(\varphi_{*p}))^\perp$ conformally onto $T_{\varphi(p)}N$, i.e., $\varphi_{*p}$ is surjective and $\varphi_*$ satisfies (1.66) for $X, Y$ vectors tangent to $\mathcal{H}_p$.

If a point $p$ is of type (i), then it is called the critical point of $\varphi$. A point $p$ of type (ii) is called regular. The number $\Lambda(p)$ is called the square dilation; it is necessarily non-negative. Its square root $\lambda(p) = \sqrt{\Lambda(p)}$ is called the dilation. We note that $dim(M) > dim(N)$ for a nonconstant horizontally weakly conformal map [20, Proposition 2.4.3]. A horizontally weakly conformal map $\varphi : M \longrightarrow N$ is said to be *horizontally homothetic* if the gradient of its dilation $\lambda$ is vertical, i.e., $\mathcal{H}(grad\lambda) = 0$ at regular points. If a horizontally weakly conformal map $\varphi$ has no critical points, then it is called a *horizontally conformal submersion*. One can conclude that the notion of horizontal conformal maps is a generalization of the concept of Riemannian submersions. Indeed, it follows that a Riemannian submersion is a horizontally conformal submersion with dilation identically being one. We note that horizontally conformal maps were introduced independently by Fuglede [120] and Ishihara [147].

## 5. Certain product structures on manifolds

In this section, we recall some important product structures that allow us to decompose Riemannian manifolds. These structures also offer useful tools to investigate the geometry of manifolds. Our aim in this section is also to collect product structures that are scattered in different papers and theses. We first recall the usual product structure from [206]. Given two manifolds $M$ and $N$, the set of all product coordinates in $M \times N$ is an atlas on $\bar{M} = M \times N$ which is called the *product manifold* of $M$ and $N$. The $dim(M \times N) = dim(M) + dim(N)$. This construction can be generalized to the product of any finite number of manifolds. The concept of product manifolds $M \times N$ has the following properties derived from the manifolds $M$ and $N$:

(a) The projections $\pi$ and $\sigma$ are smooth submersions .

(b) For each $(p, q) \in M \times N$ the subsets

$$M \times q = \{(r, q) \in M \times N : r \in M\},$$

$$p \times N = (p, r) \in M \times N : r \in N\}$$

are submanifolds of $M \times N$.

(c) For each $(p, q)$

$$\pi | M \times q \quad \text{is a diffeomorphism from} \quad M \times q \quad \text{to} \quad M,$$

$$\sigma | p \times N \quad \text{is a diffeomorphism from} \quad p \times N \quad \text{to} \quad N.$$

(d) The tangent spaces

$$T_{(p,q)}M \equiv T_{(p,q)}(M \times q) \quad \text{and} \quad T_{(p,q)}N \equiv T_{(p,q)}(p \times N)$$

are subspaces of the tangent space $T_{(p,q)}(M \times N)$, which is their respective direct sum.

Consider two Riemannian manifolds $(M_1, g_1)$ and $(M_2, g_2)$, with $\pi$ and $\sigma$ the projection maps of $M_1 \times M_2$ onto $M_1$ and $M_2$, respectively. Let

$$g = \pi^*(g_1) + \sigma^*(g_2).$$

Then, we can then see that $g$ is also a metric tensor of a Riemannian manifold ($M = M_1 \times M_2, g$). This new Riemannian manifold $(M, g)$ is called a *usual product manifold*. If $f \in C^\infty(M, \mathbb{R})$, the lift of $f$ to $M \times N$ is $f = f \circ \pi \in C^\infty(M, \mathbb{R})$. If $x \in T_p(M)$ and $q \in N$ then the lift $\tilde{x}$ of $x$ to $(p, q)$ is the unique vector in $T_{(p,q)}M$ such that $\pi_*(\tilde{x}) = x$. If $X \in \chi(M)$ the lift of $X$ to $M \times N$ is the vector field $\tilde{X}$ whose value at each $(p, q)$ is the lift of $X_p$ to $(p, q)$. The lift of $X \in \chi(M)$ to $M \times N$ is the unique element of $\tilde{X} \in \chi(M \times N)$ that is $\pi$−related to $X$ and $\sigma$ related to the zero vector field on $N$. The set of all such horizontal lifts $\tilde{X}$ is denoted by $\mathfrak{L}(M)$. Functions, tangent vectors, and vector fields on $N$ are lifted to $M \times N$ in the same way using the projection $\sigma$. We note that we will use the same notation for a vector field and for its lift. We sometimes will also use $\Gamma(TM)$ instead of $\mathfrak{L}(M)$.

In 1969, Bishop and O'Neill [34] introduced a new concept of product manifolds, called *warped product manifolds*, as follows.

**Definition 10.** [34] Let $(B, g_1)$ and $(F, g_2)$ be two Riemannian manifolds, $f : B \to (0, \infty)$ and $\pi : B \times F \to B$, $\sigma : B \times F \to F$ the projection maps given by $\pi(x, y) = x$ and $\sigma(x, y) = y$ for every $(x, y) \in B \times F$. Denote the *warped product manifold* $M = (B \times_f F, g)$, where

$$g(X, Y) = g_1(\pi_* X, \pi_* Y) + f(\pi(x, y)) g_2(\sigma_* X, \sigma_* Y)$$

for every $X$ and $Y$ of $M$ and $*$ is symbol for the tangent map.

The manifolds $B$ and $F$ are called the *base* and the *fiber* of $M$. They proved that $M$ is a complete manifold if and only if both $B$ and $F$ are complete Riemannian manifolds and also constructed a large variety of complete Riemannian manifolds of everywhere negative sectional curvature using warped product. The Levi-Civita connection $\nabla$ and curvature tensor field can be related to the base and the fiber.

**Lemma 3.** *[206] Let $M = B \times_f F$ be a warped product, and let $\nabla, \nabla^B$ and $\nabla^F$ denote the Levi-Civita connection on $M, B$ and $F$, respectively. If $X, Y \in \Gamma(TB)$ and $U, W \in \Gamma(TF)$, then:*

**(a)** $\nabla_X Y = \nabla_X^B Y$,

**(b)** $\nabla_X U = \nabla_U X = \frac{Xf}{f} U$, and

**(c)** $\nabla_U W = -\frac{g(U,W)}{f} grad_B f + \nabla_U^F W$.

From the above lemma, the following properties can be obtained.
(1) For each $y \in F$, the map $\pi|_{(B \times y)}$ is an isometry onto $B$.
(2) For each $x \in B$, the map $\sigma|_{(x \times F)}$ is a positive homothety onto $F$, with scalar factor $1/h(x)$.
(3) For each $(x, y) \in M$, the leaf $B \times y$ and the fiber $x \times F$ are orthogonal at $(x, y)$.
(4) The fibers $x \times F = \pi^{-1}(x)$ and the leaves $B \times y = \times \sigma^{-1}(y)$ of the warped product are totally geodesic and totally umbilical, respectively.

**Lemma 4.** *[206] Let $M = B \times_f F$ be a warped product with curvature $R$, If $X, Y, Z \in \Gamma(TB)$ and $U, V, W \in \Gamma(TF)$, then:*

**(a)** $R(X, Y)Z = R^B(X, Y)Z$,

**(b)** $R(V, X)Y = -\frac{H_B^f(X,Y)}{f} V$,

**(c)** $R(X, Y)V = R(V, W)X = 0$,

**(d)** $R(X, V)W = -\frac{g(V,W)}{f} \nabla_X^B grad_B f$,

**(e)** $R(V, W)U = R^F(V, W)U - \frac{|grad_B f|_B^2}{f^2}[g(W, U)V - g(V, U)W]$.

Warped product manifolds find their applications in general relativity. Indeed, generalized Robertson-Walker space-times and standard static space-times are two well-known warped product spaces [22]. We note that there are close relations between generalized Einstein manifolds (Section 1) and warped products. Indeed, in [58] the author proved that a complete $n-$ dimensional generalized quasi-Einstein manifold with harmonic Weyl tensor and with zero radial Weyl curvature is locally a warped product with $(n - 1)$-dimensional Einstein fibers. Also in [198], the author proved that a $4-$ dimensional generalized $m-$ quasi-Einstein manifold with harmonic anti-self dual Weyl tensor is locally a warped product with three-dimensional Einstein fibers under certain conditions. After the notion of warped product manifolds, many extensions of such manifolds have appeared. There are now various types of warped products in addition to the ones considered above. We now recall them and their basic equations.

**Definition 11.** [44] A *multiply warped product* $(M, g)$ is a product manifold of form $M = B \times_{b_1} F_1 \times_{b_2} F_2 \cdots \times_{b_m} F_m$ with the metric

$$g = g_B \oplus b_1^2 g_{F_1} \oplus b_2^2 g_{F_2} \cdots \oplus b_m^2 g_{F_m},$$

where for each $i \in \{1, \cdots, m\}$, $b_i : B \to (0, \infty)$ is smooth and $(F_i, g_{F_i})$ is a Riemannian manifold.

Reissner-Nördstrom space-times can be expressed as multiply warped products. Like for warped product manifolds, one can define lifts of vector fields for multiply warped product manifolds. Let $X \in \chi(B)$, then the lift of $X$ to $B \times F$ is the vector field $\tilde{X} \in \chi(B \times F)$ such that $\pi_*(\tilde{X}) = X$ and $\sigma_{*i}(\tilde{X}) = 0$ for any $i \in \{1, 2, ...m\}$. Similarly, let $V \in \chi(F_i)$, then the lift of $V$ to $B \times F$ is the vector field $\tilde{V}_i \in \chi(B \times F)$ such that $\pi_*(\tilde{V}_i) = 0$ and $\sigma_{*i}(\tilde{V}_i) = V_i$ and also $\sigma_{*j}(\tilde{V}_i) = 0$ for any $j \in \{1, 2..., m\} - i$. We will denote the set of all lifts of all vector fields of B by $\mathcal{L}(B)$ and the set of all lifts of all vector fields of $F_i$ by $\mathcal{L}(F_i)$ for any $i \in \{1, 2, ..., m\}$. Here also, we will use the same notation for a vector field and for its lift. Therefore, sometimes we just use $\Gamma(TF_i)$ instead of $\mathcal{L}(F_i)$.

**Proposition 2.** *[289] Let $M = B \times_{b_1} F_1 \times_{b_2} F_2 \cdots \times_{b_m} F_m$ be a multiply warped product and let $X, Y \in \Gamma(TB)$ and $U \in \Gamma(TF_i)$, $W \in \Gamma(TF_j)$. Then*

**(1)** $\nabla_X Y = \nabla_X^B Y,$

**(2)** $\nabla_X U = \nabla_U X = \frac{X(b_i)}{b_i} U,$

**(3)** $\nabla_U W = 0$ *if $i \neq j$, and*

**(4)** $\nabla_U W = \nabla_U^{F_i} W - \frac{g(U,W)}{b_i} \operatorname{grad}_B b_i, i = j.$

By using the above result, we list the following notes [289]:

(1) The fibers $x \times F_2 = \pi^{-1}(x)$ of a multiply warped product are totally umbilical.

(2) The leaves $B \times \sigma^{-1}(y)$ of a multiply warped product are totally geodesic, and the fibers $x \times F_2 = \pi^{-1}(x)$ are totally geodesic if $\operatorname{grad}_B(b_i)_p = 0$, for all $i \in \{1, ..., m\}$.

**Proposition 3.** *[289] Let $M = B \times_{b_1} F_1 \times_{b_2} F_2 \cdots \times_{b_m} F_m$ be a multiply warped product and let $X, Y, Z \in \Gamma(TB)$ and $V \in \Gamma(TF_i)$, $W \in \Gamma(TF_j)$, $U \in \Gamma(TF_k)$. Then*

**(1)** $R(X, Y)Z = R^B(X, Y)Z,$

**(3)** $R(V, X)Y = -\frac{H_B^{b_i}(X,Y)}{b_i} V,$

**(2)** $R(X, V)W = R(V, W)X = R(V, X)W = 0$ *if $i \neq j$,*

**(4)** $R(X, Y)V = 0,$

**(5)** $R(V, W)X = 0, i = j,$

**(6)** $R(V, W)U = 0, i = j \neq k,$

**(7)** $R(U, V)W = -g(V, W)\frac{g_B(\operatorname{grad}_B b_i, \operatorname{grad}_B b_k)}{b_i b_k} U, if \ i = j \neq k,$

**(8)** $R(X, V)W = -\frac{g(V,W)}{b_i}\nabla^B_X(\text{grad}_B b_i)$ *if* $i = j$, *and*

**(9)** $R(V, W)U = R^{F_i}(V, W)U + \frac{|\text{grad}_B b_i|^2_B}{b_i^2}(g(W, U)V - g(V, U)W)$, $i = j = k$.

We now recall the notion of a doubly warped product manifold.

**Definition 12.** [45] A *doubly warped product* $(M, g)$ is a product manifold of form $M =_f B \times_b F$, with smooth functions $b : B \to (0, \infty)$, $f : F \to (0, \infty)$ and the metric tensor

$$g = f^2 g_B \oplus b^2 g_F.$$

We have the following formulas for such structures.

**Proposition 4.** *[9] Let* $M =_{f_2} M_1 \times_{f_1} M_2$ *be a doubly warped product furnished with the metric tensor* $g = f_2^2 g_1 \oplus f_1^2 g_2$ *where* $(M_i, g_i, D_i)$ *is a Riemannian manifold,* $f_i : M_i \to (0, \infty)$ *is a positive function and* $i = 1, 2$. *Then the Levi-Civita connection D on M is given:*

**(a)** $D_{X_i} X_j = X_i (\ln f_i) X_j + X_j (\ln f_j) X_i$, *and*

**(b)** $D_{X_i} Y_i = D^i_{X_i} Y_i - \frac{f_j}{f_i^2} g_i (X_i, Y_i) (\text{grad}_j f_j$,

*where* $i \neq j$ *and* $X_i, Y_i \in \mathfrak{X}(M_i)$.

By using the above result, we have the following results:

(1) The fibers $x \times M_2 = \pi^{-1}(x)$ and the leaves $M_1 \times \sigma^{-1}(y)$ of a double warped product are totally umbilical.

(2) The leaves $M_1 \times \sigma^{-1}(y)$ of a double warped product are totally geodesic if $\text{grad}_{M_2}(f_2)_p = 0$, and the fibers $x \times M_2 = \pi^{-1}(x)$ are totally geodesic if $\text{grad}_{M_1}(f_1)_p = 0$.

**Proposition 5.** *[289] Let* $M =_f B \times_b F$ *be a doubly warped product with the metric* $g = f^2 g_B + b^2 g_F$ *and let* $X, Y, Z \in \Gamma(TB)$ *and* $V, W, U \in \Gamma(TF)$. *Let* $\bar{X} = X + V \in \chi(M)$, $\bar{Y} = Y + W \in \chi(M)$ *and* $\bar{Z} = Z + U \in \chi(M)$. *Let also* $\pi : B \times F \longrightarrow B$ *and* $\sigma : B \times F \longrightarrow F$ *be projections. Then we have the following expressions.*

**(1)** *If* $X, Y, Z \in \Gamma(TB)$, *then*

$$\begin{aligned}
R(X, Y)Z &= R^B(X, Y)Z \\
&\quad + (g_B(Y, Z)X(b) - g_B(X, Z)Y(b))\frac{f\,\text{grad}_F(f)}{b^3} \\
&\quad - + (g_B(Y, Z)X - g_B(X, Z)Y)\frac{f\,\text{grad}_F(f)}{b^2}.
\end{aligned}$$

**(2)** *If* $X, Y \in \Gamma(TB)$ *and* $V \in \Gamma(TF)$ *then*

$$R(X, V)Y \;=\; \frac{H_B^b(X, Y)}{b}V - \frac{H^{f\circ\sigma}(X, V)}{f}X$$
$$+\frac{g_B(X, Y)}{b}(f\nabla_V^F grad_F(f) - \frac{V(f)}{f}grad_B(b)).$$

**(c)** *If* $X, Y \in \Gamma(TB)$ *and* $V \in \Gamma(TF)$, *then*

$$R(X, Y)V = -\frac{H^{f\circ\sigma}(V, Y)}{f}X + \frac{H^{f\circ\sigma}(V, X)}{f}Y.$$

As a generalization of a warped product manifold, the following notion of a twisted product was first defined in [65].

**Definition 13.** [69] Let $(B, g_B)$ and $(F, g_F)$ be Riemannian manifolds of dimensions $r$ and $s$, respectively, and let $\pi : B \times F \to B$ and $\sigma : B \times F \to F$ be the canonical projections. Also let $b : B \times F \to (0, \infty)$ be a positive smooth function. Then the *twisted product* of Riemannian manifolds $(B, g_B)$ and $(F, g_F)$ with twisting function $b$ is the product manifold $B \times F$ with metric tensor

$$g = g_B \oplus b^2 g_F$$

given by

$$g = \pi^* g_B + (b \circ \pi)^2 \sigma^* g_F.$$

We denote this Riemannian manifold $(M, g)$ by $B \times_b F$. In particular, if $b$ is constant on $F$, then $B \times_b F$ is the *warped product* of $(B, g_B)$ and $(F, g_F)$ with warping function $b$. Moreover, if $b = 1$, then we obtain a usual product. If $b$ is not constant, then we have a proper twisted product.

Let $(B, g_B)$ and $(F, g_F)$ be Riemannian manifolds with Levi-Civita connection ${}^B\nabla$ and ${}^F\nabla$, respectively, and let $\nabla$ denote the Levi-Civita connection and the gradient of the twisted product manifold $(B \times_b F)$ of $(B, g_B)$ and $(F, g_F)$ with twisting function $b$. We have the following proposition.

**Proposition 6.** *[115] Let* $M = B \times_b F$ *be a twisted product manifold with the* $g = g_B \oplus b^2 g_F$ *and let* $X, Y \in \mathcal{L}(B)$ *and* $U, V \in \mathcal{L}(F)$. *Then we have*

$$\nabla_X Y \;=\; {}^B\nabla_X Y,$$
$$\nabla_X U \;=\; \nabla_U X = X(k)U, and$$
$$\nabla_U V \;=\; {}^F\nabla_U V + U(k)V + V(k)U - g_F(U, V)\nabla k,$$

*where $k = \log b$.*

Let $M = B \times_b F$ be the twisted product of $(B, g_B)$ and $(F, g_F)$ with twisting function $b$. $B \times_b F$ is called *mixed Ricci-flat* if $Ric(X, V) = 0$ for all $X \in \mathcal{L}(B)$ and $V \in \mathcal{L}(F)$. In [115], the authors showed that mixed Ricci-flat twisted products could be expressed as warped products.

**Definition 14.** [115] A *doubly twisted product* $(M, g)$ is a product manifold of form $M =_f B \times_b F$, with smooth functions $b : B \times F \to (0, \infty)$, $f : B \times F \to (0, \infty)$, and the metric tensor

$$g = f^2 g_B \oplus b^2 g_F.$$

**Proposition 7.** *[115] Let $M = B \times_b F$ be a doubly twisted product manifold with the $g = f^2 g_B \oplus b^2 g_F$ and let $X, Y \in \mathcal{L}(B)$ and $U, V \in \mathcal{L}(F)$. let $k = log(b)$ and $l = log(f)$. Then we have:*

**(1)** $\nabla_X Y = {}^B\nabla_X Y + X(l)Y + Y(l)X - \nabla l$, *and*

**(2)** $\nabla_X V = V(l)X + X(l)V$.

Now, for $X, Y \in \mathcal{L}(B)$, define $h_B^k(X, Y) = XY(k) - (\nabla_X^B Y)k$. Let $R^B$ and $R^F$ be the curvature tensors of $(B, g_B)$ and $(F, g_F)$, respectively, and let $R$ be the curvature tensor of $_f B \times_b F$. Then we have the following relation among them.

**Proposition 8.** *[115] Let $X, Y, Z \in \mathcal{L}(B)$ and $V \in \mathcal{L}(F)$. Then*

$$
\begin{aligned}
R(X, Y)Z &= R^B(X, Y)Z - g(Y, Z)(h^l(X) + X(l)(\nabla l) \\
&\quad + g(X, Z)(h^l(Y) + Y(l)(\nabla l) + (h_B^l(X, Z) - X(l)Z(l))Y \\
&\quad - (h_B^l(Y, Z) - Y(l)Z(l))X \\[6pt]
R(X, Y)V &= (H^l(X, V)Y - H^l(Y, V)X + V(l)X(l)Y - V(l)Y(l)X \\
R(X, V)Y &= + (h_B^k(X, Y) + X(k)Y(k) - X(l)Y(k) - X(k)Y(l))V \\
&\quad + (Y(k)V(l) - VY(l))X + (V(l)\nabla l + h^l(V))g(X, Y),
\end{aligned}
$$

*where $h^l$ and $H^l$ are the Hessian tensor and the Hessian form of $l$ on $_f B \times_b F$, respectively.*

As a generalization of a multiply warped product manifold and a twisted product manifold, a multiply twisted product manifold was defined as follows.

**Definition 15.** [299] A *multiply twisted product* $(M, g)$ is a product manifold of form

$M = B \times_{b_1} F_1 \times_{b_2} F_2 \cdots \times_{b_m} F_m$ with the metric

$$g = g_{_B} \oplus b_1^2 g_{_{F_1}} \oplus b_2^2 g_{_{F_2}} \cdots \oplus b_m^2 g_{_{F_m}},$$

where for each $i \in \{1, \cdots, m\}$, $b_i : B \times F_i \to (0, \infty)$ is smooth.

Here, $(B, g_{_B})$ is called the base manifold and $(F_i, g_{_{F_i}})$ is called the fiber manifold and $b_i$ is called as the twisted function. Obviously, twisted products and multiply warped products are the special cases of multiply twisted products.

**Proposition 9.** *[299] Let $M = B \times_{b_1} F_1 \times_{b_2} F_2 \cdots \times_{b_m} F_m$ be a multiply twisted product and let $X, Y \in \Gamma(TB)$ and $U \in \Gamma(TF_i)$, $W \in \Gamma(TF_j)$. Then:*

(1) $\nabla_X Y = \nabla_X^B Y,$

(2) $\nabla_X U = \nabla_U X = \frac{X(b_i)}{b_i} U,$

(3) $\nabla_U W = 0$ *if* $i \neq j$, *and*

(4) $\nabla_U W = U(\ln b_i)W + W(\ln b_i)U - \frac{g_{F_i}(U,W)}{b_i}\text{grad}_{F_i} b_i - b_i g_{F_i}(U, W)\text{grad}_B b_i + \nabla_U^{F_i} W, i = j.$

From this proposition, we conclude ([299]) that $B$ is a totally geodesic submanifold and $F_i$ is a totally umbilical submanifold.

**Proposition 10.** *[299] Let $M = B \times_{b_1} F_1 \times_{b_2} F_2 \cdots \times_{b_m} F_m$ be a multiply twisted product and let $X, Y, Z \in \Gamma(TB)$ and $V \in \Gamma(TF_i)$, $W \in \Gamma(TF_j)$, $U \in \Gamma(TF_k)$. Then:*

(1) $R(X, Y)Z = R^B(X, Y)Z,$

(3) $R(V, X)Y = -\frac{H_B^{b_i}(X,Y)}{b_i}V,$

(2) $R(X, V)W = R(V, W)X = R(V, X)W = 0$ *if* $i \neq j$,

(4) $R(X, Y)V = 0,$

(5) $R(V, W)X = VX(\ln b_i)W - WX(\ln b_i)V, i = j,$

(6) $R(V, W)U = 0, i = j \neq k$ *or* $i \neq j \neq k$,

(7) $R(U, V)W = -g(V, W)\frac{g_B(\text{grad}_B b_i, \text{grad}_B b_k)}{b_i b_k}U$, *if* $i = j \neq k$,

(8) $R(X, V)W = -\frac{g(V,W)}{b_i}\nabla_X^B(\text{grad}_B b_i) + [WX(\ln b_i)]V - g_{F_i}(W, V)\text{grad}_{F_i}(X\ln b_i)$ *if* $i = j$, *and*

(9)

$$R(V, W)U = g(V, U)\text{grad}_B(W(\ln b_i)) - g(W, U)\text{grad}_B(V(\ln b_i))$$

$$+ R^{F_i}(V, W)U - \frac{|\text{grad}_B b_i|_B^2}{b_i^2}(g(W, U)V - g(V, U)W), i = j = k.$$

We quote the following theorem from [225] for product structures discussed in this section up to now.

**Theorem 6.** *[225] Let g be a Riemannian metric tensor on the manifold $B = M \times N$ and assume that the canonical foliations $\mathcal{D}_M$ and $\mathcal{D}_N$ intersect perpendicularly everywhere. Then g is the metric tensor of:*

**(i)** *a double-twisted product $M \times_{(f,g)} N$ if and only if $\mathcal{D}_M$ and $\mathcal{D}_N$ are totally umbilic foliations,*

**(ii)** *a twisted product $M \times_f N$ if and only if $\mathcal{D}_M$ is a totally geodesic foliation and $\mathcal{D}_N$ is a totally umbilical foliation,*

**(iii)** *a warped product $M \times_f N$ if and only if $\mathcal{D}_M$ is a totally geodesic foliation and $\mathcal{D}_N$ is a spherical foliation, i.e., it is umbilical and its mean curvature vector field is parallel, and*

**(iv)** *a usual product of Riemannian manifolds if and only if $\mathcal{D}_M$ and $D_N$ are totally geodesic foliations.*

We also recall the following generalization.

**Definition 16.** [269] Let $M_i, i = 1, 2, 3$ be three Riemannian manifolds with metrics $g_i$. Let $f : M_1 \to (0, \infty)$ and $\bar{f} : M_1 \times M_2 \to (0, \infty)$ be two smooth positive functions on $M_1$ and $M_1 \times M_2$, respectively. Then the *sequential warped product manifold*, denoted by $\left(M_1 \times_f M_2\right) \times_{\bar{f}} M_3$, is the triple product manifold $(M_1 \times M_2) \times M_3$ furnished with the metric tensor

$$\bar{g} = \left(g_1 \oplus f^2 g_2\right) \oplus \bar{f}^2 g_3$$

Functions $f$ and $\bar{f}$ are called warping functions. If $(M_i, g_i), i = 1, 2, 3$ are all Riemannian manifolds, then the sequential warped product manifold $\left(M_1 \times_f M_2\right) \times_{\bar{f}} M_3$ is also a Riemannian manifold. If the warping function $\bar{f}$ is defined only on $M_1$, then we get the multiply warped product manifold $M_1 \times_f M_2 \times_{\bar{f}} M_3$ with two fibers.

All the above structures find their applications in spacetime models of exact solutions of Einstein field equation: see [274] and the references we cited in the above notes.

The above constructions of product structures are straightforward generalizations of warped structures. There are some other product structures given recently. To present them, we first recall almost contact manifolds from [36]. A $(2n + 1)$-dimensional Riemannian manifold $(\bar{M}, \bar{g})$ is called a *contact metric manifold* if there exists a tensor field $\phi$ of type $(1, 1)$, a unique vector field $\xi$, called the *characteristic vector field* and its 1-form $\eta$ satisfying

$$
\begin{aligned}
\bar{g}(\phi X, \phi Y) &= g(X, Y) - \eta(X)\eta(Y), \ \bar{g}(\xi, \xi) = 1, \\
\phi^2(X) &= -X + \eta(X)\xi, \qquad g(X, \xi) = \eta(X), \\
d\eta(X, Y) &= g(X, \phi Y), \qquad \forall X, Y \in \Gamma(TM).
\end{aligned}
$$

It follows that $\phi\xi = 0$, $\eta \circ \phi = 0$, $\eta(\xi) = 1$. Then $(\phi, \xi, \eta, \bar{g})$ is called *contact metric structure* of $\bar{M}$. We say that $\bar{M}$ has a *normal contact structure* if $N_\phi + d\eta \otimes \xi = 0$, where $N_\phi$ is the Nijenhuis tensor field of $\phi$ [36]. A normal contact metric manifold is called a *Sasakian manifold*, for which we have

$$
\begin{aligned}
\bar{\nabla}_X \xi &= \phi X, \\
(\bar{\nabla}_X \phi)Y &= -\bar{g}(X, Y)\xi + \eta(Y)X.
\end{aligned}
$$

It is known that the odd dimensional spheres are Sasakian manifolds. We now define a new product structure given in [37].

**Definition 17.** [37] Let $(M_1, g_1)$ be a Riemannian manifold and $(M_2, \phi_2, \xi_2, \eta_2, g_2)$ an almost contact metric manifold. On $M_1 \times M_2$, define a metric $g$ by

$$
g = g_1 + f g_2 + f(f - 1)\eta_2 \otimes \eta_2
$$

for a positive function $f$ on $M$. This construction is called a *D–homothetic warping structure*.

In [38], the author shows that Friedman-Lemaitre-Robertson-Walker metrics in cosmology are of this type.

We now define oblique warped products from [27]. Let $(E, h)$ and $(F, k)$ be two Riemannian manifolds. Consider the product manifold $M = E \times F$ and denote by $\pi$ and $\sigma$ the projections of $M$ onto $E$ and $F$, respectively. Now, we denote by $D_E$ and $D_F$ the distributions on $M$ that are tangent to the foliations whose leaves are $\pi^{-1}(p)_{p \in E}$ and $\sigma^{-1}(q)_{q \in F}$, respectively. As they are complementary distributions in $TM$, we put

$$
TM = D_E \oplus D_F. \tag{1.67}
$$

We denote by the same symbols $h$ and $k$ Riemannian metrics on $D_E$ and $D_F$ defined by the Riemannian metrics $h$ and $k$ on $E$ and $F$, respectively. Thus we have two complementary Riemannian distributions $(D_E, h)$, and $(D_F, k)$, on $M$.

**Definition 18.** [27] Let $f$ be a positive smooth function on $E$ and $L : D_E \times D_F \longrightarrow F(M)$ be an $F(M)$–bilinear mapping. Taking into account (1.67), we denote by $P_E$ and $P_F$ the projection morphisms of $\Gamma(TM)$ onto $\Gamma(D_E)$ and $\Gamma(D_F)$, respectively. Then we define the symmetric bilinear mapping $g : \Gamma(TM) \times \Gamma(TM) \to F(M)$ by

$$g(X, Y) = h(P_E X, P_E Y) + (f \circ \pi)^2 k(P_F X, P_F Y)$$
$$+ L(P_E X, P_F Y) + L(P_E Y, P_F X), \forall X, Y \in \Gamma(TM),$$

where $F(M)$ and $\Gamma(TM)$ denote the algebra of smooth functions on $M$ and the $F(M)$−module of smooth vector fields on $M$, respectively. If $g$ is a Riemannian metric on the product manifold $M = E \times F$, then we put $M = E \times F_{(f,L)}$ and call it a *generalized warped product*. For $L = 0$, $M$ becomes a warped product with base $E$ and fiber $F$. For a nonzero $L$ we call $M = E \times F_{(f,L)}$ an *oblique warped product* with base $E$ and fiber $F$. In this case, we also say that $(h, k, f, L)$ is an oblique warped product structure on $E \times F$.

Finally, we give the following notion.

**Definition 19.** [75], [76] Let $(N_1, g_1)$ and $(N_2, g_2)$ be two Riemannian manifolds equipped with metrics $g_1$ and $g_2$, respectively. Consider the symmetric tensor field $g_{f,h}$ of type $(0, 2)$ on the product manifold $N_1 \times N_2$ defined by

$$g_{f,h} = h^2 g_1 + f^2 g_2 + 2 f h d f \otimes d h \tag{1.68}$$

for some positive differentiable functions $f$ and $h$ on $N_1$ and $N_2$, respectively. The product manifold $N_1 \times N_2$ equipped with $g_{f,h}$ is called a *convolution manifold*, which is denoted by $h N_1 \star f N_2$. When $g_{f,h}$ is a positive-definite symmetric tensor, it defines a Riemannian metric on $N_1 \times N_2$. $g_{f,h}$ is called a *convolution metric* and the convolution manifold $h N_1 \star f N_2$ is called a *convolution Riemannian manifold*.

It is obvious that this notion is a generalization of the notion of warped products. A convolution $g_{f,h}$ of two Riemannian metric is not a Riemannian metric. Therefore Chen obtained several conditions on a convolution $g_{f,h}$ of two Riemannian metrics to be a Riemannian metric; see [76].

## 6. Geometric structures along a map

In this section, we are going to define and present various geometric objects along a map. The materials of this section are going to be very important for other chapters. Therefore, we also provide some proofs for some relatively unknown notions. We start with the notion of a vector field along a map.

**Definition 20.** [122] Let $M_1$ and $M_2$ be two manifolds and $F : M_1 \longrightarrow M_2$ a map. Then a map $X : M_1 \longrightarrow TM_2$ is called a vector field along $F$ if $X(p_1) \in T_{F(p_1)} M_2$ for each $p_1 \in M_1$.

The set of vector fields along a map $F$ is denoted by $\Gamma_F(TM_2)$ and it is a module over the ring $C^\infty(M_1, \mathbb{R})$. Observe that if $M_1 = M_2 = M$ and $F = I = identity$, then it is just the set of vector fields on $M$. It is clear that every tangent vector of a curve $\alpha$ is a vector field along $\alpha$.

Let $F : M_1 \longrightarrow M_2$ be a map. If $X \in \Gamma(TM)$, then the map $F_*(X) : M_1 \longrightarrow TM_2$ defined by $(F_*(X))(p_1) = F_{*p_1}(X(p_1)$ is a vector field along $F$. If $Y \in \Gamma(TM_2)$, then $Y \circ F$ is also a vector field along $F$.

As a generalization of vector fields along a map, we have the following notion.

**Definition 21.** [122] Let $E$ be a real vector bundle over $M$ and let $F : N \longrightarrow M$ a smooth map. Then a function $W : N \longrightarrow E$ is called a *section* of $E$ along $F$ if $\pi \circ W = F$, where $\pi : E \longrightarrow M$ is the projection.

The set of sections of $E$ along $F$ will be denoted by $\Gamma_F(E)$. $\Gamma_F(E)$ is a module over $C^\infty(M, \mathbb{R})$. We note that, if $TM = E$, then the section of $E$ along $F$ is just the vector field along $F$. The following lemma shows that the set of sections along a map and the set of sections on the induced vector bundle are isomorphic.

**Lemma 5.** *[226] Let $E$ be a real vector bundle over $M$ and $F : N \longrightarrow M$ a smooth function. Then the modules $\Gamma(F^{-1}(E))$ and $\Gamma_F(E)$ are canonically isomorphic.*

*Proof.* Let $W \in \Gamma(F^{-1}(E))$, then for each $p \in N$, $W_p = (p, v)$ where $v \in E_{F(p)}$. Let $Pr_E : N \times E \longrightarrow E$ be the projection. Then the function $V : N \longrightarrow E$ defined by $V = Pr_E \circ W$ has the property that $\pi \circ V = F$. Hence we have a map $\Gamma(F^{-1}(E)) \longrightarrow \Gamma_F(E)$. It inverse pulls $V \in \Gamma_F(E)$ to $\Gamma(F^{-1}(E))$ as defined by $F^{-1}V = (id_N, V) \in \Gamma(F^{-1}(E))$. Thus $(F^{-1}V)_p = (p, V_p)$. Linearity is obvious. $\qquad\square$

The following definition extends the notion of affine connection for a map between manifolds.

**Definition 22.** [122] Let $F : M_1 \longrightarrow M_2$ be a map. A map $\nabla : \Gamma(TM_1) \times \Gamma_F(TM_2) \longrightarrow \Gamma_F(TM_2)$ is called a connection on $M_2$ along $F$ if the following conditions are satisfied:
(i)   $\nabla^F_{X+Y}U = \nabla^F_X U + \nabla^F_Y U$,
(ii)  $\nabla^F_{hX}U = h\nabla^F_X U$, $h \in C^\infty(M_1)$,
(iii) $\nabla^F_X(U + V) = \nabla^F_X U + \nabla^F_X V$, and
(iv)  $\nabla^F_X(hU) = X(h)U + h\nabla^F_X U$
for $h \in C^\infty(M, \mathbb{R})$, $X, Y \in \Gamma(TM_1)$ and $U, V \in \Gamma_F(TM_2)$.

Observe that if $M_1 = M_2 = M$ and $F = Identity$, $\nabla$ is an affine connection on $M$. It can be shown that $(\nabla^F_X U)(p_1)$ only depends on the value of $X$ at $p_1 \in M_1$.

**Theorem 7.** *[122] Let $F : M_1 \to M_2$ be map and $\overset{2}{\nabla}$ a connection on $M_2$. Then there exists a unique connection $\overset{2}{\nabla}^F$ on $M_2$ along $F$ such that for $X \in \Gamma(TM_1)$ and $Y \in \Gamma(TM_2)$,*

$$\overset{2}{\nabla}^F{}_X(Y \circ F) = \overset{2}{\nabla}_{F_*X}Y. \tag{1.69}$$

*Proof.* Uniqueness: Let $\{Y_1, Y_2, ..., Y_{n_2}\}$ be a local frame for $TM_2$. Then if $Y \in \Gamma_F(TM_2)$ we can write

$$Y = \sum_i^{n_2} h^i(Y_i \circ F)$$

for $i = 1, 2, ..., n_2$, $h^i \in C^\infty(M_1, \mathbb{R})$. Hence if $\overset{2}{\nabla}^F$ is a connection on $M_2$ along $F$, then for $X \in \Gamma(TM_1)$ we have

$$
\begin{aligned}
\overset{2}{\nabla}^F{}_X Y &= \sum_{i=1}^{n_2} X(h^i)(Y_i \circ F) + \sum_{i=1}^{n_2} h^i \overset{2}{\nabla}^F{}_X(Y_i \circ F) \\
&= \sum_{i=1}^{n_2} X(h^i)(Y_i \circ F) + \sum_{i=1}^{n_2} h^i \nabla_{F_*X} Y_i.
\end{aligned}
$$

Thus $\overset{2}{\nabla}^F$ is completely determined by $\overset{2}{\nabla}$.

Existence: Define $\overset{2}{\nabla}^F$ locally by the above formula. It follows that

$$\overset{2}{\nabla}^F{}_X(Y \circ F) = \overset{2}{\nabla}_{F_*X}Y$$

for $X \in \Gamma(TM_1)$ ve $Y \in \Gamma(TM_2)$, thus it can be seen that the oprerator $\overset{2}{\nabla}^F$ is a connection on $M_2$ along $F$, i.e., $\overset{2}{\nabla}^F$ satisfies the conditions given in the definition. □

The unique connection $\overset{2}{\nabla}^F$ on $M_2$ along $F$ is called the *pullback connection* of $\overset{2}{\nabla}$ along $F$. The following result can be obtained by using the above notions.

**Theorem 8.** *[122] Let $M_1$ be a manifold and $(M_2, g_2)$ be a Riemannian manifold with the Levi-Civita connection $\overset{2}{\nabla}$. Suppose that $F : M_1 \longrightarrow M_2$ is a map and $\overset{2}{\nabla}^F$ is the* *pullback connection of $\overset{2}{\nabla}$ along $F$. Then we have:*

**(i)** $Xg_2(U, W) = g_2(\overset{2}{\nabla}^F{}_X U, W) + g_2(U, \overset{2}{\nabla}^F{}_X W)$,

**(ii)** $\overset{2}{\nabla}^F{}_X F_* Y - \overset{2}{\nabla}^F{}_Y F_* X = F_*[X, Y]$,

*for $X, Y \in \Gamma(TM_1)$, $U, W \in \Gamma_F(TM_2)$.*

We now define the curvature tensor field of pullback connection $\overset{2}{\nabla}$ as

$$\overset{2}{R}{}^F(X, Y)U = \overset{2}{\nabla}^F{}_X \overset{2}{\nabla}^F{}_Y U - \overset{2}{\nabla}^F{}_Y \overset{2}{\nabla}^F{}_X U - \overset{2}{\nabla}^F{}_{[X,Y]} U$$

for $X, Y \in \Gamma(TM_1)$, $U \in \Gamma_F(TM_2)$. We note that $\overset{2}{R}{}^F$ is linear and the value of $\overset{2}{R}{}^F(X, Y)U$ at $p_1 \in M_1$ only depends on the values of $X, Y$ and $U$ at $p_1$. Thus we can define $\overset{2}{R}{}^F(x, y)u$ by

$$\overset{2}{R}{}^F(x, y)u = (\overset{2}{R}{}^F(X, Y)U)(p_1),$$

where $X, Y \in \Gamma(TM_1)$, $U \in \Gamma_F(TM_2)$ with $x = X(p_1)$, $y = Y(p_1)$, $u = U(p_1)$. The following proposition establishes the relation between the curvature tensor fields of pullback connection and affine connection.

**Proposition 11.** *[122] Let $F : M_1 \longrightarrow M_2$ be a map and $\overset{2}{\nabla}$ be a connection on $M_2$ with curvature tensor $\overset{2}{R}$. Let also $\overset{2}{\nabla}^F$ be the pullback of $\overset{2}{\nabla}$ along $F$ with a curvature tensor field $\overset{2}{R}{}^F$. Then for $X, Y \in \Gamma(TM_1)$, $U \in \Gamma_F(TM_2)$*

$$\overset{2}{R}(F_* X, F_* Y)U = \overset{2}{R}{}^F(X, Y)U.$$

We now present the notion of the adjoint of a map $F$. Let $(M_1, g_1)$ and $(M_2, g_2)$ be Riemannian manifolds and $F : M_1 \longrightarrow M_2$ a map. Then the adjoint map ${}^* F_{p*}$ at $p \in M_1$ is defined by

$$g_2(F_{*p}(X), W) = g_1(X, {}^* F_{p*}(W)) \tag{1.70}$$

for $X \in T_p M_1$ and $W \in T_{F(p)} M_2$, where $F_{*p}$ is the derivative of $F$ at $p \in M_1$.

We now define the second fundamental form of a map which will be useful throughout this book.

**Definition 23.** [20] Let $F : (M_1, g_1) \to (M_2, g_2)$ be a map between manifolds $(M_1, g_1)$

and $(M_2, g_2)$. The *second fundamental form* of $F$ is the map

$$\nabla F_* : \Gamma(TM_1) \times \Gamma(TM_1) \to \Gamma_F(TM_2)$$

defined by

$$(\nabla F_*)(X, Y) = \overset{2}{\nabla}^F_X F_* Y - F_*(\overset{1}{\nabla}_X Y) \tag{1.71}$$

where $\overset{1}{\nabla}$ is a linear connection on $M_1$.

It is easy to see that $\nabla F_*$ is linear in its arguments and the value of $(\nabla F_*)(X, Y)$ at $p_1 \in M_1$ only depends on the values of $X$ and $Y$ at $p_1 \in M_1$. Thus we can define $(\nabla F_*)(x, y)$ by

$$(\nabla F_*)(x, y) = (\nabla F_*)(X, Y)(p_1)$$

with $X, Y \in \Gamma(TM_1)$, $X(p_1) = x$ and $Y(p_1) = y$. Also using Theorem 8, we see that the second fundamental form is symmetric. On the other hand, for each fixed $x \in T_{p_1} M$, we can define a linear transformation

$$\nabla_x F_* : T_{p_1} M_1 \longrightarrow T_{F(p_1)} M_2$$

by

$$(\nabla_x F_*) y = (\nabla F_*)(X, Y)(p_1).$$

Moreover, for each $p_1 \in M_1$, the map

$$F_{*p_1} : (T_{p_1} M_1, g_{1_{p_1}}) \to (T_{F(p_1)} M_2, g_{2_{F(p_1)}})$$

is a linear transformation between inner product spaces $(T_{p_1} M_1, g_{1_{p_1}})$ and $(T_{F(p_1)} M_2, g_{2_{F(p_1)}})$.

**Definition 24.** [103] A smooth map $F : M \to N$ between Riemannian manifolds $M$ and $N$ is called *totally geodesic* if for each geodesic $\alpha$ in $M$ the image $F(\alpha)$ is a geodesic in $N$. Equivalently, $\nabla F_* = 0$.

Thus, by definition, every totally geodesic immersion is an example of a totally geodesic map. Also affine maps between Euclidean spaces are examples of totally geodesic maps.

**Remark 2.** Some authors use affine maps for totally geodesic maps in literature. However, an affine map is defined for a more general case. For a map $F : M \to N$ ($M$ and $N$ are not Riemannian manifolds) with linear connections, then $F$ is called an affine map if $(\nabla F_*) = 0$. In the case of symmetric connections, a map is affine if and only if it is totally geodesic [297]. It means that these two notions are same for Riemannian manifolds.

We now give the second fundamental form of a composite map.

**Lemma 6.** *[20] The second fundamental form of the composition of two maps $F_1 :$ $M_1 \to M_2$ and $F_2 : M_2 \to M_3$ is given by*

$$\nabla(F_2 \circ F_1)_* = F_{2*}(\nabla F_{1*}) + (\nabla F_{2*})(F_{1*}, F_{1*}). \tag{1.72}$$

*Proof.* We denote linear connections on $M_1$, $M_2$ and $M_3$ by $\overset{1}{\nabla}$, $\overset{2}{\nabla}$ and $\overset{3}{\nabla}$, respectively. We also denote the pullback connection the map $F_2 \circ F_1 : M_1 \to M_3$ by $\overset{F_2 \circ F_1}{\nabla^3}$. Then for $X, Y \in \chi(M_1)$, by direct calculations we have

$$(\nabla(F_2 \circ F_1)_*)(X, Y) = \overset{F_2 \circ F_1}{\nabla^3_X} (F_2 \circ F_1)_*(Y) - (F_2 \circ F_1)_*(\overset{1}{\nabla}_X Y).$$

Since $(F_2 \circ F_1)_* = F_{2*} \circ F_{1*}$, using the linearity of $F_*$, we get

$$
\begin{aligned}
(\nabla(F_2 \circ F_1)_*)(X, Y) &= \overset{F_2 \circ F_1}{\nabla^3_X} F_{2*}(F_{1*}(Y)) - F_{2*}(F_{1*}(\overset{1}{\nabla}_X Y)) \\
&= \overset{F_2 \circ F_1}{\nabla^3}_X F_{2*}(F_{1*}(Y)) - F_{2*}(\overset{2}{\nabla}^{F_1}_X F_{1*}(Y) \\
&\quad -(\nabla F_{1*})(X, Y)).
\end{aligned}
$$

Writing the second fundamental form of $F_1$, we derive

$$
\begin{aligned}
(\nabla(F_2 \circ F_1)_*)(X, Y) &= \overset{F_2 \circ F_1}{\nabla^3_X} F_{2*}(F_{1*}(Y)) - F_{2*}(\overset{2}{\nabla}^{F_1}_X F_{1*}(Y)) \\
&\quad +F_{2*}((\nabla F_{1*})(X, Y)).
\end{aligned}
$$

Now, using the second fundamental form of $F_2$, we obtain

$$
\begin{aligned}
(\nabla(F_2 \circ F_1)_*)(X, Y) &= \overset{3}{\nabla}_{(F_2 \circ F_1)_*(X)} F_{2*}(F_{1*}(Y)) \\
&\quad -F_{2*}(\overset{2}{\nabla}_{F_{1*}(X)} F_{1*}(Y)) + F_{2*}((\nabla F_{1*})(X, Y)) \\
&= \overset{3}{\nabla}_{(F_{2*}(F_{1*}X))} F_{2*}(F_{1*}(Y)) \\
&\quad -F_{2*}(\overset{2}{\nabla}_{F_{1*}(X)} F_{1*}(Y)) + F_{2*}((\nabla F_{1*})(X, Y)).
\end{aligned}
$$

Again using (1.69), we find

$$
\begin{aligned}
(\nabla(F_2 \circ F_1)_*)(X, Y) &= \overset{3}{\nabla}^{F_2}_{F_{1*}(X)} F_{2*}(F_{1*}(Y)) \\
&\quad -F_{2*}(\overset{2}{\nabla}_{F_{1*}(X)} F_{1*}(Y)) + F_{2*}((\nabla F_{1*})(X, Y)) \\
&= (\nabla F_{2*})(F_{1*}(X), F_{1*}(Y)) + F_{2*}(\overset{2}{\nabla}_{F_{1*}(X)} F_{1*}(Y)) \\
&\quad -F_{2*}(\overset{2}{\nabla}_{F_{1*}(X)} F_{1*}(Y)) + F_{2*}((\nabla F_{1*})(X, Y)) \\
&= (F_{2*}(\nabla F_{1*}) + (\nabla F_{2*})(F_{1*}, F_{1*}))(X, Y),
\end{aligned}
$$

where $\overset{2}{\nabla}^{F_1}$ and $\overset{3}{\nabla}^{F_2}$ denote the pullback connections of the maps $F_1$ and $F_2$, respectively. □

We now define harmonic and biharmonic maps, but first we need to define the notion of a tension field.

**Definition 25.** [20] Let $F : (M_1^m, g_1) \to (M_2^n, g_2)$ be a map and $\{e_1, ..., e_{n_1}\}$ an orthonormal basis for $TM_1$. Then the *tension field* $\tau(F)$ of $F$ is defined to be the trace of the second fundamental form of $F$ with respect to $g_1$, i.e.,

$$
\tau(F) = trace(\nabla F_*) = \sum_{i=1}^{m} (\nabla F_*)(e_i, e_i). \tag{1.73}
$$

The tension field of a map $F : (M_1^m, g_1) \to (M_2^n, g_2)$ is a vector field along $F$, i.e., $\tau(F) \in \Gamma_F(TM_2)$. By using (1.72), we have the following for the tension field.

**Corollary 1.** *[20] The tension field of the composition of two maps $F_1 : M_1 \to M_2$ and $F_2 : M_2 \to M_3$ is given by*

$$
\tau(F_2 \circ F_1) = F_{2*}(\tau(F_1)) + trace\,(\nabla F_{2*})(F_{1*}, F_{1*}). \tag{1.74}
$$

A map $F$ is called a *harmonic map* if it has a vanishing tension field, i.e., $\tau(F) = 0$. Geodesics, constant maps, and identity map are examples of harmonic maps. Since the second fundamental form of an isometric immersion is the mean curvature vector field of the immersed submanifold, we have the following.

**Corollary 2.** *An isometric immersion is harmonic if and only if it is minimal.*

If $F : (M_1, g) \longrightarrow (M_2, g_2)$ is a Riemannian submersion, then from Lemma 1 we get the following result.

**Corollary 3.** *A Riemannian submersion is harmonic if and only if its fibers are minimal submanifolds.*

We now give brief information for harmonic morphism taken from [20]. We first recall the tension field of weakly conformal map.

**Proposition 12.** *[20] Let $\varphi : M \to N$ be a weakly conformal map. Then, at a regular point, the normal component of the tension field is given by*

$$\tau(\varphi)^{\perp} = (dimM)\Lambda\mu^{M},$$

*where $\mu^{M}$ denotes the mean curvature of M in N.*

**Definition 26.** [20] Let $(M^{m}, g)$ and $(N^{n}, h)$ be Riemannian manifolds and $\varphi : M \to N$ be a fixed smooth map with a $M$ compact. Then the *stress-energy tensor field* of $\varphi$ is given by

$$S(\varphi) = e(\varphi) - \varphi^{*}h, \tag{1.75}$$

where the function $e(\varphi) : M \to [0, \infty)$ is the energy density function given by

$$e(\varphi)_{x} = \frac{1}{2} \parallel \varphi_{*x} \parallel^{2}$$

for $x \in M$.

The following lemma gives relation between the tension field and the stress-energy tensor field.

**Lemma 7.** *[20] Let $\varphi : (M, g) \to (N, h)$ be a smooth map. Then*

$$h(\tau(\varphi), \varphi_{*}) = -divS(\varphi).$$

From Lemma 7, we have the following result.

**Proposition 13.** *[20] Let $\varphi : (M^{m}, g) \to (N, h)$ be weakly conformal. Then the tangential component $\tau(\varphi)^{T}$ of the tension field vanishes at all regular points if and only if either $m = 2$, or $\varphi$ has constant conformality factor (i.e., it is constant or is a homothetic immersion).*

Proposition 12 and Proposition 13 give the following characterization for the harmonicity of weakly conformal maps.

**Proposition 14.** *[20] (i) A weakly conformal map from a Riemannian manifold of dimension 2 (or conformal surface) is harmonic if and only if its image is minimal at regular points. (ii) A weakly conformal map from a Riemannian manifold of dimension not equal to 2 is harmonic if and only if it is homothetic and its image is minimal.*

For $dimM = dimN$, the minimality of the image is obvious; from Proposition 14, we have the following result for maps between equidimensional manifolds.

**Corollary 4.** *[20] (i) A weakly conformal map $\varphi : M^2 \to N^2$ between Riemannian manifolds of dimension 2 is harmonic. (ii) A weakly conformal map $\varphi : M^m \to N^n$ between Riemannian manifolds of the same dimension $m \neq 2$ is harmonic if and only if it is homothetic.*

From Corollary 4 and Definition 8, we have the following proposition.

**Proposition 15.** *[20] A weakly conformal map $\varphi : M^2 \to N^2$ between Riemannian manifolds of dimension 2 and a harmonic map $\psi : N^2 \to P$ to an arbitrary Riemannian manifold is harmonic.*

We are now ready to state the definition of a harmonic morphism and give a characterization.

**Definition 27.** [20] Let $\varphi : M \to N$ be a smooth map between Riemannian manifolds. Then $\varphi$ is called a *harmonic morphism* if, for every harmonic function $f : V \to \mathbb{R}$ defined on an open subset $V$ of $N$ with $\varphi^{-1}(V)$ non-empty, the composition $f \circ \varphi$ is harmonic on $\varphi^{-1}(V)$.

It is clear that constant maps and isometries are harmonic morphisms. From Proposition 15, we have the following immediate result.

**Proposition 16.** *[20] Weakly conformal maps between two-dimensional Riemannian manifolds are harmonic morphisms.*

From Lemma 6 and Proposition 15 we have the following result.

**Lemma 8.** *[20] Let $M = (M, g)$ and $N = (N, h)$ be Riemannian manifolds. A harmonic horizontally weakly conformal map $\varphi : M \to N$ is a harmonic morphism.*

Finally, for a harmonic morphism, we give the following characterization obtained independently by Fuglede [120] and Ishihara [147].

**Theorem 9.** *[20] A smooth map $\varphi : M \rightarrow N$ between Riemannian manifolds is a harmonic morphism if and only if $\varphi$ is both harmonic and horizontally weakly conformal.*

We note that one part of the above theorem comes from Lemma 8. However, the other part needs jet theory; for this side proof, see [20].

In the rest of this section, we are going to present the theory of biharmonic maps. We note that materials given for biharmonic maps are taken mainly from [21], [208] and [290]. The notion of biharmonic map was first suggested by Eells and Sampson [103], see also [20]. The first variation formula and, thus, the Euler-Lagrange equation associated to the bienergy was obtained by Jiang in [150] and [151]. See also [189] and [291] for surveys of recent results.

**Definition 28.** [21] Let $F$ be a smooth map from a Riemannian manifold $(M_1, g_1)$ to a Riemannian manifold $(M_2, g_2)$ and $\mathcal{U}$ a compact domain of $M_1$. Then, the *bienergy* of $F$ over $\mathcal{U}$ is defined by

$$E_{2,\mathcal{U}}(F) = \frac{1}{2} \int_{\mathcal{U}} \| \tau(F) \|^2 v_{g_1}. \tag{1.76}$$

A smooth map $F$ is said to be *biharmonic* if it is a critical point of the bienergy functional $E_{2,\mathcal{U}}(F)$.

**Proposition 17.** *[151], [21], [290]. Let $F : (M_1, g_1) \longrightarrow (M_2, g_2)$ be a smooth map and for $I = (-\varepsilon, \varepsilon)$, let $\{F_t\}_{t \in I}$ be a smooth variation of $F$ supported in $\Omega$, where $\Omega \subset M$ is compact. Then*

$$\frac{d}{dt} |_{t=0} E_{2,\mathcal{U}}(F_t) = \int_{\Omega} < \tau_2(F), V > v_{g_1}, \tag{1.77}$$

*where $V$ denotes the variation vector field of $\{F_t\}_{t \in I}$ and*

$$\tau_2(F) = -\Delta^F \tau(F) - trace R^{M_2}(F_*, \tau(F))F_* \tag{1.78}$$

*is the bi-tension field of $F$.*

*Proof.* Let $\{F_t\}_{t \in I}$ be a smooth variation of $F$. Then we have a smooth map $\Phi : I \times M \longrightarrow N$ satisfying

$$\Phi(t, p) = F_t(p), \forall(t, p) \in I \times M$$
$$\Phi(0, p) = F(p), \forall p \in M. \tag{1.79}$$

The variation vector field $V \in \Gamma(F^{-1}(TM_2))$ associated to the variation $\{F_t\}_{t \in I}$ is given by

$$V(p) = \frac{d}{dt}\Big|_{t=0} F_t(p) = d\Phi(0, p)(\frac{\partial}{\partial t}) \in T_{F(p)}N.$$

Hence we have

$$\begin{aligned}
\frac{d}{dt}\Big|_{t=0} E_{2,\mathcal{U}}(F_t) &= \frac{1}{2}\int_\Omega \frac{\partial}{\partial t} <\tau(F_t), \tau(F_t)>\Big|_{t=0} v_{g_1} \\
&= \int_\Omega <\nabla^\Phi_{\frac{\partial}{\partial t}}\tau(F_t), \tau(F_t)>\Big|_{t=0} v_{g_1}.
\end{aligned}$$

Let now $E^m_{i=1}$ be a local orthonormal frame field geodesic at $p \in \Omega$. Then with respect to $E^m_{i=1}$, we have

$$\begin{aligned}
\nabla^\Phi_{\frac{\partial}{\partial t}}\tau(F_t) &= \nabla^\Phi_{\frac{\partial}{\partial t}}\sum_{i=1}^m (\nabla\Phi_*)(E_i, E_i) \\
&= \sum_{i=1}^m \nabla^\Phi_{\frac{\partial}{\partial t}}\nabla^\Phi_{E_i}\Phi_*(E_i) - \nabla^\Phi_{\frac{\partial}{\partial t}}\Phi_*(\nabla_{E_i}E_i).
\end{aligned}$$

Since $[Z, \frac{\partial}{\partial t}] = 0$ for any $Z \in \Gamma(TM_1)$, we arrive at

$$\nabla^\Phi_{\frac{\partial}{\partial t}}\tau(F_t) = \sum_{i=1}^m \nabla^\Phi_{\frac{\partial}{\partial t}}\nabla^\Phi_{E_i}\Phi_*(E_i) - \nabla^\Phi_{\nabla_{E_i}E_i}\Phi_*(\frac{\partial}{\partial t}).$$

at $p \in \Omega$. Then the geodesic frame implies that

$$\nabla^\Phi_{\frac{\partial}{\partial t}}\tau(F_t) = \sum_{i=1}^m \nabla^\Phi_{\frac{\partial}{\partial t}}\nabla^\Phi_{E_i}\Phi_*(E_i).$$

Thus by the curvature tensor field, we get

$$\begin{aligned}
\nabla^\Phi_{\frac{\partial}{\partial t}}\tau(F_t) &= \sum_{i=1}^m \nabla^\Phi_{E_i}\nabla^\Phi_{\frac{\partial}{\partial t}}\Phi_*(E_i) + \nabla^\Phi_{[\frac{\partial}{\partial t},E_i]}\Phi_*(E_i) \\
&\quad + R^\Phi(\frac{\partial}{\partial t}, E_i)\Phi_*(E_i).
\end{aligned}$$

Then $[Z, \frac{\partial}{\partial t}] = 0$ and Proposition 11 imply that

$$\nabla^\Phi_{\frac{\partial}{\partial t}}\tau(F_t) = \sum_{i=1}^m \nabla^\Phi_{E_i}\nabla^\Phi_{\frac{\partial}{\partial t}}\Phi_*(E_i) + R^{M_2}(\Phi_*(\frac{\partial}{\partial t}), \Phi_*(E_i))\Phi_*(E_i).$$

Denote by $v$ the outward pointing unit normal of $\partial\Omega$ in $M$ and by $i : \partial\Omega \longrightarrow M$ the canonical inclusion. By using the symmetries of the Riemann-Christoffel tensor field and the divergence theorem

$$\frac{d}{dt}\Big|_{t=0} E_{2,\mathcal{U}}(F_t) = \int_\Omega < trace\nabla^2 V, \tau(F) > - < V, traceR^{M_2}(F_*, \tau(F))F_* > v_{g_1}.$$

Then the proof comes from the formula $g_2(\Delta^F V, W) = g_2(V, \Delta^F W)$. $\qquad\square$

**Theorem 10.** *[151], [21] A map $F : (M_1, g_1) \longrightarrow (M_2, g_2)$ is biharmonic if and only if $\tau_2(F) = 0$.*

We note that the harmonicity of a map is also given by a variation formula. But this is very well known in the literature, therefore we have omitted it. It is clear that every harmonic map is biharmonic map. Therefore we give the following definition.

**Definition 29.** [151], [21], [150] A map between two Riemannian manifolds is said to be *proper biharmonic* if it is a non-harmonic biharmonic map.

We now give some characterizations for biharmonic maps. First of all, we have the following result.

**Corollary 5.** *[21] Let $(M, g)$ be a Riemannian manifold and $\alpha : I \longrightarrow M$ a geodesic. Let $t : J \longrightarrow I$ be a change of parameter and consider $\beta = \alpha \circ t : J \longrightarrow M$ to be a geodesic, then $\tilde{\alpha} = \alpha \circ t$ is a proper biharmonic map if and only if $\frac{d^4 t}{ds^4} = 0$ ve $\frac{d^2 t}{ds^2} \neq 0$.*

**Corollary 6.** *[21] Let $(M, g)$ be a Riemannian manifold and $\alpha : I \longrightarrow M$ a curve parametrized by arc length from an open interval $I \subset \mathbb{R}$. In this case $T = \dot{\alpha}$ and $\tau(\alpha) = \nabla_T T$, where $\nabla$ is the Levi-Civita connection on M. Thus $\alpha$ is a biharmonic map if and only if*

$$\nabla_T^3 T - R(T, \nabla_T T)T = 0.$$

The rest of this section is devoted to the biharmonicity of submanifolds.

**Theorem 11.** *[21], [208] Let $M_1^{m_1}$ be an immersed submanifold of a spaceform $(M_2(c), g_2)$ Then the canonical inclusion $F : M_1^{m_1} \longrightarrow M_2^{m_2}(c)$ is biharmonic if and only if*

$$-\bar{\Delta}^\perp H - izh(., A_{H.}) + m_1 c H = 0 \qquad (1.80)$$

*and*

$$2iz A_{\nabla_{()}^\perp H}(.) + \frac{m_1}{2}\nabla(\| H \|^2) = 0, \qquad (1.81)$$

*where $\bar{\Delta}^\perp$ is the Laplacian on normal bundle and $\nabla(\| H \|^2)$, denotes the gradient of*

$(\parallel H \parallel^2)$ *on* $M_1^{m_1}$.

*Proof.* Let $M_2(c)$ be a space form with sectional curvature $c$ and $F : M_1^{m_1} \longrightarrow M_2(c)$ the canonical inclusion of a submanifold $M_1^{m_1}$. Then from (1.33) we have

$$iz\overset{2}{R}(F_*, \tau(F))F_* = -m_1 c\tau(F) = -m_1^2 cH,$$

hence we derive

$$\tau_2(F) = -m_1(\bar{\Delta}H - m_1 cH).$$

Thus submanifold $M_1$ is biharmonic if and only if

$$\bar{\Delta}H - m_1 cH = 0. \tag{1.82}$$

Consider now $\{e_i\}_{i=1}^{m_1}$ to be a local orthonormal frame field on $M_1$, geodesic at $p \in M_1$. Thus at $p \in M_1$, we find

$$\bar{\Delta}H = -\sum^{m_1} \overset{2}{\nabla}_{e_i} \overset{2}{\nabla}_{e_i} H.$$

Here, using (1.31), we get

$$\bar{\Delta}H = -\sum^{m_1} \overset{2}{\nabla}_{e_i}(-A_H e_i + \nabla_{e_i}^{\perp} H).$$

Then Equations (1.30) and (1.31) imply that

$$\bar{\Delta}H = -\sum^{m_1}\{-\overset{1}{\nabla}_{e_i} A_H e_i + h(e_i, A_H e_i) - A_{\nabla_{e_i}^{\perp} H} e_i + \nabla_{e_i}^{\perp} \nabla_{e_i}^{\perp} H\}.$$

Thus we arrive at

$$\begin{aligned} \bar{\Delta}H &= -\bar{\Delta}^{\perp}H + iz\overset{1}{\nabla}_{(.)} A_H(.) - izh(., A_H(.)) \\ &\quad + izA_{\nabla_{(.)}^{\perp} H}(.). \end{aligned} \tag{1.83}$$

On the other hand, by direct computations we derive

$$\sum_{i=1}^{m_1} \overset{1}{\nabla}_{e_i} A_H e_i = \sum_{i,j=1}^{m_1} e_i g_1(A_H e_i, e_j) e_j,$$

where we have used that $\overset{1}{\nabla}$ is a metric connection and $\{e_j\}_{j=1}^{m_1}$ is a geodesic at $p \in M_1$. Using (1.32), we have

$$\sum_{i=1}^{m_1} \overset{1}{\nabla}_{e_i} A_H e_i = \sum_{i,j=1}^{m_1} e_i g_2(h(e_i, e_j), H) e_j.$$

Again using (1.30) we obtain

$$\sum_{i=1}^{m_1} \overset{1}{\nabla}_{e_i} A_H e_i = \sum_{i,j=1}^{m_1} \{g_2(\overset{2}{\nabla}_{e_i}\overset{2}{\nabla}_{e_j}e_i, H) + g_2(\overset{2}{\nabla}_{e_j}e_i, \overset{2}{\nabla}_{e_i}H)\}e_j,$$

and hence we get

$$\sum_{i=1}^{m_1} \overset{1}{\nabla}_{e_i} A_H e_i = \sum_{i,j=1}^{m_1} \{g_2(\overset{2}{\nabla}_{e_i}\overset{2}{\nabla}_{e_j}e_i, H) + g_2(h(e_i, e_j), \nabla^{\perp}_{e_i}H)\}e_j.$$

Then (1.32) implies that

$$\sum_{i=1}^{m_1} \overset{1}{\nabla}_{e_i} A_H e_i = \sum_{i,j=1}^{m_1} g_2(\overset{2}{\nabla}_{e_i}\overset{2}{\nabla}_{e_j}e_i, H)e_j + \sum_{i=1}^{m_1} A_{\nabla^{\perp}_{e_i}H}e_i. \tag{1.84}$$

Considering (1.84) in the curvature tensor field, we have

$$\sum_{i,j=1}^{m_1} g_2(\overset{2}{\nabla}_{e_i}\overset{2}{\nabla}_{e_i}e_j, H) = \sum_{i,j=1}^{m_1} g_2(\overset{2}{R}(e_i, e_j)e_i + \overset{2}{\nabla}_{e_j}\overset{2}{\nabla}_{e_i}e_i + \overset{2}{\nabla}_{[e_i,e_j]}e_i, H).$$

Thus using (1.30), at $p \in M_1$, we find

$$\sum_{i,j=1}^{m_1} g_2(\overset{2}{\nabla}_{e_i}\overset{2}{\nabla}_{e_i}e_j, H) = -m_1 c g_2(e_j, H) + \sum_{i,j=1}^{m_1} g_2(\overset{2}{\nabla}_{e_j}h(e_i, e_i), H).$$

Hence we get

$$\sum_{i,j=1}^{m_1} g_2(\overset{2}{\nabla}_{e_i}\overset{2}{\nabla}_{e_i}e_j, H) = \frac{m_1}{2}e_j(\| H \|^2). \tag{1.85}$$

Putting (1.85) in (1.84) we arrive at

$$\sum_{i=1}^{m_1} \overset{1}{\nabla}_{e_i} A_H e_i = \sum_{j=1}^{m_1} \frac{m_1}{2}e_j(\| H \|^2)e_j + \sum_{i=1}^{m_1} A_{\nabla^{\perp}_{e_i}H}e_i. \tag{1.86}$$

Writing (1.86)in (1.83) we find

$$\bar{\Delta}H = -\bar{\Delta}^{\perp}H - izh(., A_H(.)) + 2izA_{\nabla^{\perp}_{()}H}(.) + \frac{m_1}{2}\nabla(\| H \|^2). \tag{1.87}$$

Also writing (1.87) in (1.81), we have

$$-\bar{\Delta}^{\perp}H - izh(., A_H(.)) + m_1 cH = 2izA_{\nabla^{\perp}_{()}H}(.) + \frac{m_1}{2}\nabla(\| H \|^2).$$

Finally, considering the tangential parts and normal parts of above equation we obtain assertions.                                                             □

In this particular case, we have the following result.

**Corollary 7.** *[21] Let M be a hypersurface of a space form $(M_2^{m_2}(c), g_2)$. Then M is biharmonic if and only if the following conditions are satisfied.*

$$-\bar{\Delta}^{\perp}H - (m_1 c - \| A \|^2)H + m_1 cH = 0,$$

$$2A(\nabla(\| H \|)) + m_1 \| H \| \nabla(\| H \|) = 0.$$

**Remark 3.** We note that Eells and Lemaire [102] suggested the notion of $k$-harmonic maps and Euler-Lagrange equations for $k$-harmonic maps were obtained by Wang [298]. We also note that $k$-harmonic submanifolds were defined and studied by Maeta in [181].

# CHAPTER 2

# Applications of Riemannian Submersions

## Contents

## Abstract

In this chapter, we give two applications of Riemannian submersions and it consists of two sections. In the first section, we give an application of Riemannian submersions in robotic theory [12]. To this aim, we first define Lie groups and give basic properties of such manifolds, then we give applications of Riemannian submersions in redundant robotic theory. Applications of Riemannian submersions in Kaluza-Klein theory have been given in detail in the book [111] and papers [112] and [40]; therefore, in the second section of this chapter, we give brief information for the Kaluza-Klein theory in terms of principal fiber bundles and Riemannian submersions.

**Keywords:** Lie group, matrix group, homogeneous space, rigid body transformation, forward kinematic map, gauge group, Hessian, divergence, Laplacian, vector bundle, pullback principal fiber bundle, Kaluza-Klein theory, connection, Yang-Mills theory

*Anyone who has never made a mistake has never tried anything new.*

*Albert Einstein*

## 1. Applications of Riemannian submersions in robotic theory

Riemannian submersions have many applications in different research areas. Indeed, Riemannian submersions have their applications in Kaluza-Klein theory [111], [40], statistical machine learning processes [317], medical imaging [186], statistical analysis on manifolds [32] and the theory of robotics [12]. Of course it is beyond the scope of this book to mention all applications of Riemannian submersions, but we will focus on applications of Riemannian submersions in redundant robotic chains given in

Reimannian Submersions, Reimannian Maps in Hermitian Geometry, and their Applications
http://dx.doi.org/10.1016/B978-0-12-804391-2.50002-6

[12] and Kaluza-Klein theory. In this section, we give an application of Riemannian submersions in robotic theory. For this aim, we first give a brief review of Lie groups. and recall some basic information from robotic theory.

## 1.1. A geometric introduction to $\mathbf{SE}(3)$

In this subsection, we recall some geometric terms which will be used in the next subsection. We start by recalling the notion of Lie Group. A *Lie group G* is a set which is both a group and a differentiable manifold such that the group operations

$$
\begin{aligned}
\bullet : \quad G \times G \quad &\longrightarrow \quad G \\
(x, y) \quad & \qquad xy
\end{aligned}
$$

and

$$
\begin{aligned}
\star : \quad G \quad &\longrightarrow \quad G \\
x \quad & \qquad x^{-1}
\end{aligned}
$$

are differentiable. The identity element of the group is denoted by $e$. The dimension of Lie group $G$ is the dimension of $G$ as a manifold. A simple example of a Lie group is the $n$−dimensional vector space $\mathbb{R}^n$ with additive operation of its vectors. We now list certain Lie groups and their Lie subgroups which are used in this section.

### The General Lie Group $GL(n, \mathbb{R})$

The general linear group $GL(n, \mathbb{R})$ is the set of non-singular matrices of dimension $n^2$:

$$
Gl(n, \mathbb{R}) = \{M \mid M \in \mathbb{R}^{n \times n}, det(M) \neq 0\}.
$$

The group manifold of $GL(n, \mathbb{R})$ lies in $\mathbb{R}^{n^2}$, where the coordinates of $\mathbb{R}^{n^2}$ are given by the $n^2$ entries $(a_{ij})$. It cannot be all of $\mathbb{R}^{n^2}$, because we must exclude the matrices with zero determinant.

Let $GL^+(n)$ denote the set of all $n \times n$ real matrices with positive determinant:

$$
GL^+(n) = \{M \mid M \in \mathbb{R}^{n \times n}, det(M) > 0\}.
$$

### The Orthogonal Group $O(n)$

This group is defined by

$$
O(n) = \{M \mid M \in \mathbb{R}^{n \times n}, MM^T = I\}.
$$

Since $det(M^T) = det(M)$, we have that $det(M)^2 = 1$ and hence $det(M) = \pm 1$. The manifold for this group consists of two disconnected components. One part consists of those matrices that have $det(M) = +1$; these include the identity element $I_n$ and can be thought of as rotations about the origin. The other part consists of those orthogonal matrices with $det(M) = -1$; these are usually thought of as reflections. In two dimensions, 0(2) consists of matrices of the form

$$\begin{pmatrix} \cos\theta & -\sin\theta \\ \sin\theta & \cos\theta \end{pmatrix}, \begin{pmatrix} \cos\theta & \sin\theta \\ \sin\theta & -\cos\theta \end{pmatrix}.$$

The first of these correspond to an anticlockwise rotation by 0 about the origin, while the second are reflections in a line through the origin at an angle of $\frac{\theta}{2}$ from the first axis.

### The Special Orthogonal Group $SO(n)$

This is the group of determinant 1 orthogonal matrices. According to the remarks above, it is the group of rotations about the origin in $n$-dimensional space. For the scope of this chapter, it is $SO(2)$ and $S0(3)$ that will be most important, since these are the rigid body rotations about a fixed center in two and three dimensions. The group $SO(2)$ consists of matrices of the form

$$\begin{pmatrix} \cos\theta & -\sin\theta \\ \sin\theta & \cos\theta \end{pmatrix}.$$

The group manifold is a circle.

### General linear group $GL(n, \mathbb{C})$

This group consists of $n \times n$ non-singular matrices of complex elements and real dim. $2n^2$.

### Unitary Group $U(n)$

This group is defined by $U(n) = \{A \in GL(n, \mathbb{C}) : A^{-1} = \bar{A}^t\}$ with real dim. $n^2$. Identifying $S^1 = \{(x, y) \in \mathbb{R}^2 \mid x^2 + y^2 = 1\}$ with the unit circle

$$U(1) = \{z \in \mathbb{C} : \mid z \mid = 1\}$$

in $\mathbb{C}$, $S^1$ also inherits a group structure, given by

$$(x_1, y_1)(x_2, y_2) := (x_1 x_2 - y_1 y_2, x_1 y_2 + x_2 y_1), (x_1, y_1)^{-1} = (x_1, -y_1).$$

We note that there is a one-to-one correspondence between complex numbers and certain real $2 \times 2$ real matrices given by

$$z = x + iy \in \mathbb{C} = \mathbb{R}^2 \leftrightarrow A = \begin{pmatrix} x & -y \\ y & x \end{pmatrix} \in M_2\mathbb{R}$$

and the ordinary scalar product on $\mathbb{R}^2$ defined by $< z_1, z_2 > = x_1 x_2 + y_1 y_2$, for $z_1(x_1, y_1)$, $z_2 = (x_2, y_2)$, can also be expressed as $\frac{1}{2}(z_1 \bar{z}_2 + \bar{z}_1 z_2)$, which corresponds to the scalar product $< A_1, A_2 > = \frac{1}{2} \text{trace}(A_1 A_2^T)$. The length of the complex number $\mid z \mid$ becomes $\mid A \mid^2 = detA$. Moreover, every element (matrix) of $U(1)$ may be written in the form $A = A(\alpha) \equiv e^{i\alpha} = \cos\alpha + i\sin\alpha, \alpha \in \mathbb{R}$. Hence it is easy to see that there are group isomorphisms (see [16] for details)

$$U(1) \cong S^1 \cong SO(2).$$

We note that $U(1)$ is called the *circle group*.

Consider a group $G$ and a manifold $X$. We say that $G$ *acts* on $X$ if there is a differentiable map

$$a : G \times X \longrightarrow X$$

that satisfies

$$a(e, x) = x, \forall x \in X \tag{2.1}$$

and

$$a(g_1, a(g_2 x)) = a(g_1 g_2, x), \forall x \in X, \forall g_1, g_2 \in G. \tag{2.2}$$

Equations (2.1) and (2.2) imply that $a$ is a diffeomorphism. Linear actions of groups on vector spaces are called *representations* that is, actions which satisfy

$$g(h_1 + h_2) = g(h_1) + g(h_2) \text{ for all } g \in G \text{ and all } h_1, h_2 \in H. \tag{2.3}$$

Suppose we have a group $G$ and a commutative group $H$ together with a linear action of $G$ on $H$. That is, a map $G \times H \longrightarrow H$ given by $g(h)$ satisfying Equations (2.1), (2.2) and (2.3). The *semi-direct product* of $G$ and $H$, written $G \ltimes H$, has the same elements as the direct product. But the product of two elements is defined as

$$(g_1, h_1)(g_2, h_2) = (g_1 g_2, h_1 + g_1(h_2)) \tag{2.4}$$

This is a group. The identity element in such a group is $(e, e)$, and the inverse of an element $(g, h)$ is given by

$$(g, h)^{-1} = (g^{-1}, -g^{-1}(h)).$$

### The Proper Euclidean Group

The group of proper rigid body transformations in $\mathbb{R}^n$ is the semi-direct product of the special orthogonal group with $\mathbb{R}^n$ itself. We will denote it by $SE(n)$ for the special Euclidean group:

$$SE(n) = SO(N) \ltimes \mathbb{R}^n.$$

There is an $(n + 1)$–dimensional representation of $SE(n)$ given by

$$(R, t) \rightarrow \begin{pmatrix} R & t \\ 0 & 1 \end{pmatrix},$$

where $R \in SO(n), t \in \mathbb{R}^n$. It is known that all proper rigid body motions in three-dimensional space, with the exception of pure translations, are equivalent to a screw motion, that is, a rotation about a line together with a translation along the line.

Lie groups are non-linear objects and their study requires quite a lot of effort. But it is possible to associate to every point of a Lie group $G$ a real vector space, which is the tangent space of the Lie group at that point. We first recall the notion of Lie algebra.

**Definition 30.** [173] A *Lie algebra* over $\mathbb{R}$ ( or more generally a field $F$) is a real vector space $\mathfrak{g}$ with a bilinear map

$$[,] : \mathfrak{g} \times \mathfrak{g} \to \mathfrak{g} ( \quad \text{called} \quad \text{the} \quad \text{bracket}),$$

such that:
**(i)** $[X, Y] = -[Y, X]$, and
**(ii)** $[[X, Y], Z] + [[Y, Z], X] + [[Z, X], Y] = 0$ (Jacoby identity) for all $X, Y, Z \in \mathfrak{g}$.

Let $M$ be an $m$−dimensional manifold. Then the vector space $\chi(M)$ of all vector fields on $M$ is a Lie algebra with respect to the Lie bracket (see: section 1 of 1) of vector fields. The set $GL(n, \mathbb{R})$ is a real vector space of dimension $n^2$. If we set

$$[A, B] = AB - BA,$$

then $GL(n, \mathbb{R})$ becomes a Lie algebra.

Let $a$ be an element of a Lie group $G$. We define the maps

$$L_a : G \to G, \ L_a(g) = ag \ (\text{left} \quad \text{translation}),$$

$$R_a : G \to G, \ R_a(g) = ga \ (\text{right} \quad \text{translation}).$$

These maps are smooth; in fact, they are diffeomorphisms. A vector field $X$ on a Lie group $G$ is left-invariant if $X \circ L_a = dL_a(X)$ for all $a \in G$. A *left-invariant vector field* is determined by its value at the identity element $e$ of the Lie group, since $X_a = dL_a(X_e)$ for all $a \in G$. Let $\mathfrak{g}$ denote the set of all left-invariant vector fields on a Lie group $G$. The usual addition of vector fields and scalar multiplication by real numbers make $\mathfrak{g}$ a vector space. Furthermore, $\mathfrak{g}$ is closed under the bracket operation on vector fields. Thus $\mathfrak{g}$ is a Lie algebra, called the Lie algebra of $G$. The Lie algebras of $SO(3)$ and $SE(3)$ are denoted by $\mathfrak{so}(3)$ and $\mathfrak{se}(3)$. In kinematics, $\mathfrak{se}(3)$ corresponds to the space of twists.

A Riemannian metric on a Lie group $G$ is called *left-invariant* if

$$g(u, v) = g((dL_a)_x u, (dL_a)_x v)_{L_a(x)}$$

for all $a, x \in G$ and $u, v \in T_x G$. Similarly, a Riemannian metric is right-invariant if each $R_a$ is an isometry. It is very well known that there is a one-to-one correspondence between left-invariant metrics on a Lie group $G$, and scalar products on its Lie algebra $\mathfrak{g}$. Moreover, a metric on $G$ that is both left-invariant and right-invariant is called *bi-invariant. One-parameter subgroup* of a Lie group $G$ is a smooth homomorphism $\phi : (\mathbb{R}, +) \to G$. Thus $\phi : (\mathbb{R}, +) \to G$ is a curve such that $\phi(s + t) = \phi(s)\phi(t)$, $\phi(0) = e$, and $\phi(-t) = \phi(t)^{-1}$. For each $X \in \mathfrak{g}$, there exists a unique one-parameter subgroup $\phi_X : R \to G$ such that $\phi'(0) = X$. We now recall the definition of the *exponential map*. The exponential map $exp : \mathfrak{g} \to G$ is defined by $exp(X) = \phi_X(1)$, where $\phi_X$ is the unique one-parameter subgroup of $X$. If we take $G = GL(n, \mathbb{R})$ with $\mathfrak{g} = M_n\mathbb{R}$, the term "exponential map" will be justified, since it will coincide with the usual exponential map for matrices.

Like left-invariant metrics on a Lie group, there is a one-to-one correspondence between bi-invariant metrics on $G$ and special scalar products on $\mathfrak{g}$. But we first need the following notion. Let $G$ be a Lie group. The map $I_g : G \to G$. $g \in G$ given by $I_g(h) = ghg^{-1}$ is called the *conjugation map*. The tangent map of the conjugation map at identity $e$, $Ad_g = d_e I_g$ is called the *adjoint action* o f $G$ on $\mathfrak{g}$, its Lie algebra. The *adjoint representation* of $g$ is the homomorphism *ad* given by $ad(X) = (dAd)_e(X)$. If $G$ is a subgroup of $GL(n, \mathbb{R})$ (like $SO(3)$ and $SE(3)$), then $Ad_g S = gS g^{-1}$ for all $g \in G$) and $S \in \mathfrak{g}$ (the multiplication being multiplication of matrices). A consequence of this notion is the following important theorem.

**Theorem 12.** *[16] The adjoint representation of $\mathfrak{g}$ satisfies $ad(X)Y = [X, Y]$ for all* $X, Y \in \mathfrak{g}$.

A Lie algebra that satisfies the property $[X, Y] = 0$, for all $X, Y \in \mathfrak{g}$, is called *Abelian*. We now state the following result from [16]: There is a one-to-one correspondence between bi-invariant metrics on $G$ and Ad-invariant scalar products on $\mathfrak{g}$, that is,

$$(Ad(g)X, Ad(g)Y) = (X, Y)$$

for all $g \in G$, $X, Y \in \mathfrak{g}$. It is known that every compact Lie group admits a bi-invariant metric. In the scope of this chapter, we are interested in the geometry of the group $SE(3)$, therefore we note the following.

**Theorem 13.** *[193] There does not exist a bi-invariant metric (positive definite) on* $SE(3)$.

The fact that a bi-invariant metric does not exist on $SE(3)$ does not mean that we cannot define a notion of length on $SE(3)$. Rather, it implies that the definition of a metric is not intrinsic. It involves a choice. Therefore, we restrict our discussion to the

Lie group $SO(3)$ of rotations in $\mathbb{R}^3$. We first note that, for a given curve

$$A(t) : [-a, a] \to SE(3), A(t) = \begin{bmatrix} R(t) & d(t) \\ 0 & 1 \end{bmatrix}$$

an element $S(t)$ of the Lie algebra $\mathfrak{se}(3)$ can be identified with the tangent vector $\dot{A}(t)$ at an arbitrary point $t$ by

$$S(t) = A^{-1}(t)\dot{A}(t) = \begin{bmatrix} \hat{\omega}(t) & R^T \dot{d} \\ 0 & 0 \end{bmatrix},$$

where $\hat{\omega}(t) = R(t)^T \dot{R}(t)$ is the corresponding element from $\mathfrak{so}(3)$. A curve on $SE(3)$ physically represents a motion of the rigid body. We now note the following result which describes the chosen metric for $SO(3)$.

**Theorem 14.** *[29] Let W be the matrix representation of a quadratic form $<,>$ defined at identity of $SO(3)$ and extended through left invariance throughout the manifold. Then the quadratic is bi-invariant if and only if W has the form: $W = \alpha I$.*

If $\alpha$ is positive, the quadratic form becomes a metric and the following is true:

**Theorem 15.** *[29] Let W be the matrix representation of a metric $<,>$ defined at identity of $SO(3)$ and extended through left invariance throughout the manifold. Then the metric is bi-invariant if and only if W has the form: $W = \alpha I, \alpha > 0$.*

We now recall some basic materials for homogeneous spaces. First of all, we recall the following proposition.

**Proposition 18.** *[16] Let G be a Lie group, and K a closed subgroup of G. Then there is a unique way to make $G/K$, the set $\{gK, g \in G\}$ of left cosets modulo K, a manifold so that the projection $\pi : G \to G/K$ is a submersion.*

On the other hand, an action ( see (2.1) and (2.2)) is called *transitive* if for any $m, n \in M$ there exists a $g \in G$ such that $g.m = n$. A *homogeneous space* is a manifold $M$ with a transitive action of a Lie group $G$. Equivalently it is a manifold of the form $G/K$, where $G$ is a Lie group and $K$ a closed subgroup of $G$.

**Definition 31.** [31] A Riemannian manifold $(M, g)$ is *homogeneous* if the group of isometries $I(M, g)$ acts transitively.

In fact, from Myers-Steenrod's theorem [194], the group $I(M, g)$ is a Lie group and if a closed group $G$ of $I(M, G)$ acts transitively, then $M$ is called $G-$ *homogeneous*. Let $(M, g)$ be a Riemannian manifold and $G$ a closed subgroup of the isometry group of $(M, g)$. By assuming $\pi$ as a submersion, there exists one and only one Riemannian metric $\hat{g}$ on $M/G$ such that $\pi$ is a Riemannian submersion. A Riemannian homogeneous space $M$ is diffeomorphic to a homogeneous space $G/K$, where $G = I(M)$ and $K$ is the isotropy subgroup $\{g \in G : g.m = m\}$ at $m \in M$. This means that any homogeneous Riemannian manifold can be written as a coset space of a connected Lie group with an invariant Riemannian metric. Let $\mathfrak{g}$ and $\mathfrak{k}$ be, respectively, the Lie algebras of $G$ and $K$. Denote the adjoint representation of $G$ on $\mathfrak{g}$ by $Ad$. For $k \in K$, $Ad(k)$ keeps the subalgebra $\mathfrak{k}$ invariant. $Ad(k)$ induces a linear map on the quotient space $\mathfrak{g}/\mathfrak{k}$. Denote this map by $Ad_{\mathfrak{g}/\mathfrak{k}}(k)$. Then there is a one-to-one correspondence between the $G-$invariant Riemannian metrics on $G/K$ and the $Ad_{\mathfrak{g}/\mathfrak{k}}(K)$ invariant inner products on $\mathfrak{g}/\mathfrak{k}$ [92]. We list some well-known examples of homogeneous space.

- A Lie group is a homogeneous space. In this case, for instance, $G \times G$ acts on $G$ by left and right translations.
- The sphere $S^n$ may be viewed as the homogeneous manifold $SO(n + 1)/SO(n)$.
- Consider $\mathbb{R}^n$ as a Riemannian manifold with the Euclidean metric. Then $E(n)$, the set of $(n + 1) \times (n + 1)$ real matrices of the form

$$\begin{bmatrix} A & b \\ 0 & 1 \end{bmatrix}, A \in O(n), b \in \mathbb{R}$$

is a closed Lie subgroup of $GL(n + 1, \mathbb{R})$. Define a map $E(n) \to \mathbb{R}^n \times \mathbb{R}^n$ by identifying $\mathbb{R}^n$ with the subset $S = \{(x, 1) \in \mathbb{R}^{n+1} : x \in \mathbb{R}^n\}$ and restricting the linear action of $E(n)$ on $\mathbb{R}^{n+1}$ to $S$. Then we can obtain that this is a smooth action of $E(n)$ on $\mathbb{R}^n$ by isometries of the Euclidean metric and $E(n)$ acts transitively on $\mathbb{R}^n$. Thus Euclidean space is a homogeneous space.

We also recall that a homogeneous space $G/K$ is called *reductive* if $\mathfrak{g}$ admits a decomposition $\mathfrak{g} = \mathfrak{k} \oplus \mathfrak{m}$ such that $Ad_K(\mathfrak{m}) \subset \mathfrak{m}$. The above definition is not very restrictive. Indeed, we quote the following result.

**Proposition 19.** *[165], [166] Any homogeneous Riemannian manifold $G/K$ is a reductive homogeneous space in the sense that there exists a direct sum decomposition (reductive decomposition) of the form*

$$\mathfrak{g} = \mathfrak{k} + \mathfrak{m},$$

*where $\mathfrak{m} \subset \mathfrak{g}$ is a vector subspace such that $Ad(K)(\mathfrak{m}) \subset \mathfrak{m}$.*

For the geometry of Lie groups and applications of such groups, see [16], [29], [31], [63], [92], [113], [142], [161], [162], [170], and [264].

## 1.2. Forward kinematics map as an example of Riemannian submersions

In this subsection, we are going to give an application of Riemannian submersions in redundant robotic theory. We first recall some notions from robotic theory from [43], [193], and [272]. A particle is an object with mass concentrated at a point. A rigid body is an object with mass and volume. A free mechanical system is a collection $P_1, ..., P_N$ of particles and $B_1, ..., B_N$ of rigid bodies which move independently of one another. To specify the location of a particle, choose an inertial reference frame $(O_{spatial}, \{s_1, s_2, s_3\})$ consisting of a spatial origin $O_{spatial}$ and an orthonormal frame $\{s_1, s_2, s_3\}$ at $O_{spatial}$. The position of the particle $P_j$ is exactly determined by a vector $r_j \in R^3$ from $O_{spatial}$ to the location of $P_j$. The configuration of a free mechanical system is specified by a point in

$$\underbrace{R^3 \times ... R^3}_{N_P copies} \times \underbrace{(SO(3) \times R^3) \times ...(SO(3) \times R^3)}_{N_B copies}.$$

Robot manipulators can be regarded as open-loop link mechanisms consisting of several links connected together by joints. Joints are typically rotary or linear. A *rotary joint* is like a hinge and allows relative rotation between two links. A *linear joint* allows a linear relative motion between two links. Each joint represents the interconnection between two links. A configuration of a manipulator is a complete specification of the location of every point on the manipulator. The set of all possible configurations is called the *configuration space*. If we know the values for the joint variables (i.e., the joint angle for revolute joints, or the joint offset for prismatic joints), then it is straightforward to infer the position of any point on the manipulator, since the individual links of the manipulator are assumed to be rigid, and the base of the manipulator is assumed to be fixed.

As an example, for a two-link manipulator, a planar joint can be built from a revolute joint attached to two independent prismatic joints. The motion of a planar joint is restricted to $SE(2)$, regarded as a three-dimensional subgroup of $SE(3)$. Here we have

$$(SO(3) \times R^3) \times (SO(3) \times R^3).$$

However, the actual configurations of the system are specified by the angles $\theta_1$ and $\theta_2$, as shown in Figure 2.1. Each angle is measured by a point on the circle:

$$S^1 = \{(x, y) \in R^2 \mid x^2 + y^2 = 1\}.$$

Thus the configurations of this simple two-link robot are specified by a point in $S^1 \times S^1$.

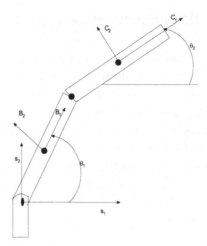

Figure 2.1 Two-link planar manipulator:

An object is said to have *n−degrees-of-freedom* (DOF) if its configuration can be minimally specified by *n* parameters. Thus, the number of DOF is equal to the dimension of the configuration space. For a robot manipulator, the number of joints determines the number DOF. A rigid object in three-dimensional space has six DOF: three for positioning and three for orientation (e.g., roll, pitch, and yaw angles). Therefore, a manipulator should typically possess at least six independent DOF. With fewer than six DOF, the arm cannot reach every point in its work environment with arbitrary orientation. Certain applications such as reaching around or behind obstacles may require more than six DOF. A manipulator having more than six links is referred to as a kinematically redundant manipulator. The workspace of a manipulator is the total volume swept out by the end-effector as the manipulator executes all possible motions. The workspace is constrained by the geometry of the manipulator as well as mechanical constraints on the joints. The workspace is often broken down into a *reachable workspace* and a *dexterous workspace*. The reachable workspace is the entire set of points reachable by the manipulator, whereas the dexterous workspace consists of those points that the manipulator can reach with an arbitrary orientation of the end-effector. Obviously the dexterous workspace is a subset of the reachable workspace; for details see [43], [193], and [272]. Consider an ordinary six-joint robot. Suppose we know the joint variables (angles or lengths) for each joint. How can we work out the position and orientation of the end-effector? This problem is called the *forward kinematic* problem for the robot. Finding possible sets of joint parameters, given the final position and orientation of the end-effector, is the problem of *inverse kinematics*.

The rest of this subsection is taken from a paper by Altafini [12], which may be consulted for the details we cannot discuss in this section. The forward kinematics of a robot arm is represented by the smooth map

$$\rho : Q \quad \rightarrow \quad SE(3)$$
$$q = [q_1, ..., q_n] \quad \rightarrow \quad g = \rho(q), \tag{2.5}$$

where $SE(3)$ is the Lie group of rigid body motions in 3– dimensional space. Each joint variable leaves on $\mathbb{S}$ or on $\mathbb{R}$ and therefore lies in the left/right-invariant translation of a one-dimensional subgroup of $SE(3)$. Hence their product $Q$ is an $n$–dimensional abelian group. The movements of the end-effector are the resulting of the composition of rototranslatins of the $n$–one parameter joints. This can be represented as a product of exponentials of the single one-degree-of-freedom screw motion. The interpretation of that method is the following. One fixes a coordinate system on $Q$ and one on $SE(3)$ and identifies them through $\rho$. Now differentiating (2.5), we have

$$\rho_*^B : TQ \quad \rightarrow \quad TSE(3)$$
$$(q, \dot{q}) \quad \rightarrow \quad (g, g X_{\rho(g)}) = (g, g J^b(q)\dot{q}), \tag{2.6}$$

where $\dot{q}$ corresponds to the velocity in coordinates $\dot{q} = \dot{q}\frac{\partial}{\partial q}$.

Using the natural parallelism of an abelian group, instead of tangent bundle $\rho_*^B$ one can consider the map between tangent space $\rho_*$, which is called differential forward kinematics:

$$\rho_* : T_q Q \quad \rightarrow \quad \mathfrak{se}(3)$$
$$\dot{q} \quad \rightarrow \quad X_{\rho(g)} = J^b(q)\dot{q}, \tag{2.7}$$

where $J^b$ is the Jacobian of the product of exponentials.

In joint space $Q$, it is possible to construct a positive definite and symmetric metric $M(q)$ by

$$M(q) = \sum_{i=1}^{9} J_i^T(q) M_i J_i(q), \tag{2.8}$$

where $M_i$ is the quadratic form representing the generalized inertia tensor of the $i$–th joint and $J_i(q)$ is the $\mathbb{R}^6$–velocity of the $j$–th joint ($j \leq i$), which is referred to the $i$–th link. Consequently, $Q$ is a Riemannian manifold.

Moreover, the Levi-Civita connection determined by $M$ is locally flat. For flat manifolds, it is possible to construct an isometry

$$\varphi : (Q, M(q)) \longrightarrow (\tilde{Q}, I) \tag{2.9}$$

such that

$$
\begin{aligned}
< \dot{\tilde{q}}, \dot{\tilde{q}} >_{\tilde{Q}} &= \varphi_*^T(q)\varphi_*(q) \\
&= \dot{q}^T N^T(q) N(q) \dot{q} \\
&= < \dot{q}, \dot{q} >_Q,
\end{aligned} \tag{2.10}
$$

where $N(q)$ is the Jacobian of the isometry. Then $M(q)$ can be written as

$$M(q) = N(q)^T N(q). \tag{2.11}$$

Thus the metric of $Q$ can be reduced to the identity.

Now assume the following condition:

**R1** Assume that at $q \in Q$, $rank(\rho_*(q)) = 6$.

Since the dimension of $SE(3)$ is 6, this assumption implies that we have a 6 degree of freedom manipulator.

It is known that in $SE(3)$, there is no ad-invariant Riemannian metric (see Theorem 13) which implies that there is no natural way of transporting vector fields between points of $SE(3)$ and that there is no natural concept of distance on $SE(3)$. The two most common approaches to tackle this obstruction are: (1) Ad-invariant pseudo Riemannian structure; (2) double geodesic. The double geodesic method is based on discarding the group structure of $SE(3)$ and considering separately the bi-invariant metric of $SO(3)$ and the Euclidean metric of $\mathbb{R}^3$. The corresponding quadratic form is

$$M_{dg} = \begin{pmatrix} I & 0 \\ 0 & I \end{pmatrix}. \tag{2.12}$$

In fact, above quadratic form is special. We can choose a more general form (see Theorem 14), but this will be enough for our aim. The Riemannian connection is defined by

$$\tilde{\nabla}_{gA_i}(gA_j) = g\tilde{\nabla}_{A_i}A_j = \Gamma_{ij}^k gA_k \tag{2.13}$$

for all $g \in SE(3)$ and $A_i$ the elements of an orthonormal basis of left-invariant vector fields. Since $Q$ is a Riemannian manifold, at each $q \in Q$, we have the following decomposition:

$$T_q Q = \mathcal{H}_q \oplus \mathcal{V}_q,$$

where $\mathcal{V}_q$ is the kernel space of $\rho_*(q)$ and $\mathcal{H}_q$ is the orthogonal complementary subspace to $\mathcal{V}_q$ in $T_qQ$.

**R2** Assume that at $q \in Q$, $dim\mathcal{H}_q = 6$ and $dim\mathcal{V}_q = n - 6$.

Now we have the following result.

**Proposition 20.** *[12] Under the assumption* **R2**, *the forward kinematic map is a Riemannian submersion.*

*Proof.* For the orthogonal subalgebra of $\mathfrak{se}(3) = \mathfrak{so}(3) \circledS \mathbb{R}^3$, $\mathfrak{so}(3)$ is isomorphic to $(\mathbb{R}^3, \times)$. The cross product induces a Lie algebra structure on $\mathbb{R}^3$ which is compatible with the Euclidean inner product, i.e., for $x, y \in \mathbb{R}^3$, $< x, y >_{\mathbb{R}^3} = < \tilde{x}, \tilde{y} >_{\mathfrak{so}(3)}$. Now for $\dot{q}_X, \dot{q}_Y \in \mathcal{H}_q$, from (2.10) we have

$$< \dot{q}_X, \dot{q}_Y >_Q = \dot{q}_X^T N^T(q)N(q)\dot{q}_Y.$$

Using Equations (2.11) and (2.8), we get

$$< \dot{q}_X, \dot{q}_Y >_Q = \dot{q}_X^T J_i^T(q_X)M_iJ_i(q_Y)\dot{q}_Y. \tag{2.14}$$

On the other hand, from Equations (2.7) and (2.12), we have

$$< \dot{q}_X, \dot{q}_Y >_{\mathfrak{se}(3)} = \dot{q}_X^T J^b(q_X)M_{dg}J_b(q_Y)\dot{q}_Y. \tag{2.15}$$

Then, since the metric of $Q$ is reduced to the identity by the isometry $\varphi$, the proof is complete due to Equations (2.14) and (2.15). $\qquad\square$

In this construction, since $Q$ is abelian, the horizontal distribution is also integrable. Also note that, in robotics, $\rho^{-1}(g)$ is the set of joint movements that do not affect the end effector. For details, we refer the reader to [11] and [12].

## 2. Kaluza-Klein theory

In this section, we give another application of Riemannian submersions from mathematical physics: the Kaluza-Klein theory. Since applications of Riemannian submersions in Kaluza-Klein theory were also given in [111] in detail, here we give a brief review of five-dimensional Kaluza-Klein theory and describe it in terms of the principal fiber bundle. Therefore, we begin this section by recalling the principal fiber bundle.

**Definition 32.** Let $M$ be a manifold and $G$ a Lie group. A *principal fiber bundle* over $M$ with group $G$ consists of a manifold $P$ and an action of $G$ on $P$ satisfying the following conditions:

**(i)** $G$ acts freely on $P$ on the right

$$P \times G \;\to\; P$$
$$(p,g) \;\to\; pg = R_{gp} \in P, \; (pg = p \Rightarrow g = e).$$

**(ii)** $M$ is the quotient space of $P$ by the equivalence relation $\tilde{R}$ induced by $G$

$$\tilde{R} = \{(p_1, p_2) \mid \exists g : p_1 g = p_2\},$$

which is written $M = P/G$, and the canonical projection $\pi : P \longrightarrow M$ is differentiable. $\pi^{-1}(u) = \{pg \mid \pi(u) = p$ denotes the fiber through $u \in M$, which is a closed submanifold of $P$ and which coincides with an equivalence of $\tilde{R}$.

**(iii)** $P$ is locally trivial, that is, every point $x$ of $M$ has a neighborhood $U$ such that $\pi^{-1}(U)$ is isomorphic with $U \times G$ in the sense that there is a diffeomorphism $\psi :$ $\pi^{-1}(U) \longrightarrow U \times G$ such that $\psi(u) = (\pi(u), \varphi(u))$ where $\varphi$ is a mapping of $\pi^{-1}(U)$ into $G$ satisfying $\varphi(ua) = \varphi(u)a$ for all $u \in \pi^{-1}(U)$ and $a \in G$.

A principal fiber bundle will be denoted by $P(M, G)$. We call $P$ the total space or the bundle space, $M$ the base space, $G$ the structure group, and $\pi$ the projection. (ii) implies that every fiber is diffeomorphic to $G$. Let a Lie group $G$ act on a manifold $M$ on the right, then one can assign to each element $A \in \mathfrak{g}$ a vector field $A*$ on $M$, where $\mathfrak{g}$ is the Lie algebra of $G$. Principal fiber bundles find their applications in physics. For instance, a gauge group is principal fiber bundle over the base (generally Minkowski space or the underlying Euclidean space $\mathbb{R}^4$) with structure group $G$, a compact Lie group, [99].

**Definition 33.** Let $P(M, G)$ be a principal fiber bundle on a manifold $M$. For each $p \in P$, let $T_p(P)$ be tangent space of P at $p$ and $\mathcal{V}_p$ the subspace of $T_p(P)$ consisting of vectors tangent to the fiber through $p$. A *connection* $\Gamma$ in $P$ is an assignment

$$\mathcal{H} : p \longrightarrow \mathcal{H}_p \subset T_p(P)$$

of a subspace of $T_p P$ to each $p \in P$ such that

**(i)** $T_p P = \mathcal{V}_p \oplus \mathcal{H}_p$,

**(ii)** $dR_g \mathcal{H}_p = \mathcal{H}_{pg}, \forall p \in P$ and $\forall g \in G$, and

**(iii)** $\mathcal{H}$ is a $C^\infty$ distribution on $P$.

The subspaces $\mathcal{V}_p$ and $\mathcal{H}_p$ are called vertical space and the horizontal space of the connection at $p \in P$, respectively.

A vector field $X \in T_pP$ is called vertical (resp. horizontal) if it lies in $\mathcal{V}_p$ (resp. $\mathcal{H}_p$). From Definition 33, every vector $X \in T_pP$ can be uniquely written as

$$X = Y + Z, \text{ where } Y \in \mathcal{V}_p \text{ and } Z \in \mathcal{H}_p.$$

We call $Y$ ( resp. $Z$) the vertical component of $X$ and denote it by $vX$ (resp. $hX$). Given a connection $\Gamma$ in $P$, we define a $1$–form $\omega$ on $P$ with values in the Lie algebra $\mathfrak{g}$ of $G$ as follows. The form $\omega$ is called the connection form of the given connection $\Gamma$.

**Proposition 21.** *[303] Given a connection $\Gamma$ in $P(M, G)$, $G$ is the structural group of $P$. Then there exist a $\mathfrak{g}$–valued $1$–form $\omega$ on $P$ which satisfies the following conditions:*
**(a)** $\omega(X)$ *is vertical, that is,* $\forall X \in \chi(P)$

$$\begin{aligned}
\omega(X) &= \omega(Y) = \hat{X}_v, Y \in \mathcal{V}_p \subset T_pP, \hat{X}_v \in \mathfrak{g} \\
\omega(X) &= 0 \quad \Leftrightarrow \quad X \in \mathcal{H}_p.
\end{aligned}$$

**(b)** $R_g^*\omega(X) = \omega(dR_gX) = (adg^{-1})\omega(X), \forall g \in \mathfrak{g}.$
*Conversely, given a $\mathfrak{g}$–valued $1$–form $\omega$ on $P$ satisfying conditions (a) and (b), there is a unique connection $\Gamma$ in $P$ whose connection form is $\omega$.*

We note that connection form has its applications in mathematical physics. Indeed, the electromagnetic vector potential and Yang-Mills potential are interpreted as one form with values in the Lie algebra of the gauge group. More precisely, a gauge field is connection in the principal fiber bundle $P$, to which the vector bundle of the particle fields is associated [99, Pages 44-58]. For a nice survey on gauge fields for both mathematical and physical aspects, see [195].

The theory of Kaluza-Klein unifies gravity and eloctromagnetism. The original Kaluza-Klein theory was formulated as a $5$–dimensional unified theory after the gravitation theory of Einstein, and such a theory was derived with one extra spatial dimension. The idea was first proposed by Kaluza [155] in 1919 by assuming that all the metric components were independent of the fifth coordinate (cylindrical condition). In fact, Kaluza had basically taken Einstein's theory and added an extra dimension to it by expanding five dimensions. In this way, he was able to take the two forces known at time, gravity, and electromagnetism, and combine them into a single, unified force. The main aspect of Kaluza's theory was that nature can be explained through pure geometry, and physics only depends on the first four coordinates. However, there was no any information as to why it does not depend on the extra dimensions. In 1926, Klein [159] reexamined Kaluza's theory and provided an explanation for Kaluza's fifth dimension by proposing it to have a circular topology so that there is a little circle at each point in four-dimensional spacetime. Thus the global space has topology $R^4 \times S^1$. He also made several important improvements that also seemed to have application to the then-emerging quantum theory.

We now give a brief construction for the Kaluza Klein theory, taken from [185]. The five-dimensional metric tensor is given by

$$ds^2 = g_{ij}dx^i dx^j + I^2(B_i dx^i + d\theta)(B_j dx^j + d\theta), i, j = 1, .., 4, \qquad (2.16)$$

where $ds^2$ is the five-dimensional metric, $g_{ij}$ are the components of the four-dimensional metric, $I$ is the scalar potential, and $B_i$ the electromagnetic potential. $g_{ij}$, $I$ and $B_i$ depend only on $x_1, ..., x_4$, but not on $x_5$. Once we have a metric, we can construct the Christoffel symbols $\Gamma^k_{ij}$, the Riemann-Christoffel curvature tensor $R^l_{ijk}$, the Ricci tensor $R_{ij}$, the curvature invariant $R$, and then the field equations, for such calculations of Kaluza-Klein metric, see [167]. It is easy to check that a coordinate transformation of the fifth dimension $\theta \longrightarrow \theta + \Lambda(x^i)$, i.e., a local transformation of the group $U(1)$ in $M^5$, is equivalent to a gauge transformation of the four electromagnetic potential $B_i \longrightarrow B_i + \partial_i \Lambda$, because of the transformation rule. This means that the compact manifold is providing the internal symmetry space for the (abelian) gauge group, and internal symmetry has now to be interpreted as just another spacetime symmetry, but associated with the extra spatial dimension. The field equations can be deduced from the Einstein-Hilbert action in five dimensions:

$$S = -\frac{1}{16\pi G_k} \int d^5x \sqrt{-\hat{g}_5}\hat{R}, \qquad (2.17)$$

where $\hat{g}_5$ is the determinant of the metric components $\hat{g}_{IJ}$ and $\hat{R}$ is the five-dimensional curvature scalar. If we substitute the metric (2.16) into (2.17) and integrate over the $\theta$ coordinate, we get the four-dimensional action

$$S = -\frac{2}{16\pi G_k} \int d^4x \sqrt{-g_4}I[R + \frac{1}{4}I^2 B_{ij}B^{ij}]. \qquad (2.18)$$

Here, $g_4$ is the determinant of the four-dimensional metric, $R$ the four-dimensional curvature scalar, and $B_{ij}$ the Maxwell tensor $B_{ij} = B_{i,j} - B_{j,i}$. Variation of (2.18) with respect to the metric yields the Einstein's equations coupled with the Faraday stress tensor for $B_{ij}$ and a scalar stress tensor for $I$ as sources. Variation with respect to $B_i$ gives the Maxwell equations for the potential $B_i$ and with respect to $I$, one finds a field equation for the scalar potential $I$ where the currents are electromagnetic and gravitational. When $I$ is constant for some $x^i$, one recovers the Einstein-Maxwell theory.

The massless vector field can be identified as the gauge potential of the electro magnetic $U(1)$ theory. Summing up, the five-dimensional Kaluza-Klein theory appears to be a unified theory of gravity with electro magnetism and $U(1)$ gauge symmetry stems from the isometry of the internal space $S^1$.

A modern version of Kaluza-Klein theory is constructed as follows. One picks an invariant metric on the circle $S^1$ that is the fiber of $U(1)$–bundle of electro magnetism. An invariant metric $g$ is one that is invariant under rotations of the circle. Suppose this metric gives the circle $S^1$ a total length $I$, then consider metrics on the bundle $P$ that are consistent with both the fiber metric and the metric underlying 4–dimensional spacetime $M$ so that $\pi : P \longrightarrow M$ is a (semi) Riemannian submersion. Thus the Kaluza-Klein theory can be given in terms of the bundle $P(M^4, U(1))$ associated with a connection in $P(M^4, U(1))$. According to [80] once the above geometrical interpretation of Kaluza-Klein theories are understood, we can extend the theory to higher dimensions; see [100] for a full discussion of the Kaluza-Klein theory for higher dimensions. Such generalizations are called Yang-Mills theories [305]. This interesting geometrical approach constitutes the base for constructing the modern gauge theories which are used to describe the physics of electromagnetic, weak, and strong interactions.

The Kaluza-Klein theory was ignored back then, the reason being that quantum physics was a popular topic at that time. However, this theory has opened a way for unification. Indeed, string theory borrows from Kaluza-Klein the general notion that extra dimensions are required for unified theory. If we were to follow a Kaluza-Klein approach and ask how many dimensions are needed to combine all four forces within a single framework with five covering gravity and electromagnetism, a couple more for the weak force, and a few more for the strong, we would need a minimum of 11 dimensions. But it is known that the exact number of dimension is ten. For string theory and its relation with the Kaluza-Klein theory; see [302, Chapter 10], [318, Chapter 17], and a popular science book [312].

Finally, we note that there are some different applications based on Riemannian submersions. We recall two of them. One is Watson's paper on Riemannian submersions and instantons [301], and the other paper is written by Hogan [143], which takes Riemannian submersions as a starting point to investigate Bergman's approach to the Kaluza-Klein theory in the special case of five dimensions.

# CHAPTER 3

# Riemannian submersions From Almost Hermitian Manifolds

Contents

## Abstract

In this chapter, we introduce various new Riemannian submersions from almost Hermitian manifolds on to Riemannian manifolds. In section 1, we first review almost Hermitian manifolds and their submanifolds, and give brief information about holomorphic submersions and invariant Riemannian submersions. In this section, we also present three important maps, namely almost complex or holomorphic maps, pseudo-horizontally weakly conformal maps, and pluriharmonic maps, defined between almost complex manifolds and Riemannian manifolds. Then, in section 2, we introduce anti-invariant Riemannian submersions from almost Hermitian manifolds to Riemannian manifolds, and show that such submersions are useful to investigate the geometry of total space. In sections 3 and 4, as a generalization of anti-invariant submersions and invariant submersions, we also introduce semi-invariant submersions and slant submersions. In section 5, as a generalization of slant submersions we consider point-wise slant submersions and show that such submersions have rich geometric properties. In section 6, we define and study generic submersions as a generalization of semi-invariant submersions. In sections 7 and 8, we introduce semi-slant submersions and hemi-slant submersions as generalizations of semi-invariant submersions and slant submersions. In section 9, we check the Einstein conditions of the base space for anti-invariant submersions. In the last section, we investigate various submersions in this chapter in terms of Clairaut's relation.

**Keywords:** almost Hermitian manifold, Kähler manifold, complex space form, almost complex map, holomorphic map, pseudo-horizontally weakly conformal map, Pluriharmonic map, CR-submanifold,, slant submanifold, generic submanifold, semi-slant submanifold, hemi-slant submanifold, point-wise slant submanifold, skew CR-submanifold, holomorphic submersion, invariant submersion, anti-invariant submersion, Lagrangian submersion, tangent bundle, Sasaki metric, semi-invariant submersion, pointwise slant submersion, generic submersion, semi-slant submersion, hemi-slant submersion, Einstein manifold, Clairaut submersion

*Character consists of what you do on the third and fourth tries.*

*James A. Michener*

## 1. Almost Hermitian manifolds

In this section, we review almost Hermitian manifolds and submanifolds of almost Hermitian manifolds due to the fact that the construction of Riemannian submersions from almost Hermitian manifolds has been inspired from such submanifolds. Let $M$ be a $C^\infty$ real $2n$-dimensional manifold, covered by coordinate neighborhoods with coordinates $(x^i)$, where $i$ runs over $1, 2, \ldots, n, \bar{1}, \bar{2}, \ldots, \bar{n}$. $M$ can be considered as a *complex manifold* of dimension $n$ if we define complex coordinates $(z^a = x^a + iy^a)$ on a neighborhood of $z \in M$ such that the intersection of any two such coordinate neighborhoods is regular. Let there exist an endomorphism $J$ (a tensor field $J$ of type $(1, 1)$) of the tangent space $T_p(M)$, at each point $p$ of $M$, such that

$$J(\partial /\partial x^a) = \partial /\partial y^a, \qquad J(\partial /\partial y^a) = -(\partial /\partial x^a),$$

and hence $J^2 = -I$, where $I$ is the identity morphism of $T_p M$. Alternatively, $M$ considered as an $n$-dimensional complex manifold admits a globally defined tensor field,

$$J = i\partial /\partial z^a \otimes dz^a - i\partial /\overline{\partial z^a} \otimes \overline{dz^a},$$

which retains the property $J^2 = -I$ and remains as a real tensor with respect to any of the $z^a$-coordinate charts. Moreover, the tensor field $J$ remind the real manifold that it has a complex structure.

The above property is not sufficient for the existence of a complex structure on $M$. Indeed Fukami-Ishihara [121] proved that the six-dimensional sphere $S^6$ has no complex structure but its tangent bundle admits such an endomorphism $J$. A differentiable manifold $M$ is said to be an *almost complex manifold* if there exists a linear map $J : TM \longrightarrow TM$ satisfying $J^2 = -id$ and $J$ is said to be an *almost complex structure* of $M$. As we have seen from the above discussion, a complex manifold $M$ is an almost complex manifold. Since $J^2 = -id$, for a suitable basis of the tangent bundle, we can construct $J^2$ as

$$\begin{pmatrix} -1 & 0 & \dots & 0 \\ 0 & -1 & \dots & 0 \\ \cdot & \cdot & & \cdot \\ 0 & 0 & 0 & -1 \end{pmatrix}$$

which implies that $(-1)^n = det(J^2) = (det J)^2 \geq 0$. Thus an almost complex manifold is even-dimensional. The tensor field $N$ of type $(1, 2)$ defined by

$$N_J(X, Y) = [JX, JY] - [X, Y] - J([X, JY] + [JX, Y]), \tag{3.1}$$

for any $X, Y \in \Gamma(TM)$, is called the *Nijenhuis tensor field* of $J$. Then, $J$ defines a complex structure [199] on $M$ if and only if $N$ vanishes on $M$.

Let $M$ be an almost complex manifold and $T_pM$ tangent space at $p \in M$. The complexification of $T_pM$ is denoted by $T_p^C M$ and it is defined by
1. $X_p + iY_p \in T_p^C M$ if and only if $X_p, Y_p \in T_pM$,
2. $(X_p + iY_p) + (Z_p + iW_p) = (X_p + Z_p + i(Y_p + W_p))$, $X_p, Y_p, Z_p, W_p \in T_pM$,
3. $(a + ib)(X_p + iY_p) = (aX_p - bY_p) + i(bX_p + aY_p)$ for $a, b \in \mathbb{R}$.
With these algebraic operations, $T_p^C M$ becomes a complex vector space. Complex conjugation in $T_p^C M$ is defined by $\overline{X_p + iY_p} = X_p - iY_p$. Identifying $T_p(M)$ with $\{X_p + iO_p \mid X_p \in T_pM\}$, $T_pM$ is a subspace of $T_p^C M$. It is easy to see that $dim_{\mathbb{C}} T_p^C M = dim_{\mathbb{R}} T_pM$.

Let $(M, J)$ be an almost complex manifold with a complex structure $J$. Then $J_p$ can be extended as an isomorphism of $T_p^C M$. We define $T_p^{(0,1)} M$ and $T_p^{(1,0)} M$, respectively, by

$$T_p^{(0,1)} M = \{X_p + iJ_pX_p \mid X_p \in T_p(M)\}$$

and

$$T_p^{(1,0)} M = \{X_p - iJ_pX_p \mid X_p \in T_p(M)\}.$$

It follows that $T_p^{(0,1)} M$ and $T_p^{(1,0)} M$ are subspaces of $T_p^C M$. If $Z_p \in T_p^{(0,1)} M$, then $\overline{Z_p} \in T_p^{(1,0)} M$. $T_p^{(0,1)} M$ and $T_p^{(1,0)} M$ are said to be complex conjugates of each other. Let

$$T^C M = \bigcup_{p \in M} T_p^C M, \ T^{(1,0)} M = \bigcup_{p \in M} T_p^{(1,0)} M, \ T^{(0,1)} M = \bigcup_{p \in M} T_p^{(0,1)} M,$$

then we have the following decomposition.

**Theorem 16.** $T^C M = T^{(0,1)} M \oplus T^{(1,0)} M$.

Thus the complexified tangent space $T_p^C M$ has the property of being expressible as the direct sum of two subspaces that are complex conjugates of each other. Moreover, vectors in $T_p^{(0,1)}M$ (resp., $T_p^{(1,0)}M$ ) are said to be of type $(0,1)$ (resp., $(1,0)$). Furthermore, $Z_p \in T_p^{(0,1)}M$ if and only if $J_p Z_p = -iZ_p$ and that $Z_p \in T_p^{(1,0)}M$ if and only if $J_p Z_p = iZ_p$. Above spaces $T_p^{(0,1)}M$ and $T_p^{(1,0)}M$ also give an important method for the integrability of almost complex structures of almost complex manifolds.

**Theorem 17.** $T^{(0,1)}M$ and $T^{(1,0)}M$ are involutive if and only if the Nijenhuis tensor $N$ vanishes identically.

For more information about above decomposition, we refer to [97], [162], and [311].

Now consider a Riemannian metric $g$ on an almost complex manifold $(M, J)$. We say that the pair $(J, g)$ is an *almost Hermitian structure* on $M$, and $M$ is an *almost Hermitian manifold* if

$$g(JX, JY) = g(X, Y), \qquad \forall X, Y \in \Gamma(M). \qquad (3.2)$$

Moreover, if $J$ defines a complex structure on $M$, then $(J, g)$ and $M$ are called a *Hermitian structure* and *Hermitian manifold*, respectively. The *fundamental 2-form* $\Omega$ of an almost Hermitian manifold is defined by

$$\Omega(X, Y) = g(X, JY), \quad \forall X, Y \in \Gamma(M). \qquad (3.3)$$

A Hermitian metric on an almost complex $M$ is called a *Kähler metric* and then $M$ is called a *Kähler manifold* if $\Omega$ is closed, i.e.,

$$d\Omega(X, Y, Z) = 0, \quad \forall X, Y \in \Gamma(M), \qquad (3.4)$$

It is known (see Kobayashi-Nomizu [162]) that the Kählerian condition (3.4) is equivalent to

$$(\nabla_X J)Y = 0, \forall X, Y \in \Gamma(M), \qquad (3.5)$$

where $\nabla$ is the Riemannian connection of $g$. An almost Hermitian manifold $M$ with almost complex structure $J$ is called a *nearly Kähler manifold* if

$$(\nabla_X J)Y + (\nabla_Y J)X = 0 \qquad (3.6)$$

for any vector fields $X$ and $Y$ on $M$. If we have

$$(\nabla_X J)Y + (\nabla_{JX} J)JY = 0 \qquad (3.7)$$

for any vector fields $X$ and $Y$ on $M$, then $M$ is called a *quasi-Kähler manifold*. Let

$\{e_1, ..., e_n, Je_1, ..., Je_n\}$ be a basis of $M$. Then the codifferential of $\Omega$, for $X \in \chi(M)$, is

$$(\delta\Omega)(X) = -\sum_{i=1}^{n}(\nabla_{e_i}\Omega)(e_i, X) + (\nabla_{Je_i}\Omega)(Je_iX).$$

If $\delta\Omega = 0$, then $M$ is called *almost semi-Kähler manifold*. If $N = 0$, (see (3.1)), $M$ is called a *semi-Kähler manifold*.

## Certain maps between (or to) Kähler Manifolds

In this paragraph, we will recall holomorphic maps between almost Hermitian manifolds, pluriharmonic maps from almost Hermitian manifolds to Riemannian manifolds, and pseudo-harmonic maps from Riemannian manifolds to Kähler manifolds. A smooth map $\phi : M \longrightarrow N$ between almost complex manifolds $(M, J)$ and $(N, \bar{J})$ is called *almost complex* (or *holomorphic*) map if

$$\phi_*(JX) = \bar{J}\phi_*(X) \tag{3.8}$$

for $X \in \Gamma(TM)$, where $J$ and $\bar{J}$ are complex structures of $M$ and $N$, respectively, see: [20], [111], [129], and [275] for other versions of this notion. For Kähler manifolds we have the following strong result.

**Theorem 18.** *[103] Every holomorphic map between Kähler manifolds is harmonic.*

*Proof.* Let $F$ be a holomorphic map between Kähler manifolds $(M, J)$ and $(N, J')$. Then from (1.71) we have

$$(\nabla F_*)(X, JY) = \overset{N}{\nabla}{}^F_X F_*(JY) - F_*(\overset{M}{\nabla}_X JY).$$

Then Equations (3.8) and (3.5) imply that

$$(\nabla F_*)(X, JY) = J'(\nabla F_*)(X, Y).$$

Hence by using symmetry of the second fundamental form of $F$, we get

$$(\nabla F_*)(JX, JY) = -(\nabla F_*)(X, Y). \tag{3.9}$$

We now choose a local orthonormal frame $\{e_1, ..., e_m, \bar{e}_1, ..., \bar{e}_m\}$ for $M$ such that $Je_i = \bar{e}_i$ for all $i$. Then from (3.9), we obtain

$$\tau(F) = \sum_{i=1}^{m}(\nabla F_*)(e_i, e_i) + (\nabla F_*)(\bar{e}_i, \bar{e}_i) = 0$$

which completes the proof.    □

We also recall the notion of a pluriharmonic map. Let $M_1$ be a Kähler manifold with complex structure $J$ and $M_2$ a Riemannian manifold. A smooth map $F : M_1 \longrightarrow M_2$ is called *pluriharmonic* if the second fundamental form $\nabla F_*$ of the map $F$ satisfies

$$(\nabla F_*)(X, Y) + (\nabla F_*)(JX, JY) = 0 \tag{3.10}$$

for any $X, Y \in \Gamma(TM_1)$, [203]. If $M_1$ and $M_2$ are Kähler manifolds and $F : M_1 \longrightarrow M_2$ is a holomorphic map, then (3.9) implies that $F$ is pluriharmonic. It is also clear that a pluriharmonic map is harmonic. For more details for pluriharmonic maps see [203].

Let $(M, g_M)$ be a Riemannian manifold of real dimension $m$ and $(N, J, g_N)$ a Hermitian manifold of complex dimension $n$. Let $\phi : M \to N$ be a map between them. For any point $p \in M$, we consider ${}^*\phi_{*p} : T_{\phi(p)}N \to T_pM$ the adjoint of the tangent map $\phi_{*p} : T_pM \to T_{\phi(p)}N$. Denoting the horizontal space of $\phi_*$ at $p \in M$ by $\mathcal{H}_p$, if $im(\phi_{*p})$ is $J-$ invariant, then we can define an almost complex structure $J_{\mathcal{H},p}$ on the space of $\mathcal{H}_p$ by

$$J_{\mathcal{H},p} = \phi_{*p}^{-1} \circ J_{\phi(p)} \circ \phi_{*p}.$$

**Definition 34.** [177], [13] Let $\phi$ be a map from a Riemannian manifold $(M, g_M)$ to a Hermitian manifold $(N, J, g_N)$. Then the map $\phi$ is called *pseudo-horizontally weakly conformal* (PHWC) at $p \in M$ if and only if $im(\phi_{*p})$ is invariant and $g \mid_{im(\phi_{*p})}$ is $J_{\mathcal{H},p}$-Hermitian. The map is called pseudo-horizontally weakly conformal if and only if it is PHWC at any point of $p$. We note that it is easy to see that

$$g_{\mathcal{H}}(J_{\mathcal{H},p}-, J_{\mathcal{H},p}-) = g_{\mathcal{H}}(-, -)$$

if and only if

$$\phi_{*p} \circ {}^*\phi_{*p} \circ J = J \circ \phi_{*p} \circ {}^*\phi_{*p}, \quad \text{simply} \quad [\phi_{*p} \circ {}^*\phi_{*p}, J] = 0. \tag{3.11}$$

We also note that the notion of pseudo-horizontally weakly conformal map was first defined by Loubeau in [177]. The origin of condition (3.11) comes from a paper [46] on the stability of harmonic maps into irreducible Hermitian symmetric space. It is also easy to see that the condition (3.11) is equivalent to the condition that $\phi_{*p} \circ {}^*\phi_{*p}$ map the holomorphic bundle $T^{(1,0)}N$ of $N$ itself (see: [177, Lemma 1]. A harmonic map which is pseudo-horizontally weakly is called a *pseudo-harmonic morphism*. Pseudo-harmonic morphisms were also defined in [177] by Loubeau. He also obtained that a smooth map $F$ from a Riemannian manifold $M$ to a Kähler manifold $N$ is a pseudo-harmonic morphism if and only if $F$ pulls back local pluriharmonic functions on $N$ to local harmonic functions on $M$. PHWC maps have been also studied in [15], [14], and [178]. In [15], the authors introduced the notion of pseudo-horizontally homothetic maps as follow.

**Definition 35.** [15], [13] Let $(M, g_M)$ and $(N, g_N, J)$ be a Riemannian manifold and a Kähler manifold, respectively. Then the map $\phi : M \to N$ is called a *pseudo-horizontally homothetic (PHH) map* at $p \in M$ if and only if

**(a)** $\phi$ is PHWC, and

**(b)** $J_{\mathcal{H}}$ is parallel in horizontal directions, i.e., $\nabla_X^M J_{\mathcal{H}} = 0$ for $X \in \Gamma(\mathcal{H})$.

More precisely, if $\phi$ is PHWC at $p \in M$, then we say that $\phi$ is PHH at $p \in M$ if

$$\phi_{*p}((\nabla_X^M{}^* \phi_{*p}(JY)_p) = J_{\phi(p)} \circ \phi_{*p}((\nabla_X^M{}^* \phi_{*p}(Y)_p)) \tag{3.12}$$

for any horizontally $X \in T_p M$ and any vector field $Y$, locally defined in a neighborhood at $\phi(p)$. And $\phi$ is a PHH if and only if it is PHH at every $p \in M$. Also $\phi$ is called a *strongly pseudo-horizontally homothetic map* if (3.12) is satisfied for every vector $X \in T_p M$. For instance, any holomorphic map between Kähler manifolds is strongly pseudo-horizontally homothetic.

## Complex space forms

We now recall curvature relations for Kähler manifolds. First we note the following identities.

**Theorem 19.** *[311] For a Kähler manifold we have the following properties*

$$R(X, Y) JZ = J R(X, Y)Z, R(JX, JY) = R(X, Y), \tag{3.13}$$

$$Ric(X, Y) = \frac{1}{2}\{Tr.J \circ R(X, JY)\}, \, Q J = J Q \tag{3.14}$$

$$Ric(JX, JY) = Ric(X, Y), \tag{3.15}$$

*where Q is the Ricci operator.*

The next theorem shows that the notion of constant curvature for Kähler manifolds does not work for Kähler manifolds.

**Theorem 20.** [311] *Let M be a real 2n−dimensional Kähler manifold. If M is of constant curvature, then M is flat provided $n > 1$.*

Therefore, the notion of constant holomorphic sectional curvature was introduced for Kähler manifolds. Consider a vector $U$ at a point $p$ of a Kähler manifold $M$. Then the pair $(U, JU)$ determines a plane $\pi$ (since $JU$ is obviously orthogonal to $U$) element called a *holomorphic section*, whose curvature $K$ is given by

$$K = \frac{g(R(U, JU)JU, U)}{(g(U, U))^2},$$

and is called the *holomorphic sectional curvature* with respect to $U$. If $K$ is independent of the choice of $U$ at each point, then $K = c$, an absolute constant. A simply connected complete Kähler manifold of constant sectional curvature $c$ is called a *complex space-form*, denoted by $M(c)$, which can be identified with the complex projective space $P_n(c)$, the open ball $D_n$ in $C^n$ or $C^n$ according as $c > 0$, $c < 0$ or $c = 0$. The curvature tensor of $M(c)$ is

$$R(X, Y)Z = \frac{c}{4}[g(Y, Z)X - g(X, Z)Y + g(JY, Z)JX$$
$$- g(JX, Z)JY + 2g(X, JY)JZ]. \tag{3.16}$$

A contraction of the above equation shows that $M(c)$ is Einstein, that is,

$$Ric(X, Y) = \frac{n+1}{2}cg(X, Y).$$

## Symplectic manifolds

We now give brief information for symplectic manifolds and their relations with complex structures. Our references for this paragraph are [36] and [270]. A *symplectic manifold* is an even dimensional differentiable manifold $M$ with a global 2−form $\Omega$ which is closed $d\Omega = 0$ and of maximal rank $\Omega^n \neq 0$. In fact the last condition comes from the non-degenerate condition of $\Omega$ and it shows that $(M, \Omega)$ is canonically oriented. A complex Euclidean space with coordinates $z_1, ..., z_n$ and the form $\Omega_0 = \frac{1}{2}\sum dz_k \wedge d\bar{z}_k$ is a symplectic manifold. A Kähler manifold $M$ with its fundamental 2−form is a symplectic manifold. An almost complex structure $J$ on a symplectic manifold $(M, \Omega)$ is *compatible* if the map that assigns the each point $p \in M$ the bilinear pairing $g_p : T_pM \times T_pM \longrightarrow \mathbb{R}$, $g_p(u, v) = \Omega_p(u, J_pv)$ is a Riemannian metric on $M$. It is known that any symplectic manifold has compatible almost complex structures. In fact, the closeness of $\Omega$ is not used to construct compatible almost complex structures. The construction holds for an *almost symplectic manifold* $(M, \Omega)$, that is, a pair of a manifold $M$ and non-degenerate form $\Omega$, not necessarily closed. However, there are symplectic manifolds that do not admit any complex structures [114]. We now define the Lagrangian submanifold. Let $i : M \to \bar{M}$ be an immersion into a symplectic manifold $(\bar{M}^{2n}, \Omega)$. We say that $M$ is *Lagrangian submanifold* if the dimension of $M$ is $n$ and $i^*\Omega = 0$. We note that since $i^*\Omega = 0$, there is no induced structure. Therefore the geometry of such submanifolds is very difficult to investigate.

## Riemann submanifolds of almost Hermitian manifolds

According to the behavior of the tangent bundle of a submanifold with respect to the action of the almost complex structure $\bar{J}$ of the ambient manifold, there are several popular classes of submanifolds, namely, Kähler submanifolds, totally real submanifolds, CR-submanifolds, slant submanifolds, generic submanifolds, semi-slant submanifolds, hemi-slant submanifolds, and pointwise slant submanifolds. A submanifold of a Kähler manifold is called a *complex submanifold* if each of its tangent space is invariant under the almost complex structure of the ambient manifold. A complex submanifold of a Kähler manifold is a Kähler manifold with induced metric structure which is called a Kähler submanifold. It is well known that a Kähler submanifold of a Kähler manifold is always minimal: [311], [202]. A *totally real submanifold M* is a submanifold such that the almost complex structure $\bar{J}$ of the ambient manifold $\bar{M}$ carries a tangent space of $M$ into the corresponding normal space of $M$. A totally real submanifold is called *Lagrangian* if $dim_R M = dim_C \bar{M}$. Real curves of Kähler manifolds are examples of totally real submanifolds. The first contribution to the geometry of totally real submanifolds was given in the early 1970's [83]. For details, see [311].

In 1978, generalizing these two types of submanifolds, Bejancu [24] defined Cauchy-Riemann (CR) submanifolds. By a *CR submanifold* we mean a real submanifold $M$ of an almost Hermitian manifold $(\bar{M}, J, \bar{g})$, carrying a $J$-invariant distribution $D$ (i.e., $JD = D$) and whose $\bar{g}$-orthogonal complement is $J$-anti-invariant (i.e., $JD^{\perp} \subseteq T(M)^{\perp}$), where $T(M)^{\perp} \to M$ is the normal bundle of $M$ in $\bar{M}$. Basic details on these may be seen in [24], [25], [28], [66], [67], and [309].

Let $(\bar{M}, J)$ be an almost complex manifold with almost complex structure J. Let $N$ be a submanifold of $\bar{M}$. For each point $x \in N$, we put $H_x = T_x N \cap J(T_x N)$. Then $H_x$ is the maximal complex subspace of the tangent space $T_x \bar{M}$ which is contained in $T_x M$. If the dimension of $H_x$ is constant along $N$, $N$ is called a *generic submanifold* of $(\tilde{M}, J)$ [69]. It follows that a CR-submanifold is generic submanifold with complementary orthogonal distribution to $H_x$ at $x \in N$.

Another generalization of holomorphic and totally real submanifolds, named slant submanifolds, was introduced by Chen as follows: let $\bar{M}$ be a Kähler manifold and $M$ be a real submanifold of $\bar{M}$. For any non-zero vector $X$ tangent to $M$ at a point $p \in M$, the angle $\theta(X)$ between $JX$ and the tangent space $T_p M$ is called the *Wirtinger angle of X*. A submanifold $M$ of $\bar{M}$ is called a *slant submanifold* [71] if the Wirtinger angle $\theta(X)$ is constant, i.e., it is independent of the choice of $p \in M$ and $X \in T_p M$. The Wirtinger angle of a slant submanifold is called the *slant angle* of the slant submanifold. By the above definition, it follows that holomorphic (resp. totally real) submanifolds are slant submanifolds with $\theta = 0$ (resp., $\theta = \frac{\pi}{2}$). A slant submanifold is called proper if it is neither totally real nor holomorphic submanifold of $\bar{M}$. It is known that every proper slant submanifold is even-dimensional. We note that there is no any inclusion relation between proper CR-submanifolds and proper slant submanifolds.

Therefore two other classes of submanifolds of almost Hermitian manifolds were introduced as follows. The submanifold $M$ is called *semi-slant* [214] if it is endowed with two orthogonal distributions $D$ and $D'$, where $D$ is invariant with respect to $J$ and $D'$ is slant, i.e., $\theta(X)$ between $JX$ and $D'_p$ is constant for $X \in D'_p$. Let $\bar{M}$ be a Kähler manifold and $M$ a real submanifold of $\bar{M}$. Then we say that $M$ is a *hemi-slant submanifold* [238] (or anti-slant submanifold [53]) if there exist two orthogonal distributions $D^\perp$ and $D^\theta$ on $M$ such that (a) $TM$ admits the orthogonal direct decomposition $TM = D^\perp \oplus D^\theta$, (b) the distribution $D^\perp$ is an anti-invariant distribution, i.e., $JD^\perp \subset TM^\perp$, and (c) the distribution $D^\theta$ is slant with slant angle $\theta$ [53].

We also note that the notion of skew CR-submanifolds was introduced in [234] as follows. Let $M$ be a submanifold of a Kähler manifold $\bar{M}$. For any $X_p$ and $Y_p$ in $T_pM$ we have $g(TX_p, Y_p) = -g(X_p, TY_p)$. Hence, it follows that $T^2$ is a symmetric operator on the tangent space $T_pM$, for all $p$. Therefore, its eigenvalues are real and its diagonalizable. Moreover, its eigenvalues are bounded by $-1$ and $0$. For each $p \in M$, we may set

$$\mathcal{D}_p^\lambda = Ker\{T^2 + \lambda^2(p)I\}_p,$$

where $I$ is the identity transformation and $\lambda(p)$ belongs to the closed real interval $[0, 1]$ such that $-\lambda^2(p)$ is an eigenvalue of $T^2(p)$. Notice that $\mathcal{D}_p^1 = KerF$ and $\mathcal{D}_p^0 = KerT$. $\mathcal{D}_p^1$ is the maximal $J-$ invariant subspace of $T_pM$ and $\mathcal{D}_p^0$ is the maximal anti $J-$ invariant subspace of $T_pM$. From now on, we denote the distributions $\mathcal{D}^1$ and $\mathcal{D}^0$ by $\mathcal{D}^T$ and $\mathcal{D}^\perp$, respectively. Since $T_p^2$ is symmetric and diagonalizable, if $-\lambda_1^2(p), ..., -\lambda_k^2(p)$ are the eigenvalues of $T^2$ at $p \in M$, then $T_pM$ can be decomposed as the direct sum of the mutually orthogonal eigenspaces, i.e.,

$$T_pM = \mathcal{D}_p^{\lambda_1} \oplus ... \oplus \mathcal{D}_p^{\lambda_k}.$$

Each $\mathcal{D}_p^{\lambda_i}$, $1 \leq i \leq k$, is a $T-$ invariant subspace of $T_pM$. Moreover, if $\lambda_i \neq 0$ then $\mathcal{D}_p^{\lambda_i}$ is even dimensional. Let $M$ be a submanifold of a Kähler manifold $\bar{M}$. $M$ is called a *generic submanifold* if there exists an integer $k$ and functions $\lambda_i$, $1 \leq i \leq k$ defined on $M$ with values in $(0, 1)$ such that

**1.** Each $-\lambda_i^2(p)$, $1 << i \leq k$ is a distinct eigenvalue of $T^2$ with

$$T_pM = \mathcal{D}_p^T \oplus \mathcal{D}_p^\perp \oplus \mathcal{D}_p^{\lambda_1} \oplus ... \oplus \mathcal{D}_p^{\lambda_k}$$

for $p \in M$, and

**2.** the dimensions of $\mathcal{D}_p^T$, $\mathcal{D}_p^\perp$ and $\mathcal{D}^{\lambda_i}$, $1 \leq i \leq k$ are independent of $p \in M$.

Moreover, if each $\lambda_i$ is constant on $M$, then $M$ is called a *skew CR-submanifold*. Thus, we observe that CR-submanifolds are a particular class of skew CR-submanifolds with $k = 0$, $\mathcal{D}^T \neq \{0\}$ and $\mathcal{D}^\perp \neq \{0\}$. And slant submanifolds are also a particular class of skew CR-submanifolds with $k = 1$, $\mathcal{D}^T = \{0\}$, $\mathcal{D}^\perp = \{0\}$ and $\lambda_1$ is constant. Moreover, if $\mathcal{D}^\perp = \{0\}$, $\mathcal{D}^T \neq \{0\}$ and $k = 1$, then $M$ is a semi-slant submanifold. Skew CR-submanifolds have been also studied in [140], [176], [234], [242] and [287].

Finally, we recall the notion of pointwise (or quasi) slant submanifolds. Let $\bar{M}$ be a Kähler manifold and $M$ be a real submanifold of $\bar{M}$. If at each given point $p \in M$, the Wirtinger angle $\theta(X)$ is independent of the choice of the nonzero tangent vector $X \in \Gamma(TM)$, then $M$ is called a *pointwise slant submanifold* of $\bar{M}$. In this case, $\theta$ can be regarded as a function on $M$, which is called the slant function of the pointwise slant submanifold. We note that pointwise slant submanifolds have been defined by Etayo [109] and such submanifolds have been studied in details by Chen and Garay [81]. The notion of a pointwise semi-slant submanifold was also introduced in [252] as follows: let $\bar{M}$ be a Kähler manifold and $M$ a real submanifold of $\bar{M}$. Then we say that $M$ is a *pointwise semi-slant submanifold* if there exist two orthogonal distributions $D^T$ and $D^\theta$ on $M$ such that (a) $TM$ admits the orthogonal direct decomposition $TM = D^T \oplus D^\theta$. (b) The distribution $D^T$ is a holomorphic distribution, i.e., $JD^T = D^T$. (c) The distribution $D^\theta$ is pointwise slant with slant function $\theta$. See [97] for CR-submanifolds of complex projective spaces. Also see [74]. and [188] for a good survey on the submanifolds of almost Hermitian manifolds.

## 2. Holomorphic submersions and invariant submersions

In this section, we give brief notes about holomorphic submersions, define invariant Riemannian submersions, and show that invariant submersions may not be a holomorphic submersions.

Let $M$ be a complex $m-$dimensional almost Hermitian manifold with Hermitian metric $g_M$ and almost complex structure $J_M$ and $N$ be a complex $n-$dimensional almost Hermitian manifold with Hermitian metric $g_N$ and almost complex structure $J_N$. In [300], Watson considered the holomorphic Riemannian submersion, called the almost Hermitian submersion, and obtained the fundamental properties of this map. Later, this topic was studied by many authors; see [111] for different total spaces and base spaces. More precisely, a Riemannian submersion $F : M \longrightarrow N$ is called an almost Hermitian submersion if $F$ is an almost complex mapping. The main result of this notion is that the vertical and horizontal distributions are $J_M-$invariant. We note that in most cases, the base manifold and each fibre have the same kind of structure as the total space. Indeed, we have the following theorem.

**Theorem 21.** *[300], [311] Let $\pi : (M, J_M, g_M) \longrightarrow (B, J_B, g_B)$ be an almost Hermitian submersion. If $M$ is quasi-Kähler, nearly Kähler, Kähler, or Hermitian, then $B$ has the same property.*

*Proof.* Let $\Omega_M$ and $\Omega_B$ be the fundamental 2$-$forms of $M$ and $B$, respectively. We first claim $\Omega_M = \pi^*\Omega_B$ on basic vector fields. If $X$ and $Y$ are basic vector fields on $M$, and $X_*$ and $Y_*$ are their associated vector fields on $B$, then

$$\Omega_M(X, Y) = g_M(X, J_M Y) = g_B(X_*, J_B Y_*) \circ \pi = \pi^*\Omega_B(X_*, Y_*).$$

Since $\pi^*$ commutes with the exterior derivative $d_M$ on differential forms, we also see that $d_M\Omega = \pi^* d_B\Omega_B$. If $M$ is almost Kähler, then $\pi^* d_B\Omega_B = 0$. Since $\pi^*$ is a linear isometry, we obtain $d_B\Omega_B = 0$ and therefore $B$ is almost Kähler. Suppose that $M$ is nearly Kähler. It is easy to see that the basic vector field associated to $\nabla^B_{X_*} J_B X_*$ for any vector field $X_*$ on $B$ is $\mathcal{H}\nabla^M_X J_M X$, which vanishes on $M$. Thus from (3.6), we conclude that $B$ is nearly Kähler. Similarly, we see that if $M$ is quasi-Kähler, then $B$ is quasi-Kähler. Moreover, the basic vector field on $M$ associated to the Nijenhuis tensor $N^B(X_*, Y_*)$ is $\mathcal{H}N^M(X.Y)$. Therefore, if $M$ is Hermitian, $B$ is also Hermitian. Furthermore, when $M$ is Kähler, $B$ is Kähler.                                  □

Watson also found the following result.

**Theorem 22.** *[300] Let $F : (M, g_M, J_M) \longrightarrow (N, g_N, J_N)$ be a holomorphic submersion from an almost Hermitian manifold $(M, g_M, J_M)$ onto an almost Hermitian manifold $(N, g_N, J_N)$. Then:*

**(1)** *If the total space is almost semi-Kähler, then the base manifold is almost semi-Kähler if and only if each fiber is minimal.*

**(2)** *A quasi-Kähler submersion is curvature decreasing.*

**(3)** *An almost semi-Kähler submersion is Betti number decreasing.*

In the rest of this section, we are going to show that an invariant submersion may not be a holomorphic submersion. We first give the following definition.

**Definition 36.** [243] Let $F$ be a Riemannian submersion from an almost Hermitian manifold $(M, g_M, J_M)$ onto an almost Hermitian manifold $(N, g_N, J_N)$. Then we say that $F$ is an invariant Riemannian submersion if the vertical distribution is invariant with respect to $J_M$, i.e.,

$$J_M(ker F_*) = ker F_*.$$

From the above definition, we have:

**Corollary 8.** *[243] Let $F$ be a Riemannian submersion from an almost Hermitian manifold $(M, g_M, J_M)$ onto an almost Hermitian manifold $(N, g_N, J_N)$. Then the horizontal distribution is invariant with respect to $J_M$.*

**Example 4.** [243] Let $F : \mathbb{R}^8 \longrightarrow \mathbb{R}^4$ be a map defined by

$$F(x_1, x_2, x_3, x_4, x_5, x_6, x_7, x_8) = (\frac{x_1 - x_5}{\sqrt{2}}, \frac{x_2 - x_6}{\sqrt{2}}, \frac{x_3 - x_7}{\sqrt{2}}, \frac{x_4 - x_8}{\sqrt{2}}).$$

Then, by direct calculations,

$$ker F_* = span\{Z_1, Z_2, Z_3, Z_4\},$$

where

$$Z_1 = \partial x_1 + \partial x_5 \quad, \quad Z_2 = \partial x_2 + \partial x_6$$
$$Z_3 = \partial x_3 + \partial x_7 \quad, \quad Z_4 = \partial x_4 + \partial x_8$$

and

$$(ker\, F_*)^\perp = span\{H_1, H_2, H_3, H_4\},$$

where

$$H_1 = \partial x_1 - \partial x_5 \quad, \quad H_2 = \partial x_2 - \partial x_6$$
$$H_3 = \partial x_3 - \partial x_7 \quad, \quad H_4 = \partial x_4 - \partial x_8.$$

Then considering complex structures on $\mathbb{R}^8$ and $\mathbb{R}^4$ by

$$J^8(a_1, a_2, a_3, a_4, a_5, a_6, a_7, a_8) = (-a_2, a_1, -a_4, a_3, -a_6, a_5, -a_8, a_7)$$

and

$$J^4(a_1, a_2, a_3, a_4) = (-a_3, -a_4, a_1, a_2),$$

it is easy to see that $ker\, F_*$ and $(ker\, F_*)^\perp$ are invariant with respect to $J^8$. Thus $F$ is an invariant submersion.

However, $F$ is not a holomorphic submersion. Indeed, for $H_1$, we have $J^4 F_*(H_1) = \sqrt{2}\partial y_3$ and $F_* J^8(H_1) = \sqrt{2}\partial y_2$, i.e., $J^4 F_*(H_1) \neq F_*(J^8 H_1)$.

Thus we can state the following.

**Proposition 22.** *[243] Every holomorphic submersion is an invariant submersion. However, an invariant submersion may not be a holomorphic submersion.*

For an invariant Riemannian submersion, we also have:

**Proposition 23.** *[243] Let $F$ be an invariant submersion from an almost Hermitian manifold onto a Riemannian manifold. Then the fibers are minimal submanifolds.*

## 3. Anti-invariant Riemannian submersions

In this section, we consider a Riemannian submersion from an almost Hermitian manifold under the assumption that the fibers are anti-invariant with respect to the complex structure of the almost Hermitian manifold. This assumption implies that the horizontal distribution is also anti-invariant. Roughly speaking, almost Hermitian submersions are useful for describing the geometry of base manifolds; however, anti-invariant submersions serve to determine the geometry of total manifolds. Also note that the geometry of anti-invariant Riemannian submersions is quite different from the geometry of almost Hermitian submersions. For example, from Theorem 18, we know that every holomorphic map between Kähler manifolds is harmonic, thus it follows that any Hermitian (holomorphic) submersion between Kähler manifolds is harmonic. However, this result is not valid for anti-invariant Riemannian submersions.

**Definition 37.** [239] Let $M$ be a complex $m-$dimensional almost Hermitian manifold with Hermitian metric $g_M$ and almost complex structure $J$ and $N$ be a Riemannian manifold with Riemannian metric $g_N$. Suppose that there exists a Riemannian submersion $F : M \longrightarrow N$ such that $ker F_*$ is anti-invariant with respect to $J$, i.e., $J(ker F_*) \subseteq (kerF_*)^\perp$. Then we say that $F$ is an *anti-invariant Riemannian submersion*.

First of all, since the distribution $ker F_*$ is integrable, the above definition implies that the integral manifold (fiber) $F^{-1}(q)$, $q \in M_2$, of $ker F_*$ is a totally real submanifold of $M_1$. From Definition 37, we have $J(ker F_*)^\perp \cap ker F_* \neq \{0\}$. We denote the complementary orthogonal subbundle to $J(ker F_*)$ in $(kerF_*)^\perp$ by $\mu$. Then we have

$$(ker F_*)^\perp = Jker F_* \oplus \mu. \tag{3.17}$$

It is easy to see that $\mu$ is an invariant subbundle of $(ker F_*)^\perp$, under the endomorphism $J$. Thus, for $X \in \Gamma((ker F_*)^\perp)$, we have

$$JX = BX + CX, \tag{3.18}$$

where $BX \in \Gamma(ker F_*)$ and $CX \in \Gamma(\mu)$.

**Example 5.** [239] Let $F : \mathbb{R}^4 \longrightarrow \mathbb{R}^2$ be a map defined by

$$F(x_1, x_2, x_3, x_4) = (\frac{x_1 + x_4}{\sqrt{2}}, \frac{x_2 + x_3}{\sqrt{2}}).$$

Then, by direct calculations,

$$ker F_* = span\{Z_1 = \partial x_1 - \partial x_4, Z_2 = \partial x_2 - \partial x_3\}$$

and

$$(ker F_*)^\perp = span\{X_1 = \partial x_1 + \partial x_4, X_2 = \partial x_2 + \partial x_3\}.$$

Then it is easy to see that $F$ is a Riemannian submersion. Moreover, $JZ_1 = X_2$ and $JZ_2 = X_1$ imply that $J(ker F_*) = (ker F_*)^\perp$. As a result, $F$ is an anti-invariant Riemannian submersion.

We now give a characterization for a Riemannian submersion to be invariant or anti-invariant.

**Proposition 24.** *[284] Let $\pi : M(c) \to N$ be a Riemannian submersion from a complex space form $M(c)$ with $c \neq 0$ to a Riemannian manifold $N$. Then the fibers of $M(c)$ are invariant or anti-invariant with respect to the almost complex structure $J$ of $M(c)$ if and only if*

$$g((\nabla_U T)_V W, X) = g((\nabla_V T)_U W, X), \tag{3.19}$$

*where $U, V$ and $W$ are vertical vector fields and $X$ is a horizontal vector field on $M(c)$.*

*Proof.* Let $U, V$ and $W$ be vertical vector fields and $X$ be a horizontal vector field on $M(c)$. Then from (3.16), we have

$$R(U, V)W = \tfrac{c}{4}\{g(V, W)U - g(U, W)V + g(JV, W)JU \\ -g(JU, W)JV + 2g(U, JV)JW\}. \tag{3.20}$$

From (3.20), we see that $R(U, V)W$ is vertical, if the fibers are invariant or anti-invariant with respect to the almost complex structure $J$ of $M(c)$. Thus we have $R(U, V, W, X) = 0$. Therefore, (3.19) follows from the formula (1.60).

Conversely, assume that (3.19) holds. Then for $U, V$, and $W$, it is not difficult to see that $R(U, V)W$ is vertical from (1.60). If we put $W = U$ in (3.20), then we have

$$R(U, V)U = \frac{c}{4}\{g(V, U)U - g(U, U)V + g(U, JV)JU\}. \tag{3.21}$$

Thus, we see that $g(U, JV)JU$ is vertical from (3.21), since $R(U, V)U$ is vertical. So, we conclude that either $JU$ is vertical or $g(U, JV) = 0$. It means that either $J(ker\pi_*) \subseteq ker\pi_*$ or $J(ker\pi_*) \subseteq (ker\pi_*)^\perp$, i.e., either the fibers are invariant or anti-invariant with respect to the almost complex structure $J$ of $M(c)$. □

Thus we have the following result for an anti-invariant submersion.

**Corollary 9.** *[284] Let $\pi : M(c) \to N$ be an anti-invariant Riemannian submersion from a complex space form $M(c)$ with $c \neq 0$ to a Riemannian manifold $N$. Then the equality (3.19) holds.*

**Lemma 9.** *[239] Let $F$ be an anti-invariant Riemannian submersion from a Kähler manifold $(M, g_M, J)$ to a Riemannian manifold $(N, g_N)$. Then we have*

$$g_M(CY, JV) = 0 \tag{3.22}$$

*and*

$$g_M(\nabla_X CY, JV) = -g_M(CY, J\mathcal{A}_X V) \tag{3.23}$$

*for $X, Y \in \Gamma((ker\, F_*)^\perp)$ and $V \in \Gamma(ker F_*)$.*

*Proof.* For $Y \in \Gamma((ker\, F_*)^\perp)$ and $V \in \Gamma(ker\, F_*)$, using (3.2) we have

$$g_M(CY, JV) = g_M(JY - BY, JV) = g_M(JY, JV)$$

due to $BY \in \Gamma(ker\, F_*)$ and $JV \in \Gamma((ker F_*)^\perp)$. Hence $g_M(JY, JV) = g_M(Y, V) = 0$ which is (3.22). Now, using Equations (3.22) and (3.5), we get

$$g_M(\nabla_X CY, JV) = -g_M(CY, J\nabla_X V)$$

for $X, Y \in \Gamma((ker\, F_*)^\perp)$ and $V \in \Gamma(ker F_*)$. Then using (1.49) we have

$$g_M(\nabla_X CY, JV) = -g_M(CY, J\mathcal{A}_X V) - g_M(CY, J\mathcal{V}\nabla_X V).$$

Since $J\mathcal{V}\nabla_X V \in \Gamma(Jker\, F_*)$, we obtain (3.23). □

We now investigate the geometry of leaves of the vertical distribution and the horizontal distribution. But we first study the integrability of distributions $(ker\, F_*)^\perp$.

**Theorem 23.** *[239] Let $F$ be an anti-invariant Riemannian submersion from a Kähler manifold $(M, g_M, J)$ to a Riemannian manifold $(N, g_N)$. Then the following assertions are equivalent to each other:*

**(a)** *$(ker\, F_*)^\perp$ is integrable.*

**(b)**

$$\begin{aligned} g_N((\nabla F_*)(Y, BX), F_* JV) &= g_N((\nabla F_*)(X, BY), F_* JV) \\ &+ g_M(CY, J\mathcal{A}_X V) - g_M(CX, J\mathcal{A}_Y V). \end{aligned}$$

**(c)** *$g_M(\mathcal{A}_Y BX - \mathcal{A}_X BY, JV) = g_M(CY, J\mathcal{A}_X V) - g_M(CX, J\mathcal{A}_Y V)$, for $X, Y \in \Gamma((ker\, F_*)^\perp)$ and $V \in \Gamma(ker\, F_*)$.*

*Proof.* For $Y \in \Gamma((ker\, F_*)^\perp)$ and $V \in \Gamma(ker\, F_*)$, we see from Definition 37, $JV \in \Gamma((ker\, F_*)^\perp)$ and $JY \in \Gamma(ker\, F_* \oplus \mu)$. Thus using (3.2), for $X \in \Gamma((ker\, F_*)^\perp)$, we get

$$g_M([X, Y], V) = g_M(\nabla_X JY, JV) - g_M(\nabla_Y JX, V).$$

Then from (3.17) we have

$$\begin{aligned} g_M([X, Y], V) &= g_M(\nabla_X BY, JV) + g_M(\nabla_X CY, JV) \\ &- g_M(\nabla_Y BX, JV) - g_M(\nabla_Y CX, JV). \end{aligned}$$

Since $F$ is a Riemannian submersion, we obtain

$$\begin{aligned} g_M([X, Y], V) &= g_N(F_*\nabla_X BY, F_*JV) + g_M(\nabla_X CY, JV) \\ &\quad - g_N(F_*\nabla_Y BX, F_*JV) - g_M(\nabla_Y CX, JV). \end{aligned}$$

Thus, from (1.71) and (3.23), we have

$$\begin{aligned} g_M([X, Y], V) &= g_N(-(\nabla F_*)(X, BY) + (\nabla F_*)(Y, BX), F_*JV) \\ &\quad - g_M(CY, J\mathcal{A}_X V) + g_M(CX, J\mathcal{A}_Y V) \end{aligned}$$

which proves (a) $\Leftrightarrow$ (b). On the other hand, using (1.71) we get

$$(\nabla F_*)(Y, BX) - (\nabla F_*)(X, BY) = F_*(\nabla_Y BX - \nabla_X BY).$$

Then (1.49) implies that

$$(\nabla F_*)(Y, BX) - (\nabla F_*)(X, BY) = F_*(\mathcal{A}_Y BX - \mathcal{A}_X BY).$$

Since $\mathcal{A}_Y BX - \mathcal{A}_X BY \in \Gamma((ker F_*)^\perp)$, this shows that (b) $\Leftrightarrow$ (c).    $\square$

We say that an anti-invariant Riemannian submersion is a *Lagrangian Riemannian submersion* if $J(ker F_*) = (ker F_*)^\perp$. We note that there is a close relation between Lagrangian submersions and Lagrangian fibration in symplectic geometry. Let $(X, \omega)$ be a smooth symplectic $2n$–dimensional manifold, $B$ a smooth $n$–dimensional manifold and $\triangle \subset B$ a closed subset with $B_0 = B - \triangle$ dense in $B$. A lagrangian fibration on $X$ is a smooth map $f : X \to B$ such that the fibers of $f$ over $B_0$ are Lagrangian submanifolds and $f$ restricted to $B_0$ is a submersion; see [55] and [56].

**Remark 4.** We recall some notes for mirror symmetry in both mathematics and physics. We first give brief information about De Rham cohomology taken from [184]. Consider the vector subspace (of $\wedge^p M$) $Z^p(M)$ of all closed $p$-forms on a smooth manifold $M$ of dimension $n$. The set of all exact $p$-forms on $M$ is another vector subspace $B^p M$ of $\wedge^p M$ and, since every exact form is closed, $B^p M \subseteq Z^p(M)$. Two closed p-forms $\sigma$, $\acute{\sigma}$ are said to be cohomologous if their difference is an exact form. The equivalence classes are known as cohomology classes and the set of all these classes is denoted by $H^p M$. $H^p(M) = Z^p(M)/B^p(M)$, the quotient space of the vector space of closed forms modulo the subspace of exact forms, can be easily seen to be a vector space when addition and scalar multiplication of equivalence classes are defined in the natural way; if $\omega_1 \sim \acute{\omega}_1$ and $\omega_2 \sim \acute{\omega}_2$, then $\omega_1 + \omega_2 \sim \acute{\omega}_1 + \acute{\omega}_2$ and $c\omega \sim c\acute{\omega}$, $c \in \mathbb{R}$. The zero element of $H^p(M)$ is the set of all exact $p$-forms on $M$. Being a vector space, $H^p(M)$ is an abelian group and is called the $p$ th *De Rham cohomology* group of $M$. De Rham has proved that the dimension of $H^p(M)$ is the $p$ th *Betti number* $B_p$ of $M$ as defined in algebraic topology. The *Euler-Poincare characteristic* is defined by

$\chi = \sum_{p=0}^{n}(-1)^{p}B_{p}$. Also $\chi = 2 - 2g$, where $g$ is the number of holes in $M$. See [184] for cohomology groups of some well-known geometric objects. For a full discussion of cohomology groups; see [41].

For mirror symmetry in physics, we quote the following notes from [232]: In string theory, 10–dimensional spacetime is written as a product space of the form $M^4 \times K^6$, where $M^4$ is (Minkowski) flat space and $K^6$ is some real six dimensional compact spaces. For the compact part, Calabi-Yau manifolds were found to be suitable geometric objects. *Calabi-Yau manifolds* are compact spaces satisfying the condition of Ricci flatness and Kählericity. Ricci flatness means that the metric is a solution of the vacuum Einstein equations for general relativity. The selection of a single such space is a difficult task because there are a huge number of Calabi-Yau spaces (in six) meeting the required conditions. In string theory, the topological and complex structure of the compact manifold determines the low energy physics in the real, four non-compact dimensions. What was required by the string theorists was a Calabi-Yau space with an Euler characteristic $\chi$ of $\mp 6$. These were found by Yau himself. However, there is an entire family of "mirror" Calabi-Yau spaces with opposite Euler number. These look distinct from a topological and complex structure perspective, but they are descriptions of the same physical world, for more details; see [232] and [233].

On the other hand, in mathematics, mirror symmetry predicts that there exist pairs of Calabi-Yau manifolds $(M, N)$ such that symplectic geometry on $N$ mirrors complex geometry on $M$. Set $n = dim_{\mathbb{C}}M = dim_{\mathbb{C}}N$. A concrete prediction of mirror symmetry is that $H^q(M, \Omega_M^p) = H^q(N, \Omega_N^{n-p})$; this means that the Hodge diamond of $N$ is the reflection of the Hodge diamond of $M$ about a diagonal, here the Hodge diamond is an arrangement of *Hodge numbers* $h^{p,q}$ and the Betti and Hodge numbers are related by $B_n = \sum_{p+q=n} h^{p,q}$. Like Betti numbers, Hodge numbers of $M$ are also topological invariants, for homological mirror symmetries and their relations with Lagrangian fibrations; see [57] and [232].

An almost Hermitian structure $(J, g)$ defines an almost symplectic structure $\omega(X, Y) = g(JX, Y)$. The following lemma shows that both notions coincide.

**Lemma 10.** *[284] Let* $\pi : (M, J, g) \to N$ *be a submersion from an almost Hermitian manifold to a manifold. Then the following conditions are equivalent:*
**(1)** *The fibers of* $\pi$ *are Lagrangian submanifolds.*
**(2)** $J(ker\pi_*) = (ker\pi_*)^{\perp}$.
*Moreover, the horizontal distribution* $(ker\pi_*)^{\perp}$ *is also Lagrangian.*

*Proof.* (1) $\Rightarrow$ (2). Let $X$ and $Y$ be vertical, that is; $X, Y \in ker\pi_*$. Then $g(JX, Y) = \omega(X, Y) = 0$, which proves (2). The converse is similar.    □

In order to have a Lagrangian submersion $\pi : (M, J, g) \to N$ dimensions must be related in the following way: $dim(M) = 2dim(N)$. The most natural examples of manifolds having this relation are given by the tangent (resp. cotangent) bundle of $M = TN \to N$ (resp. $M = T^*N \to N$). We recall the tangent bundle endowed with Sasaki metric; for the geometry of tangent bundle we refer to [127] and [128]. The study of the geometry of the tangent bundle was initiated by by Sasaki in [263], and was further developed in [88] and [98]. Let $(M, g)$ be a Riemannian manifold. Then we have a direct sum decomposition

$$T_u TM = T_u^H TM \oplus T_u^V TM \qquad (3.24)$$

for $p \in M$, $u \in T_pM$. The elements of $T_u^H TM$ is called horizontal vectors. Similarly, the elements of $T_u^V TM$ is called vertical vectors. For $X \in T_pM$, the horizontal lift of $X$ to a point $(p, u) \in TM$ is the unique vector $X^H \in T_u^H TM$ such that $\pi_*(X^H) = X \circ \pi$. The vertical lift of $X$ to a point $(p, u) \in TM$ is the unique vector $X^V$ such that $X^V(df) = X(f)$ for all functions $f \in C^\infty(M, \mathbb{R})$, where $df$ is the function defined by $(df)(p, u) = u(f)$.

**Definition 38.** [263] Let $(M, g)$ be a Riemannian manifold. Then the Sasaki metric $\hat{g}$ on the tangent bundle $TM$ of M is the Riemannian metric that satisfies:
**(i)** $\hat{g}_{(p,u)}(X^H, Y^H) = g_p(X, Y)$,
**(ii)** $\hat{g}_{(p,u)}(X^V, Y^H) = 0$, and
**(iii)** $\hat{g}_{(p,u)}(X^V, Y^V) = g_p(X, Y)$
for all vector fields $X, Y \in \chi(M)$, $(p, u) \in TM$.

By using the above metric components, the Levi-Civita connection $\hat{\nabla}$ of the Sasaki metric on $TM$ and its Riemann curvature tensor $\hat{R}$ are calculated by Kowalski in [164]. There are interesting connections between the geometric properties of $(M, g)$ and $(TM, \hat{g})$. For instance, in [164], the author showed that the tangent bundle of an $(M, g)$, with metric $\hat{g}$, is locally symmetric ($\hat{\nabla}\hat{R} = 0$) only if $M$ is locally Euclidean, a useful survey on the geometry of tangent bundle including also the Cheeger-Gromoll metric and natural metric; see [127]. On the other hand, in the seminal paper [98], Dombrowski defines the almost complex structure $J$ on the tangent bundle $TM$ of a manifold $N$ having a linear connection, which is given by the conditions $J(X^H) = X^V$; $J(X^V) = -X^H$, H and V being the horizontal and vertical lifts. The tangent bundle $(TM, J, \hat{g})$ of a Riemannian manifold $(M, g)$ is an almost Hermitian manifold. Then we have the following result.

**Lemma 11.** *[284] With the above notation, $\pi : (TN, J, \hat{g}) \to (N, g)$ is a Lagrangian submersion.*

**Remark 5.** We note that the natural almost complex structure on the tangent bundle was first studied by Nagano in [196]. Tachibana and Okumura [279] showed that the almost Hermitian structure $(J, \hat{g})$ is an almost Kähler structure which is not Kähler unless the base manifold $M$ is locally flat. Therefore it has been an important problem to find Kähler structure on tangent bundle by searching suitable almost complex structures; see [144], [163], [191], [192], [204], [210], [262], [278], [281], [280] and [295]. We also note that all these modified almost complex structures imply that the projection $\pi : TM \to M$ is a Lagrangian submersion. Thus there are many examples of Lagrangian submersions in the theory of tangent bundle. We add that tangent bundle techniques have been very useful in different areas. For instance, it has been used in mathematical physics as a gauge theories of gravity and supergravity on a supergroup manifold in [118] and gauge transformations of symmetries of elementary particles on the fiber [23]. It has been applied to Lagrangian dynamics in [86]. Very recently, it found its application in visual curve completion in [313].

For the totally geodesicity of leaves of the horizontal distribution, we have the following result.

**Theorem 24.** *[239] Let $F$ be an anti-invariant Riemannian submersion from a Kähler manifold $(M, g_M, J)$ to a Riemannian manifold $(N, g_N)$. Then the following assertions are equivalent to each other:*
**(a)** *$(ker\, F_*)^\perp$ defines a totally geodesic foliation on $M$.*
**(b)** *$g_M(\mathcal{A}_X BY, JV) = g_M(CY, J\mathcal{A}_X V)$.*
**(c)** *$g_N((\nabla F_*)(X, JY), F_* JV) = g_M(CY, J\mathcal{A}_X V).$ for $X, Y \in \Gamma((ker\, F_*)^\perp)$.*

*Proof.* From (3.5) and (1.49) we get

$$g_M(\nabla_X Y, V) = g_M(\nabla_X BY, JV) + g_M(\nabla_X CY, JV)$$

for $X, Y \in \Gamma((ker\, F_*)^\perp)$ and $V \in \Gamma(ker\, F_*)$. Then, using (3.23) we obtain

$$g_M(\nabla_X Y, V) = g_M(\mathcal{A}_X BY, JV) - g_M(CY, J\mathcal{A}_X V)$$

which shows (a) $\Leftrightarrow$ (b). On the other hand, from (1.49) and (1.71), we have

$$g_M(\mathcal{A}_X BY, JV) = g_N(-(\nabla F_*)(X, BY), F_* JV)$$

which tells us that (b) $\Leftrightarrow$ (c).    □

We now show that Lagrangian submersions have a very strong property for the horizontal distribution. However, we first give the following lemmas, which are crucial for the next corollary.

**Lemma 12.** *Let $\pi : M \to N$ be a Lagrangian submersion from a Kähler manifold $M$ to a Riemannian manifold $N$. Then we have*

$$\textbf{(a)}\, \mathcal{A}_X JY = J\mathcal{A}_X Y, \quad \textbf{(b)}\, \mathcal{T}_V JW = J\mathcal{T}_V W,$$

*where $X$ and $Y$ are horizontal vector fields, and $V$ and $W$ are vertical vector fields on $M$.*

*Proof.* We prove only (a). (b) can be obtained in a similar way. From (3.5) we have

$$\nabla_X JY = J\nabla_X Y.$$

Using (1.49) and (1.50) we get

$$\mathcal{A}_X JY + \mathcal{V}\nabla_X JY = J\mathcal{H}\nabla_X Y + J\mathcal{A}_X Y.$$

Now, taking the vertical and horizontal parts of this equation, we find

$$\mathcal{A}_X JY = J\mathcal{A}_X Y$$

and

$$\mathcal{V}\nabla_X JY = J\mathcal{H}\nabla_X Y,$$

which gives us the assertion.                                    □

**Lemma 13.** *[284] Let $\pi : M \to N$ be a Lagrangian submersion from a Kähler manifold $M$ to a Riemannian manifold $N$. Then we have*

$$\mathcal{A}_X JY = -\mathcal{A}_Y JX,$$

*where $X$ and $Y$ are horizontal vector fields on $M$.*

*Proof.* Since the tensor $\mathcal{A}$ has the alternation property, from Lemma 12 (a), we have $\mathcal{A}_X JY = -J\mathcal{A}_Y X$. Applying Lemma 12 (a) again, we get $\mathcal{A}_X JY = -J\mathcal{A}_Y X = -\mathcal{A}_Y JX$.                                    □

**Theorem 25.** *[284] Let $\pi : M \to N$ be a Lagrangian submersion from a Kähler manifold $M$ to a Riemannian manifold $N$. Then the horizontal distribution is integrable and it defines a totally geodesic foliation on $M$.*

*Proof.* Considering (1.50), it will be enough to show that $\mathcal{A}_X Y = 0$ for horizontal vector fields $X$ and $Y$ on $M$. Since $\mathcal{A}$ is skew-symmetric with respect to $g$, using Lemma

12 (a) and Lemma 13, for a horizontal vector field $Z$, we have

$$
\begin{aligned}
g(\mathcal{A}_X JY, Z) &= g(J\mathcal{A}_X Y, Z) \\
&= -g(\mathcal{A}_X Y, JZ) \\
&= g(\mathcal{A}_Y X, JZ) \\
&= -g(X, \mathcal{A}_Y JZ) \\
&= g(X, \mathcal{A}_Z JY) \\
&= -g(\mathcal{A}_Z X, JY) \\
&= g(\mathcal{A}_X Z, JY) \\
&= -g(Z, \mathcal{A}_X JY)
\end{aligned}
$$

which gives $\mathcal{A}_X Z = 0$. $\qquad\qquad\qquad\qquad\qquad\qquad\qquad\qquad$ □

For the vertical distribution, we have the following result.

**Theorem 26.** *[239] Let $F : (M, g_M, J) \longrightarrow (N, g_N)$ be a Lagrangian Riemannian submersion, where $(M, g_M)$ is a Kähler manifold and $(N, g_N)$ is a Riemannian manifold. Then the following assertions are equivalent to each other:*
**(a)** *$(ker\, F_*)$ defines a totally geodesic foliation on $M$.*
**(b)** *$g_N((\nabla F_*)(Z, JX), F_* JV) = 0$ for $X \in \Gamma((ker\, F_*)^\perp)$ and $Z, W \in \Gamma(ker\, F_*)$.*
**(c)** *$\mathcal{T}_Z JW = 0$.*

*Proof.* Using (3.5), we get $g_M(\nabla_Z W, X) = g_M(\nabla_Z JW, JX)$. Hence we have $g_M(\nabla_Z W, X) = -g_M(JW, \mathcal{H}\nabla_Z JX)$. Then Riemannian submersion $F$ and (1.71) imply that

$$
g_M(\nabla_Z W, X) = g_N(F_* JW, (\nabla F_*)(Z, JX)),
$$

which is (a) $\Leftrightarrow$ (b). By direct calculation, we derive

$$
g_N(F_* JW, (\nabla F_*)(Z, JX)) = g_M(\nabla_Z JW, JX).
$$

Then from (1.48), we have

$$
g_N(F_* JW, (\nabla F_*)(Z, JX)) = g_M(\mathcal{T}_Z JW, JX).
$$

Since $\mathcal{T}_Z JW \in \Gamma(ker\, F_*)$, this shows that (b) $\Leftrightarrow$ (c). $\qquad\qquad$ □

We have the following characterization for a Lagrangian Riemannian submersion to be totally geodesic.

**Theorem 27.** *[239] Let $F$ be a Lagrangian Riemannian submersion from a Kähler manifold $(M, g_M, J)$ to a Riemannian manifold $(N, g_N)$. Then $F$ is a totally geodesic* *map if and only if*

$$\mathcal{T}_W JV = 0, \forall W, V \in \Gamma(ker\, F_*). \tag{3.25}$$

*Proof.* First of all, Lemma 1 and (1.71) imply that the second fundamental form of a Riemannian submersion satisfies

$$(\nabla F_*)(X, Y) = 0, \forall X, Y \in \Gamma((ker F_*)^\perp). \tag{3.26}$$

For $W, V \in \Gamma(ker\, F_*)$, by using (3.2), (1.48), and (1.71), we obtain

$$(\nabla F_*)(W, V) = F_*(J\mathcal{T}_W JV). \tag{3.27}$$

On the other hand, from (3.2), (3.5) and (1.71) we get

$$(\nabla F_*)(X, W) = F_*(J\nabla_X JW)$$

for $X \in \Gamma((ker\, F_*)^\perp)$. Then using (1.50), we obtain

$$(\nabla F_*)(X, W) = F_*(J\mathcal{A}_X JW). \tag{3.28}$$

Since $J$ is non-singular, proof comes from (3.26), (3.27), (3.28) and Theorem 25.  □

**Corollary 10.** *[284] Let $F$ be a Lagrangian Riemannian submersion from a Kähler manifold $(M, g_M, J)$ onto a Riemannian manifold $(N, g_N)$. Then $F$ is a totally geodesic map if and only if it has totally geodesic fibers.*

*Proof.* From Lemma 12 (b), we see that $\mathcal{T}_W JV = 0$ is equivalent to $\mathcal{T}_V W = 0$ due to $J$ being non-singular.  □

We also have the following necessary and sufficient condition for an anti-invariant Riemannian submersion to be harmonic.

**Theorem 28.** *Let $F$ be an anti-invariant Riemannian submersion from a Kähler manifold $(M, g_M, J)$ to a Riemannian manifold $(N, g_N)$. Then $F$ is harmonic if and only if*

$$trace\{J\mathcal{T}_{(.)}J(.) + C\mathcal{H}\nabla_{(.)}(.)\} = 0.$$

*Proof.* From (1.47), (3.5), (3.18), and (1.48), we have

$$\mathcal{T}_V W = -J\mathcal{T}_V JW - B\mathcal{H}\nabla_V JW - C\mathcal{H}\nabla_V JW - \hat{\nabla}_V W$$

for vertical vector fields $V$ and $W$ on $M$. Taking the horizontal parts of the above equation we get,

$$\mathcal{T}_V W = -J\mathcal{T}_V JW - C\mathcal{H}\nabla_V JW,$$

which gives the assertion.    □

For the Lagrangian case, we have the following result.

**Theorem 29.** *[239] Let F be a Lagrangian Riemannian submersion from a Kähler manifold $(M, g_M, J)$ to a Riemannian manifold $(N, g_N)$. Then F is harmonic if and only if $Trace J\mathcal{T}_V = 0$ for $V \in \Gamma(ker F_*)$.*

*Proof.* We know that $F$ is harmonic if and only if $F$ has minimal fibers. Thus $F$ is harmonic if and only if $\sum_{i=1}^{m_1} \mathcal{T}_{e_i} e_i = 0$. On the other hand, from (3.5), (1.48), and (1.47), we have

$$\mathcal{T}_V JW = J\mathcal{T}_V W \qquad (3.29)$$

for any $V, W \in \Gamma(ker F_*)$. Using (3.29), we get

$$\sum_{i=1}^{m_1} g_M(\mathcal{T}_{e_i} Je_i, V) = -\sum_{i=1}^{m_1} g_M(\mathcal{T}_{e_i} e_i, JV)$$

for any $V \in \Gamma(ker F_*)$. Thus skew-symmetric $\mathcal{T}$ implies that

$$\sum_{i=1}^{m_1} g_M(Je_i, \mathcal{T}_{e_i} V) = \sum_{i=1}^{m_1} g_M(\mathcal{T}_{e_i} e_i, JV).$$

Then, using (1.45), we have

$$\sum_{i=1}^{m_1} g_M(Je_i, \mathcal{T}_V e_i) = \sum_{i=1}^{m_1} g_M(\mathcal{T}_{e_i} e_i, JV).$$

Thus our assertion comes from (3.2).    □

In the rest of this section, we obtain decomposition theorems by using the existence of anti-invariant Riemannian submersions. From Theorem 6, Theorem 24, and Theorem 26, we have the following decomposition theorem.

**Theorem 30.** *[239] Let F be a Lagrangian Riemannian submersion from a Kähler manifold $(M, g_M, J)$ to a Riemannian manifold $(N, g_N)$. Then M is a locally product manifold if and only if $\mathcal{A}_X JY = 0$ and $\mathcal{T}_V JW = 0$ for $X, Y \in \Gamma((ker F_*)^\perp)$ and $V, W \in \Gamma(ker F_*)$.*

**Theorem 31.** *[239] Let F be a Lagrangian Riemannian submersion from a Kähler manifold $(M, g_M, J)$ to a Riemannian manifold $(N, g_N)$. Then M is a locally twisted product manifold of the form $M_{(ker F_*)^\perp} \times_f M_{kerF_*}$ if and only if*

$$\mathcal{T}_V JX = -g_M(X, \mathcal{T}_V V) \parallel V \parallel^{-2} JV$$

*for $X \in \Gamma((ker F_*)^\perp)$ and $V \in \Gamma(ker F_*)$, where $M_{(ker F_*)^\perp}$ and $M_{kerF_*}$ are integral manifolds of the distributions $(ker F_*)^\perp$ and $ker F_*$.*

*Proof.* From (3.5) and (1.48), we get

$$g_M(\nabla_V W, X) = g_M(\mathcal{T}_V JW, JX)$$

for $V, W \in \Gamma(ker F_*)$ and $X \in \Gamma((ker F_*)^\perp)$. Since $\mathcal{T}_V$ is skew-symmetric, we have

$$g_M(\nabla_V W, X) = -g_M(JW, \mathcal{T}_V JX).$$

This implies that $ker F_*$ is totally umbilical if and only if

$$\mathcal{T}_V JX = -X(\lambda)JV.$$

Then, by direct computations, it easy to see that this is equivalent to

$$\mathcal{T}_V JX = -g_M(X, \mathcal{T}_V V) \parallel V \parallel^{-2} JV.$$

Then, the proof follows from Theorem 26 and Theorem 6. □

However, in the sequel we show that the notion of anti-invariant Riemannian submersion puts some restrictions on the source manifold.

**Theorem 32.** *[239] Let $(M, g_M)$ be a Kähler manifold and $(N, g_N)$ be a Riemannian manifold. Then there do not exist anti-Riemannian submersion from M onto N such that M is a locally proper twisted product manifold of the form $M_{kerF_*} \times_f M_{(ker F_*)^\perp}$.*

*Proof.* Suppose that $F : (M, g_M) \longrightarrow (N, g_N)$ is an anti-Riemannian submersion and $M$ is a locally twisted product of the form $M_{ker F_*} \times_f M_{(kerF_*)^\perp}$. Then from Theorem 6, we know that $M_{kerF_*}$ is a totally geodesic foliation and $M_{(ker F_*)^\perp}$ is a totally umbilical foliation. We denote the second fundamental form of $M_{(ker F_*)^\perp}$ by $h$. Then we get $g_M(\nabla_X Y, V) = g_M(h(X, Y), V)$ for $X, Y \in \Gamma((ker F_*)^\perp)$ and $V \in \Gamma(ker F_*)$. Since $M_{(kerF_*)^\perp}$ is totally umbilical, we have

$$g_M(\nabla_X Y, V) = g_M(H, V)g_M(X, Y), \tag{3.30}$$

where $H$ is the mean vector field of $M_{(kerF_*)^\perp}$. On the other hand, from (3.5), we derive

$g_M(\nabla_X Y, V) = -g_M(JY, \nabla_X JV)$. Then, using (1.50), we obtain

$$g_M(\nabla_X Y, V) = -g_M(JY, \mathcal{A}_X JV). \tag{3.31}$$

Thus, from (3.30) and (3.31), we have

$$\mathcal{A}_X JV = -g_M(H, V)JX.$$

Hence, we arrive at

$$g_M(\mathcal{A}_X JV, JX) = -g_M(H, V) \parallel X \parallel^2 .$$

Then, using (1.50), we get

$$g_M(\nabla_X JV, JX) = -g_M(H, V) \parallel X \parallel^2 .$$

Thus (3.5) implies that

$$g_M(\nabla_X V, X) = -g_M(H, V) \parallel X \parallel^2 .$$

Hence, we obtain

$$g_M(V, \nabla_X X) = g_M(H, V) \parallel X \parallel^2 .$$

Then, using (1.46), we have $\mathcal{A}_X X = 0$, which implies that $g_M(H, V) \parallel X \parallel^2 = 0$. Since $g_M$ is a Riemannian metric and $H \in \Gamma(ker F_*)$, we conclude that $H = 0$. This shows that $(ker F_*)^\perp$ is totally geodesic, so $M$ is the usual product of Riemannian manifolds. Thus, the proof is complete. □

For more decomposition theorems determined by the existence of anti-invariant submersions; see [8].

## 4. Semi-invariant submersions

In this section, we introduce semi-invariant submersions from almost Hermitian manifolds onto Riemannian manifolds as a generalization of invariant Riemannian submersions and anti-invariant Riemannian submersions. We give examples, investigate the geometry of foliations which have arisen from the definition of a Riemannian submersion, and find necessary-sufficient conditions for a total manifold to be a locally product Riemannian manifold. We also find necessary and sufficient conditions for a semi-invariant submersion to be totally geodesic. Moreover, we obtain a classification for semi-invariant submersions with totally umbilical fibers and show that such submersions put some restrictions on total manifolds.

**Definition 39.** [249] Let $M_1$ be a complex $m$-dimensional almost Hermitian manifold with Hermitian metric $g_1$ and almost complex structure $J$ and $M_2$ be a Riemannian manifold with Riemannian metric $g_2$. A Riemannian submersion $F : M_1 \longrightarrow M_2$ is called a semi-invariant submersion if there is a distribution $\mathcal{D}_1 \subseteq ker\, F_*$ such that

$$ker\, F_* = \mathcal{D}_1 \oplus \mathcal{D}_2 \tag{3.32}$$

and

$$J(\mathcal{D}_1) = \mathcal{D}_1, \; J(\mathcal{D}_2) \subseteq (ker F_*)^\perp, \tag{3.33}$$

where $\mathcal{D}_2$ is orthogonal complementary to $\mathcal{D}_1$ in $ker\, F_*$.

We note that it is known that the distribution $ker\, F_*$ is integrable. Hence, the above definition implies that the integral manifold (fiber) $F^{-1}(q)$, $q \in M_2$, of $ker\, F_*$ is a CR-submanifold of $M_1$. We now give some examples of semi-invariant submersions.

**Example 6.** Every invariant Riemannian submersion from almost Hermitian manifolds onto Riemannian manifolds is a semi-invariant submersion with $\mathcal{D}_2 = \{0\}$.

**Example 7.** Every anti-invariant Riemannian submersion from an almost Hermitian manifold onto a Riemannian manifold is a semi-invariant submersion with $\mathcal{D}_1 = \{0\}$.

**Example 8.** Every Hermitian submersion from an almost Hermitian manifold onto an almost Hermitian manifold is a semi-invariant submersion with $\mathcal{D}_2 = \{0\}$.

**Example 9.** [249] Let $F$ be a submersion defined by

$$F: \qquad \mathbb{R}^6 \qquad \longrightarrow \qquad \mathbb{R}^3$$
$$(x_1, x_2, x_3, x_4, x_5, x_6) \qquad (\tfrac{x_1+x_2}{\sqrt{2}}, \tfrac{x_3+x_5}{\sqrt{2}}, \tfrac{x_4+x_6}{\sqrt{2}}).$$

Then it follows that

$$ker\, F_* = span\{V_1 = -\partial x_1 + \partial x_2, V_2 = -\partial x_3 + \partial x_5, V_3 = -\partial x_4 + \partial x_6\}$$

and

$$(ker\, F_*)^\perp = span\{X_1 = \partial x_1 + \partial x_2, X_2 = \partial x_3 + \partial x_5, X_3 = \partial x_4 + \partial x_6\}.$$

Hence we have $JV_2 = V_3$ and $JV_1 = -X_1$. Thus it follows that $\mathcal{D}_1 = span\{V_2, V_3\}$ and $\mathcal{D}_2 = span\{V_1\}$. Moreover we can see that $\mu = span\{X_2, X_3\}$. By direct computations, we also have

$$g_{\mathbb{R}^6}(JV_1, JV_1) = g_{\mathbb{R}^3}(F_*(JV_1), F_*(JV_1)), \; g_{\mathbb{R}^6}(X_2, X_2) = g_{\mathbb{R}^3}(F_*(X_2), F_*(X_2))$$

and

$$g_{\mathbb{R}^6}(X_3, X_3) = g_{\mathbb{R}^3}(F_*(X_3), F_*(X_3)),$$

which show that $F$ is a Riemannian submersion. Thus $F$ is a semi-invariant submersion.

Let $F$ be a semi-invariant submersion from a Kähler manifold $(M_1, g_1, J)$ onto a Riemannian manifold $(M_2, g_2)$. We denote the complementary distribution to $J\mathcal{D}_2$ in $(ker F_*)^\perp$ by $\mu$. Then for $V \in \Gamma(ker F_*)$, we write

$$JV = \phi V + \omega V, \tag{3.34}$$

where $\phi V \in \Gamma(\mathcal{D}_1)$ and $\omega V \in \Gamma(J\mathcal{D}_2)$. Also, for $X \in \Gamma((ker F_*)^\perp)$, we have

$$JX = \mathcal{B}X + CX, \tag{3.35}$$

where $\mathcal{B}X \in \Gamma(\mathcal{D}_2)$ and $CX \in \Gamma(\mu)$. Then, by using (3.34), (3.35), (1.47), and (1.48), we get

$$(\nabla_V \phi)W = \mathcal{B}\mathcal{T}_V W - \mathcal{T}_V \omega W \tag{3.36}$$

$$(\nabla_V \omega)W = C\mathcal{T}_V W - \mathcal{T}_V \phi W, \tag{3.37}$$

for $V, W \in \Gamma(ker F_*)$, where

$$(\nabla_V \phi)W = \hat{\nabla}_V \phi W - \phi \hat{\nabla}_V W$$

and

$$(\nabla_V \omega)W = \mathcal{H}\nabla_V^1 \omega W - \omega \hat{\nabla}_V W.$$

We now investigate the integrability of the distributions $\mathcal{D}_1$ and $\mathcal{D}_2$. Since the fibers of semi-invariant submersions from Kähler manifolds are CR-submanifolds and $\mathcal{T}$ is the second fundamental form of the fibers, the following results can be deduced from Theorem 1.1 of [26, page.39]. However, we give a proof for completeness.

**Lemma 14.** *[249] Let $F$ be a semi-invariant submersion from a Kähler manifold $(M_1, g_1, J_1)$ onto a Riemannian manifold $(M_2, g_2)$. Then:*
**1.** *the distribution $\mathcal{D}_2$ is always integrable, and*
**2.** *the distribution $\mathcal{D}_1$ is integrable if and only if*

$$g_1(T_X JY - T_Y JX, JZ) = 0$$

*for $X, Y \in \Gamma(\mathcal{D}_1)$ and $Z \in \Gamma(\mathcal{D}_2)$.*

*Proof.* (a) Since $M_1$ is a Kähler manifold, its fundamental 2−form $\Omega$ is closed. Thus for $X \in \Gamma(\mathcal{D}_1)$ and $V, W \in \Gamma(\mathcal{D}_2)$, from (3.4) we get

$$3d\Omega(X, Y, Z) = -g_1([V, W], JX) = 0,$$

which shows that $\mathcal{D}_2$ is integrable.

(b) We note that the distribution $\mathcal{D}_1$ is integrable if and only if $g_1([X, Y], Z) = g_1([X, Y], W) = 0$ for $X, Y \in \Gamma(\mathcal{D}_1)$, $Z \in \Gamma(\mathcal{D}_2)$ and $W \in \Gamma((\ker F_*)^\perp)$. Since $\ker F_*$ is integrable, $g_1([X, Y], W) = 0$. Thus $\mathcal{D}_2$ is integrable if and only if $g_1([X, Y], Z) = 0$. By using (3.5), we have

$$g_1([X, Y], Z) = g_1(\nabla_X JY, JZ) - g_1(\nabla_Y JX, JZ).$$

Then from (1.47), we obtain

$$g_1([X, Y], Z) = g_1(\mathcal{T}_X JY, JZ) - g_1(\mathcal{T}_Y JX, JZ).$$

Thus the proof is complete.    □

The proof of the following proposition can be deduced from Theorem 5.1 of [26, page.63]. Again, we give a proof for completeness.

**Proposition 25.** *Let $F$ be a semi-invariant submersion from a Kähler manifold $(M_1, g_1, J)$ onto a Riemannian manifold $(M_2, g_2)$. Then the fibers of $F$ are locally product Riemannian manifolds if and only if $(\nabla_V \phi)W = 0$ for $V, W \in \Gamma(\ker F_*)$.*

*Proof.* For $V_2, W_2 \in \Gamma(\mathcal{D}_2)$, if $(\nabla_{V_2} \phi)W_2 = 0$, then we have $\phi(\hat{\nabla}_{V_2} W_2) = 0$. This implies that $\hat{\nabla}_{V_2} W_2 \in \Gamma(\mathcal{D}_2)$. Thus $\mathcal{D}_2$ defines a totally geodesic foliation on the fiber. On the other hand, for $X, Y \in \Gamma(\mathcal{D}_1)$, (3.36) and $(\nabla_X \phi)Y = 0$ imply that that $\mathcal{B}\mathcal{T}_X Y = 0$, which shows that

$$\mathcal{T}_X Y \in \Gamma(\mu) \tag{3.38}$$

Moreover, from (3.37) we have

$$-\omega \hat{\nabla}_X Y = C\mathcal{T}_X Y - \mathcal{T}_X \phi Y$$

for $X, Y \in \Gamma(\mathcal{D}_1)$. Then from (3.38) we conclude that $C\mathcal{T}_X Y - \mathcal{T}_X \phi Y \in \Gamma(\mu)$ while $\omega \hat{\nabla}_X Y \in \Gamma(J\mathcal{D}_2)$. Hence we have $\omega \hat{\nabla}_X Y = 0$, which implies that $\hat{\nabla}_X Y \in \Gamma(\mathcal{D}_1)$. Thus $\mathcal{D}_1$ defines a totally geodesic foliation on the fibers.    □

We now obtain necessary and sufficient conditions for a semi-invariant submersion to be totally geodesic.

**Theorem 33.** *[249] Let $F$ be a semi-invariant submersion from a Kähler manifold $(M_1, g_1, J)$ onto a Riemannian manifold $(M_2, g_2)$. Then $F$ is a totally geodesic map if and only if:*

1. *$\hat{\nabla}_X \phi Y + \mathcal{T}_X \omega Y$, and $\hat{\nabla}_X \mathcal{B}Z + \mathcal{T}_X CZ$ belong to $\mathcal{D}_1$, and*
2. *$\mathcal{H}\nabla_X^1 \omega Y + T_X \phi Y$ and $\mathcal{T}_X \mathcal{B}Z + \mathcal{H}\nabla_X^1 CZ$ belong to $J\mathcal{D}_2$*

*for $X, Y \in \Gamma(\ker F_*)$ and $Z \in \Gamma((\ker F_*)^\perp)$.*

*Proof.* First of all, since $F$ is a Riemannian submersion, we have

$$(\nabla F_*)(Z_1, Z_2) = 0, \forall Z_1, Z_2 \in \Gamma((ker F_*)^\perp). \tag{3.39}$$

For $X, Y \in \Gamma(ker F_*)$, we get $(\nabla F_*)(X, Y) = -F_*(\nabla_X^1 Y)$. Then from (3.5) we get

$$(\nabla F_*)(X, Y) = F_*(J\nabla_X^1 JY).$$

Using (3.34), we have $(\nabla F_*)(X, Y) = F_*(J\nabla_X^1 \phi Y + J\nabla_X^1 \omega Y)$. Then from (1.47) and (1.48), we arrive at

$$(\nabla F_*)(X, Y) = F_*(J(\hat{\nabla}_X \phi Y + \mathcal{T}_X \phi Y + \mathcal{H}\nabla_X^1 \omega Y + \mathcal{T}_X \omega Y)).$$

Using (3.34) and (3.35) in the above equation, we obtain

$$\begin{aligned}
(\nabla F_*)(X, Y) &= F_*(\phi\hat{\nabla}_X \phi Y + \omega\hat{\nabla}_X \phi Y + \mathcal{B}\mathcal{T}_X \phi Y \\
&+ \mathcal{C}\mathcal{T}_X \phi Y + \mathcal{B}\mathcal{H}\nabla_X^1 \omega Y + \mathcal{C}\mathcal{H}\nabla_X^1 \omega Y \\
&+ \phi\mathcal{T}_X \omega Y + \omega\mathcal{T}_X \omega Y).
\end{aligned}$$

Since $\phi\hat{\nabla}_X \phi Y + \mathcal{B}\mathcal{T}_X \phi Y + \phi\mathcal{T}_X \omega Y + \mathcal{B}\mathcal{H}\nabla_X^1 \omega Y \in \Gamma(ker F_*)$, we derive

$$\begin{aligned}
(\nabla F_*)(X, Y) &= F_*(\omega\hat{\nabla}_X \phi Y + \mathcal{C}\mathcal{T}_X \phi Y \\
&+ \mathcal{C}\mathcal{H}\nabla_X^1 \omega Y + \omega\mathcal{T}_X \omega Y).
\end{aligned}$$

Then, since $F$ is a linear isometry between $(ker F_*)^\perp$ and $TM_2$, $(\nabla F_*)(X, Y) = 0$ if and only if $\omega\hat{\nabla}_X \phi Y + \mathcal{C}\mathcal{T}_X \phi Y + \mathcal{C}\mathcal{H}\nabla_X^1 \omega Y + \omega\mathcal{T}_X \omega Y = 0$. Thus $(\nabla F_*)(X, Y) = 0$ if and only if

$$\omega(\hat{\nabla}_X \phi Y + \mathcal{T}_X \omega Y) = 0, \quad \mathcal{C}(\mathcal{T}_X \phi Y + \mathcal{H}\nabla_X^1 \omega Y) = 0. \tag{3.40}$$

In a similar way, for $X \in \Gamma(ker F_*)$ and $Z \in \Gamma((ker F_*)^\perp)$, $(\nabla F_*)(X, Z) = 0$ if and only if

$$\omega(\hat{\nabla}_X \mathcal{B}Z + \mathcal{T}_X \mathcal{C}Z) = 0, \quad \mathcal{C}(\mathcal{T}_X \mathcal{B}Z + \mathcal{H}\nabla_X^1 \mathcal{C}Z) = 0. \tag{3.41}$$

Then, the proof follows from (3.39)-(3.41).    □

We now investigate the geometry of leaves of the distribution $(ker F_*)^\perp$.

**Proposition 26.** *[249] Let $F$ be a semi-invariant submersion from a Kähler manifold $(M_1, g_1, J)$ onto a Riemannian manifold $(M_2, g_2)$. Then the distribution $(ker F_*)^\perp$ defines a totally geodesic foliation if and only if*

$$\mathcal{A}_{Z_1} \mathcal{B}Z_2 + \mathcal{H}\nabla_{Z_1}^1 \mathcal{C}Z_2 \in \Gamma(\mu), \quad \mathcal{A}_{Z_1}\mathcal{C}Z_2 + \mathcal{V}\nabla_{Z_1}^1 Z_2 \in \Gamma(\mathcal{D}_2)$$

*for $Z_1, Z_2 \in \Gamma((ker F_*)^\perp)$.*

*Proof.* From (3.2) and (3.5) we have $\nabla^1_{Z_1} Z_2 = -J\nabla^1_{Z_1} JZ_2$ for $Z_1, Z_2 \in \Gamma((\ker F_*)^\perp)$. Using (3.35), (1.49), and (1.50), we obtain

$$\nabla^1_{Z_1} Z_2 = -J(A_{Z_1}\mathcal{B}Z_2 + \mathcal{V}\nabla^1_{Z_1}\mathcal{B}Z_2) - J(\mathcal{H}\nabla^1_{Z_1}CZ_2 + \mathcal{A}_{Z_1}CZ_2).$$

Then, by using (3.34) and (3.35), we get

$$
\begin{aligned}
\nabla^1_{Z_1} Z_2 &= -\mathcal{B}A_{Z_1}\mathcal{B}Z_2 - CA_{Z_1}\mathcal{B}Z_2 + \phi\mathcal{V}\nabla^1_{Z_1}\mathcal{B}Z_2 \\
&\quad - \omega\mathcal{V}\nabla^1_{Z_1}\mathcal{B}Z_2 - \mathcal{B}\mathcal{H}\nabla^1_{Z_1}CZ_2 - C\mathcal{H}\nabla^1_{Z_1}CZ_2 \\
&\quad - \phi\mathcal{A}_{Z_1}CZ_2 - \omega\mathcal{A}_{Z_1}CZ_2.
\end{aligned}
$$

Hence, we have $\nabla^1_{Z_1} Z_2 \in \Gamma((\ker F_*)^\perp)$ if and only if

$$-\mathcal{B}A_{Z_1}\mathcal{B}Z_2 - \phi\mathcal{V}\nabla^1_{Z_1}\mathcal{B}Z_2 - \mathcal{B}\mathcal{H}\nabla^1_{Z_1}CZ_2 - \phi\mathcal{A}_{Z_1}CZ_2 = 0.$$

Thus $\nabla^1_{Z_1} Z_2 \in \Gamma((\ker F_*)^\perp)$ if and only if

$$\mathcal{B}(A_{Z_1}\mathcal{B}Z_2 + \mathcal{H}\nabla^1_{Z_1}CZ_2) = 0, \quad \phi(\mathcal{V}\nabla^1_{Z_1}\mathcal{B}Z_2 + \mathcal{A}_{Z_1}CZ_2) = 0$$

which completes the proof. $\qquad\qquad\qquad\qquad\qquad\qquad\qquad\qquad\qquad\square$

In a similar way, we have the following result.

**Proposition 27.** *[249] Let F be a semi-invariant submersion from a Kähler manifold $(M_1, g_1, J)$ onto a Riemannian manifold $(M_2, g_2)$. Then the distribution $\ker F_*$ defines a totally geodesic foliation if and only if*

$$\mathcal{T}_{X_1}\phi X_2 + \mathcal{H}\nabla^1_{X_1}\omega X_2 \in \Gamma(J\mathcal{D}_2), \quad \hat{\nabla}_{X_1}\phi X_2 + \mathcal{T}_{X_1}\omega X_2 \in \Gamma(\mathcal{D}_1)$$

*for $X_1, X_2 \in \Gamma(\ker F_*)$.*

From Proposition 27, we have the following result.

**Corollary 11.** *[249] Let F be a semi-invariant Riemannian submersion from a Kähler manifold $(M_1, g_1, J)$ onto a Riemannian manifold $(M_2, g_2)$. Then the distribution $\ker F_*$ defines a totally geodesic foliation if and only if*

$$
\begin{aligned}
g_2((\nabla F_*)(X_1, X_2), F_*(JZ)) &= 0, \\
g_2((\nabla F_*)(X_1, \omega X_2), F_*(W)) &= -g_1(\mathcal{T}_{X_1}W, \phi X_2)
\end{aligned}
$$

*for $X_1, X_2 \in \Gamma(\ker F_*)$, $Z \in \Gamma(\mathcal{D}_2)$. and $W \in \Gamma(\mu)$.*

*Proof.* For $X_1, X_2 \in \Gamma(\ker F_*)$, $\hat{\nabla}_{X_1}\phi X_2 + \mathcal{T}_{X_1}\omega X_2 \in \Gamma(\mathcal{D}_1)$ if and only if $g_1(\hat{\nabla}_{X_1}\phi X_2 +$

$\mathcal{T}_{X_1}\omega X_2, Z) = 0$ for $Z \in \Gamma(\mathcal{D}_2)$. Skew-symmetric $\mathcal{T}$ and (1.47) imply that

$$g_1(\hat{\nabla}_{X_1}\phi X_2 + \mathcal{T}_{X_1}\omega X_2, Z) = g_1(\nabla^1_{X_1}\phi X_2, Z)$$
$$- g_1(\omega X_2, \mathcal{T}_{X_1}Z).$$

Hence we have

$$g_1(\hat{\nabla}_{X_1}\phi X_2 + \mathcal{T}_{X_1}\omega X_2, Z) = -g_1(\phi X_2, \nabla^1_{X_1}Z)$$
$$- g_1(\omega X_2, \mathcal{T}_{X_1}Z).$$

Using again (1.47), we get

$$g_1(\hat{\nabla}_{X_1}\phi X_2 + \mathcal{T}_{X_1}\omega X_2, Z) = -g_1(JX_2, \hat{\nabla}_{X_1}Z)$$
$$- g_1(\omega X_2, \mathcal{T}_{X_1}Z).$$

Hence we have

$$g_1(\hat{\nabla}_{X_1}\phi X_2 + \mathcal{T}_{X_1}\omega X_2, Z) = -g_1(JX_2, \nabla^1_{X_1}Z).$$

Then from (3.5) we derive

$$g_1(\hat{\nabla}_{X_1}\phi X_2 + \mathcal{T}_{X_1}\omega X_2, Z) = g_1(X_2, \nabla^1_{X_1}JZ).$$

Thus we have

$$g_1(\hat{\nabla}_{X_1}\phi X_2 + \mathcal{T}_{X_1}\omega X_2, Z) = -g_1(\nabla^1_{X_1}X_2, JZ).$$

Then Riemannian submersion $F$ implies that

$$g_1(\hat{\nabla}_{X_1}\phi X_2 + \mathcal{T}_{X_1}\omega X_2, Z) = -g_2(F_*(\nabla^1_{X_1}X_2), F_*(JZ)).$$

Using (1.71), we get

$$g_1(\hat{\nabla}_{X_1}\phi X_2 + \mathcal{T}_{X_1}\omega X_2, Z) = g_2((\nabla F_*)(X_1, X_2), F_*(JZ)). \tag{3.42}$$

On the other hand, for $X_1, X_2 \in \Gamma(\ker F_*)$, $\mathcal{T}_{X_1}\phi X_2 + \mathcal{H}\nabla^1_{X_1}\omega X_2 \in \Gamma(J\mathcal{D}_2)$ if and only if $g_1(\mathcal{T}_{X_1}\phi X_2 + \mathcal{H}\nabla^1_{X_1}\omega X_2, W) = 0$ for $W \in \Gamma(\mu)$. Since $\mathcal{T}$ is skew-symmetric, we have

$$g_1(\mathcal{T}_{X_1}\phi X_2 + \mathcal{H}\nabla^1_{X_1}\omega X_2, W) = -g_1(\phi X_2, \mathcal{T}_{X_1}W)$$
$$+ g_1(\nabla^1_{X_1}\omega X_2, W).$$

Since $F$ is a Riemannian submersion, we get

$$g_1(\mathcal{T}_{X_1}\phi X_2 + \mathcal{H}\nabla^1_{X_1}\omega X_2, W) = -g_1(\phi X_2, \mathcal{T}_{X_1}W)$$
$$+ g_2(F_*(\nabla^1_{X_1}\omega X_2), F_*W).$$

Then, from (1.71), we arrive at

$$g_1(\mathcal{T}_{X_1}\phi X_2 + \mathcal{H}\nabla^1_{X_1}\omega X_2, W) = -g_1(\phi X_2, \mathcal{T}_{X_1}W)$$
$$+ \ g_2(-(\nabla F_*)(X_1, \omega X_2), F_*W). \qquad (3.43)$$

Thus, the proof follows from (3.42), (3.43) and Proposition 27          □

From Proposition 25 and Proposition 26, we have the following.

**Theorem 34.** *[249] Let F be a semi-invariant submersion from a Kähler manifold $(M_1, g_1, J)$ onto a Riemannian manifold $(M_2, g_2)$. Then $M_1$ is locally a product Riemannian manifold $M_{\mathcal{D}_1} \times M_{\mathcal{D}_2} \times M_{(ker\,F_*)^\perp}$ if and only if*

$$(\nabla \phi) = 0 \quad on \quad ker\,F_*$$

*and*

$$\mathcal{A}_{Z_1}\mathcal{B}Z_2 + \mathcal{H}\nabla^1_{Z_1}CZ_2 \in \Gamma(\mu), \mathcal{A}_{Z_1}CZ_2 + \mathcal{V}\nabla^1_{Z_1}Z_2 \in \Gamma(\mathcal{D}_2)$$

*for $Z_1, Z_2 \in \Gamma((ker F_*)^\perp)$, where $M_{\mathcal{D}_1}$, $M_{\mathcal{D}_2}$ and $M_{(ker\,F_*)^\perp}$ are integral manifolds of the distributions $\mathcal{D}_1$, $\mathcal{D}_2$ and $(ker\,F_*)^\perp$.*

Also from Corollary 11 and Proposition 26, we have the following result.

**Theorem 35.** *[249] Let F be a semi-invariant submersion from a Kähler manifold $(M_1, g_1, J)$ onto a Riemannian manifold $(M_2, g_2)$. Then $M_1$ is locally a product Riemannian manifold $M_{ker\,F_*} \times M_{(ker\,F_*)^\perp}$ if and only if*

$$g_2((\nabla F_*)(X_1, X_2), F_*(JZ)) = 0,$$
$$g_2((\nabla F_*)(X_1, \omega X_2), F_*(W)) = -g_1(\mathcal{T}_{X_1}W, \phi X_2)$$

*and*

$$\mathcal{A}_{Z_1}\mathcal{B}Z_2 + \mathcal{H}\nabla^1_{Z_1}CZ_2 \in \Gamma(\mu), \mathcal{A}_{Z_1}CZ_2 + \mathcal{V}\nabla^1_{Z_1}Z_2 \in \Gamma(\mathcal{D}_2)$$

*for $X_1, X_2 \in \Gamma(ker\,F_*)$, $W \in \Gamma(\mu)$, $Z \in \Gamma(\mathcal{D}_2)$ and $Z_1, Z_2 \in \Gamma((ker\,F_*)^\perp)$, where $M_{ker\,F_*}$ and $M_{(ker F_*)^\perp}$ are integral manifolds of the distributions $ker\,F_*$ and $(ker\,F_*)^\perp$.*

We now give two theorems on semi-invariant submersions with totally umbilical fibers. The first result shows us that a semi-invariant submersion puts some restrictions on total manifolds. Also we obtain a classification for such submersions.

By using (1.65), (3.16), and (1.60), as in CR-submanifolds, ( see Theorem 1.2 of [26, page.78]), we have the following result.

**Theorem 36.** *[249] Let F be a semi-invariant submersion with totally umbilical fibers from a complex space form $(M_1(c), g_1, J)$ onto a Riemannian manifold $(M_2, g_2)$. Then $c = 0$.*

*Proof.* For $X, Y \in \Gamma(\mathcal{D}_1)$ and $Z \in \Gamma(\mathcal{D}_2)$, from (3.16) we get

$$g_1(R^1(X, Y)Z, JZ) = \frac{c}{2} g_1(X, JY) g_1(Z, Z). \tag{3.44}$$

On the other hand, from (1.60) we have

$$g_1(R^1(X, Y)Z, JZ) = g_1((\nabla_Y \mathcal{T})_X Z, JZ) - g_1((\nabla_X \mathcal{T})_Y Z, JZ).$$

Hence we have

$$
\begin{aligned}
g_1(R^1(X, Y)Z, JZ) &= g_1(\nabla_Y \mathcal{T}_X Z - \mathcal{T}_{\nabla_Y X} Z - \mathcal{T}_X(\nabla_Y Z), JZ) \\
&\quad - g_1(\nabla_X \mathcal{T}_Y Z - \mathcal{T}_{\nabla_X Y} Z - \mathcal{T}_Y(\nabla_X Z), JZ).
\end{aligned}
$$

Since $\mathcal{T}_X Z = \mathcal{T}_Y Z = 0$ for $X, Y \in \Gamma(\mathcal{D}_1)$ and $Z \in \Gamma(\mathcal{D}_2)$ and $\mathcal{T}$ is skew symmetric, we get

$$
\begin{aligned}
g_1(R^1(X, Y)Z, JZ) &= g_1(\mathcal{T}_{[X,Y]} Z, JZ) + g_1(\nabla_Y Z, \mathcal{T}_X JZ) \\
&\quad - g_1(\nabla_X Z, \mathcal{T}_Y JZ).
\end{aligned}
$$

Using (1.47), we get

$$
\begin{aligned}
g_1(R^1(X, Y)Z, JZ) &= g_1(\mathcal{T}_{[X,Y]} Z, JZ) + g_1(\hat{\nabla}_Y Z, \mathcal{T}_X JZ) \\
&\quad - g_1(\hat{\nabla}_X Z, \mathcal{T}_Y JZ).
\end{aligned}
$$

Then skew-symmetric $\mathcal{T}$ implies that

$$
\begin{aligned}
g_1(R^1(X, Y)Z, JZ) &= g_1(\mathcal{T}_{[X,Y]} Z, JZ) - g_1(\mathcal{T}_X \hat{\nabla}_Y Z, JZ) \\
&\quad + g_1(\mathcal{T}_Y \hat{\nabla}_X Z, JZ).
\end{aligned}
$$

Then from (1.65) we arrive at

$$
\begin{aligned}
g_1(R^1(X, Y)Z, JZ) &= g_1(\mathcal{T}_{[X,Y]} Z, JZ) - g_1(X, \hat{\nabla}_Y Z) g_1(H, JZ) \\
&\quad + g_1(Y, \hat{\nabla}_X Z) g_1(H, JZ).
\end{aligned}
$$

Hence, using (1.47), we have

$$
\begin{aligned}
g_1(R^1(X, Y)Z, JZ) &= g_1(\mathcal{T}_{[X,Y]} Z, JZ) + g_1(\nabla_Y X, Z) g_1(H, JZ) \\
&\quad - g_1(\nabla_X Y, Z) g_1(H, JZ).
\end{aligned}
$$

Thus we get

$$g_1(R^1(X, Y)Z, JZ) = g_1(\mathcal{T}_{[X,Y]} Z, JZ) + g_1([Y, X], Z) g_1(H, JZ).$$

Since *ker F*∗ is integrable, using (1.65) we have

$$g_1(R^1(X, Y)Z, JZ) = g_1([X, Y], Z)g_1(H, JZ) + g_1([Y, X], Z)g_1(H, JZ).$$

Hence

$$g_1(R^1(X, Y)Z, JZ) = 0. \tag{3.45}$$

Then from (3.44) and (3.45) we have

$$cg_1(X, JY)g_1(Z, Z) = 0.$$

Since $\mathcal{D}_1$ is an invariant distribution, putting $Y = JX$ in the above equation, we obtain $cg_1(X, X)g_1(Z, Z) = 0$. Hence Riemannian metric $g_1$ implies that $c = 0$, which completes the proof. □

We now give a classification theorem for semi-invariant submersions with totally umbilical fibers. But we need the following result, which shows that the mean curvature vector field of semi-invariant submersions has a special form.

**Lemma 15.** *[249] Let F be a semi-invariant submersion with totally umbilical fibers from a Kähler manifold $(M_1, g_1, J)$ onto a Riemannian manifold $(M_2, g_2)$. Then $H \in \Gamma(J\mathcal{D}_2)$.*

*Proof.* Using (3.2), (3.5), (1.47), (3.34), and (3.35), we get

$$\mathcal{T}_{X_1} JX_2 + \hat{\nabla}_{X_1} JX_2 = \mathcal{B}\mathcal{T}_{X_1} X_2 + \mathcal{C}\mathcal{T}_{X_1} X_2 + \phi\hat{\nabla}_{X_1} X_2 + \omega\hat{\nabla}_{X_1} X_2$$

for $X_1, X_2 \in \Gamma(\mathcal{D}_1)$. Thus, for $W \in \Gamma(\mu)$, we obtain

$$g_1(\mathcal{T}_{X_1} JX_2, W) = g_1(\mathcal{C}\mathcal{T}_{X_1} X_2, W).$$

Using (1.65), we derive

$$g_1(X_1, JX_2)g_1(H, W) = g_1(J\mathcal{T}_{X_1} X_2, W).$$

Hence we have

$$g_1(X_1, JX_2)g_1(H, W) = -g_1(\mathcal{T}_{X_1} X_2, JW),$$

where $H$ is the mean curvature vector field of the fiber. Again using (1.65), we arrive at

$$g_1(X_1, JX_2)g_1(H, W) = -g_1(X_1, X_2)g_1(H, JW). \tag{3.46}$$

Interchanging the role of $X_1$ and $X_2$, we obtain

$$g_1(X_2, JX_1)g_1(H, W) = -g_1(X_2, X_1)g_1(H, JW). \tag{3.47}$$

Thus from (3.46) and (3.47), we derive

$$g_1(X_1, X_2)g_1(H, JW) = 0,$$

which shows that $H \in \Gamma(J\mathcal{D}_2)$ due to $\mu$ being invariant distribution.    $\square$

We now give a classification theorem for a semi-invariant submersion with totally umbilical fibers which is similar to that Theorem 6.1 of [310, age.96]. We note that Lemma 15 implies that we can use the method which was used in the proof of Theorem 6.1 of [310].

**Theorem 37.** *[249] Let F be a semi-invariant submersion with totally umbilical fibers from a Kähler manifold $(M_1, g_1, J)$ onto a Riemannian manifold $(M_2, g_2)$. Then either $\mathcal{D}_2$ is one-dimensional or the fibers are totally geodesic.*

*Proof.* For $Z_1, Z_2, \in \Gamma(\mathcal{D}_2)$, using (1.47) and (1.65) we have

$$g_1(\nabla_{Z_1} Z_2, JZ_2) = g_1(Z_1, Z_2)g_1(H, JZ_2).$$

Hence, we get

$$g_1([Z_1, Z_2] + \nabla_{Z_2} Z_1, JZ_2) = g_1(Z_1, Z_2)g_1(H, JZ_2).$$

Since $ker F_*$ is integrable, we derive

$$g_1(\nabla_{Z_2} Z_1, JZ_2) = g_1(Z_1, Z_2)g_1(H, JZ_2).$$

Thus, using (3.3), we get

$$-g_1(\nabla_{Z_2} JZ_1, Z_2) = g_1(Z_1, Z_2)g_1(H, JZ_2).$$

Hence, we have

$$g_1(JZ_1, \nabla_{Z_2} Z_2) = g_1(Z_1, Z_2)g_1(H, JZ_2).$$

Using (1.47) and (1.65), we arrive at

$$g_1(Z_2, Z_2)g_1(JZ_1, H) = g_1(Z_1, Z_2)g_1(H, JZ_2). \tag{3.48}$$

Interchanging the role of $Z_1$ and $Z_2$, we also have

$$g_1(Z_1, Z_1)g_1(JZ_2, H) = g_1(Z_2, Z_1)g_1(H, JZ_1). \tag{3.49}$$

Thus from (3.48) and (3.49), we obtain

$$g_1(JZ_1, H) = \frac{g_1(Z_1, Z_2)}{g_1(Z_1, Z_1)g_1(Z_2, Z_2)} g_1(H, JZ_1). \tag{3.50}$$

Then from (3.50) and Lemma 15, we conclude that either $Z_1$ and $Z_2$ are linearly dependent or $H = 0$, which completes the proof.                                    □

**Remark 6.** We can see that the construction of semi-invariant submersions comes from the notion of CR-submanifolds. However, we have avoided using the notion of CR-submersions instead of the notion of semi-invariant submersions. The reason is that there is the notion of CR-submersion which was introduced by Kobayashi [160] and it is different from our setting. In fact, it considers a Riemannian submersion from CR-submanifold of a Kähler manifold onto a Hermitian manifold. More precisely, let $\bar{M}$ be a Kähler manifold and $M$ a CR submanifold of $\bar{M}$. A submersion from a CR-submanifold $M$ onto an almost-Hermitian manifold $N$ is a Riemannian submersion [182] $\pi : M \longrightarrow N$ with the following conditions:

**(1)** $D^{\perp}$ is the kernel of $\pi_*$, and

**(2)** at each point $p \in M$, $\pi_*$ is a complex isometry of $D_p$ to $T_{\pi(p)}N$.

This definition is given by Kobayashi for the case $J(D^{\perp}) = TM^{\perp}$. Notice that for a Riemannian submersion, the vertical distribution is always integrable. For a CR-submanifold of a Kähler manifold, the distribution $D^{\perp}$ is also integrable [39]. So this definition is meaningful.See also [94], [95], [197], [266], and [294].

## 5. Generic Riemannian submersions

In this section, we define generic Riemannian submersions from an almost Hermitian manifold onto a Riemannian manifold as a generalization of semi-invariant submersions. We find the integrability conditions for the distributions and investigate the leaves of such distributions arising from a definition of generic submersions. We also obtain necessary and sufficient conditions for such submersions to be a totally geodesic map and give decomposition theorems for the total manifold of such submersions.

**Definition 40.** [6] Let $M$ be a complex $m$-dimensional almost Hermitian manifold with Hermitian metric $g$, and an almost complex structure $J$ and $B$ be a Riemannian manifold with Riemannian metric $g_B$. A Riemannian submersion $\pi : M \to B$ is called a generic Riemannian submersion if there is a distribution $\mathcal{D}_1 \subset \Gamma(ker\pi_*)$ such that

$$(ker\pi_*) = \mathcal{D}_1 \oplus \mathcal{D}_2, \quad J\mathcal{D}_1 = \mathcal{D}_1,$$

where $\mathcal{D}_2$ is the orthogonal complement of $\mathcal{D}_1$ in $\Gamma(ker\pi_*)$, and is purely real distribution on the fibers of the submersion $\pi$.

It is known that the distribution $(ker\pi_*)$ is integrable. Hence the above definition implies that the integral manifold (fiber) $\pi^{-1}(q)$, $q \in B$, of $(ker\pi_*)$ is a generic submanifold of $M$. For the generic submanifold, we refer to [68].

**Example 10.** Every semi-invariant submersion is a generic submersion with a totally real distribution $\mathcal{D}_2$.

Since invariant Riemannian submersions and anti-invariant submersions are special semi-invariant submersions, it follows that invariant Riemannian submersions and anti-invariant Riemannian submersions are examples of generic Riemannian submersions.

For any $V \in \Gamma(ker\pi_*)$ we write

$$JV = \phi V + \omega V, \tag{3.51}$$

where $\phi V \in \Gamma(ker\,F_*)$ and $\omega V \in \Gamma(ker\pi_*)^\perp$. We denote the complementary distribution to $\omega\mathcal{D}_2$ in $(\ker\pi_*)^\perp$ by $\mu$. Then we have

$$(ker\pi_*)^\perp = \omega\mathcal{D}_2 \oplus \mu, \tag{3.52}$$

and that $\mu$ is invariant under $J$. Thus, for any $X \in \Gamma(ker\pi_*)^\perp$ we have

$$JX = BX + CX, \tag{3.53}$$

where $BX \in \Gamma(\mathcal{D}_2)$ and $CX \in \Gamma(\mu)$.

From (3.51), (3.52), and (3.53), we have the following identities.

**Lemma 16.** *Let $\pi : M \to B$ be a generic submersion from an almost Hermitian manifold onto a Riemannian manifold. Then we have*
**(i)** $\phi\mathcal{D}_1 = \mathcal{D}_1$, $\omega\mathcal{D}_1 = 0$, $\phi\mathcal{D}_2 \subset \mathcal{D}_2$, $B(ker\pi_*)^\perp = \mathcal{D}_2$,
**(ii)** $\phi^2 + B\omega = -id$, $C^2 + \omega B = -id$, *and*
**(iii)** $\omega\phi + C\omega = 0$, $BC + \phi B = 0$.

We define the covariant derivative of $\phi$ and $\omega$ as follows:

$$(\nabla_V\phi)W = \hat{\nabla}_V\phi W - \phi\hat{\nabla}_V W$$

$$(\nabla_V\omega)W = \mathcal{H}(\nabla_V\omega W) - \omega\hat{\nabla}_V W.$$

Then by using (1.47), (1.48), (3.51), and (3.53), we get

$$(\nabla_V\phi)W = B\mathcal{T}_V W - \mathcal{T}_V\omega W,$$

$$(\nabla_V\omega)W = C\mathcal{T}_V W - \mathcal{T}_V\phi W,$$

for any $V,\ W \in \Gamma(ker\pi_*)$.

The next lemma establish the main properties of the O'Neill tensor fields.

**Lemma 17.** *[6] Let $\pi$ be generic Riemannian submersion from a Kähler manifold $(M, g, J)$ onto a Riemannian manifold $(B, g_B)$. Then*

**(i)** $\mathcal{T}_V \phi W + \mathcal{A}_{\omega W} V = C \mathcal{T}_V W + \omega \hat{\nabla}_V W$ and $\hat{\nabla}_V \phi W + \mathcal{T}_V \omega W = B \mathcal{T}_V W + \phi \hat{\nabla}_V W$

**(ii)** $\mathcal{A}_X BY + \mathcal{H}(\nabla_X CY) = C(\mathcal{H}(\nabla_X Y)) + \omega \mathcal{A}_X Y$ and $\mathcal{V}(\nabla_X BY) + \mathcal{A}_X CY = B(\mathcal{H}(\nabla_X Y)) + \phi \mathcal{A}_X Y,$

**(iii)** $\mathcal{A}_X \phi V + \mathcal{H}(\nabla_V \omega W) = C \mathcal{A}_X V + \omega(\mathcal{V}(\nabla_X V))$ and $\mathcal{V}(\nabla_X \phi V) + \mathcal{A}_X \omega V = B \mathcal{A}_X V + \phi(\mathcal{V}(\nabla_X V)),$

*for any $X, Y \in \Gamma(\ker\pi_*)^\perp$ and $V, W \in \Gamma(\ker\pi_*)$.*

*Proof.* Since $M$ is a Kähler manifold, for any $V, W \in \Gamma(\ker\pi_*)$ using (3.5) and (3.51) we have

$$J\nabla_V W = \nabla_V \phi W + \nabla_V \omega W.$$

Further, on using (1.47) and (1.48), we get

$$\mathcal{T}_V \phi W + \hat{\nabla}_V \phi W + \mathcal{A}_{\omega W} V + \mathcal{T}_V \omega W = J(\mathcal{T}_V W + \hat{\nabla}_V W).$$

Since $\mathcal{T}_V W$ and $\hat{\nabla}_V W$ are the horizontal and vertical, therefore again using (3.51) and (3.53), we get

$$\mathcal{T}_V \phi W + \hat{\nabla}_V \phi W + \mathcal{A}_{\omega W} V + \mathcal{T}_V \omega W = B\mathcal{T}_V W + C\mathcal{T}_V W + \phi\hat{\nabla}_V W + \omega\hat{\nabla}_V W. \quad (3.54)$$

By comparing the vertical and horizontal parts in (3.54), we get the result. Proof of (*ii*) and (*iii*) follows on similar lines as in (*i*).  □

**Lemma 18.** *[6] Let $\pi : M \to B$ be a generic Riemannian submersion from a Kähler manifold $(M, g, J)$ onto a Riemannian manifold $(B, g_B)$. Then*

$$g(J\mathcal{T}_U V, \xi) = g(\mathcal{T}_U JV, \xi),$$

*for any $U \in \Gamma(\ker\pi_*)$, $V \in \Gamma(\mathcal{D}_1)$ and $\xi \in \Gamma(\mu)$.*

*Proof.* Since $M$ is a Kähler manifold, for any $V \in \Gamma(\mathcal{D}_1)$ and $U \in \Gamma(\ker\pi_*)$ using (3.5) we have

$$J\nabla_U V = \nabla_U JV.$$

On using (1.47), we get

$$J(\mathcal{T}_U V + \hat{\nabla}_U V) = \mathcal{T}_U JV + \hat{\nabla}_U JV.$$

Taking the inner product with $\xi \in \Gamma(\mu)$, we get

$$g(J\mathcal{T}_U V, \xi) + g(J\hat{\nabla}_U V, \xi) = g(\mathcal{T}_U JV, \xi) + g(\hat{\nabla}_U JV, \xi). \quad (3.55)$$

Since $\mu$ is invariant under $J$, then the result follows from (3.55).  □

Now, we investigate the integrability of the distributions $\mathcal{D}_1$ and $\mathcal{D}_2$. Since we have seen that the fibers of generic submersions from Kähler manifolds are generic Riemannian submanifolds and $\mathcal{T}$ is the second fundamental form of the fibers, we have the following theorem:

**Theorem 38.** *[6] Let $\pi$ be a generic Riemannian submersion from a Kähler manifold $(M, g, J)$ onto a Riemannian manifold $(B, g_B)$. Then the distribution $\mathcal{D}_1$ is integrable if and only if*

$$g(\mathcal{T}_V JW, \omega U) = g(\mathcal{T}_W JV, \omega U), \tag{3.56}$$

*for any $V,\ W \in \Gamma(\mathcal{D}_1)$ and $U \in \Gamma(\mathcal{D}_2)$.*

*Proof.* Since $M$ is a Kähler manifold, for any $V,\ W \in \Gamma(\mathcal{D}_1)$, (3.5) and (1.47) give

$$
\begin{aligned}
J[V, W] &= J\nabla_V W - J\nabla_W V \\
&= \nabla_V JW - \nabla_W JV \\
&= \mathcal{T}_V JW - \mathcal{T}_W JV + \hat{\nabla}_V JW - \hat{\nabla}_W JV.
\end{aligned}
$$

Thus we have

$$\mathcal{T}_V JW - \mathcal{T}_W JV = J[V, W] + \hat{\nabla}_W JV - \hat{\nabla}_V JW. \tag{3.57}$$

Now if $\mathcal{D}_1$ is integrable, then $J[V, W] \in \Gamma(\mathcal{D}_1)$ as $[V, W] \in \Gamma(\mathcal{D}_1)$. Hence in (3.57), the right-hand side is vertical while the left-hand side is horizontal. On comparing the horizontal and vertical parts, we get

$$\mathcal{T}_V JW = \mathcal{T}_W JV,$$

for any $V,\ W \in \Gamma(\mathcal{D}_1)$. In particular, we get

$$g(\mathcal{T}_V JW, \omega U) = g(\mathcal{T}_W JV, \omega U).$$

Conversely, suppose that (3.57) holds, i.e.,

$$g(\mathcal{T}_V JW - \mathcal{T}_W JV, \omega U) = 0,$$

which shows us that

$$\mathcal{T}_V JW - \mathcal{T}_W JV \in \Gamma(\mu).$$

Now for any $\xi \in \Gamma(\mu)$, using (1.45), we have

$$g(\mathcal{T}_V JW - \mathcal{T}_W JV, \xi) = g(J\mathcal{T}_V W - J\mathcal{T}_W V, \xi) = 0,$$

which implies that $\mathcal{T}_V JW - \mathcal{T}_W JV = 0$, for any $V,\ W \in \Gamma(\mathcal{D}_1)$. Thus from (3.57), we get

$$J[V, W] = \hat{\nabla}_V JW - \hat{\nabla}_W JV.$$

Since $\hat{\nabla}_V JW - \hat{\nabla}_W JV$ lies in $\Gamma(ker\pi_*)$, this implies that $[V, W]$ lies in $\Gamma(\mathcal{D}_1)$ and hence $\mathcal{D}_1$ is integrable. $\qquad\square$

**Theorem 39.** *[6] Let $\pi$ be a generic Riemannian submersion from a Kähler manifold $(M, g, J)$ onto a Riemannian manifold $(B, g_B)$. Then the distribution $\mathcal{D}_2$ is integrable if and only if*

$$\hat{\nabla}_W \phi V - \hat{\nabla}_V \phi\omega + \mathcal{T}_W \omega V - \mathcal{T}_V \omega W \in \Gamma(\mathcal{D}_2), \tag{3.58}$$

*for any $V, W \in \Gamma(\mathcal{D}_2)$.*

*Proof.* For any $V, W \in \gamma(\mathcal{D}_2)$, using (3.5), (1.47), (1.48), (3.51), and (3.53) we have

$$
\begin{aligned}
\nabla_V W &= -J\nabla_V JW \\
&= -J(\nabla_V \phi W + \nabla_V \omega W) \\
&= -J(\mathcal{T}_V \phi W + \hat{\nabla}_V \phi W + \mathcal{A}_\omega W V + \mathcal{T}_V \omega W) \\
&= -(B\mathcal{T}_V \phi W + C\mathcal{T}_V \phi W + \phi\hat{\nabla}_V \phi W + \omega\hat{\nabla}_V \phi \\
&\quad + B\mathcal{A}_{\omega W} V + C\mathcal{A}_{\omega W} V + \phi\mathcal{T}_V \omega W + \omega\mathcal{T}_V \omega W). 
\end{aligned}
\tag{3.59}
$$

Similarly, we get

$$
\begin{aligned}
\nabla_W V &= -(B\mathcal{T}_W \phi V + C\mathcal{T}_W \phi V + \phi\hat{\nabla}_W \phi V + \omega\hat{\nabla}_W \phi V \\
&\quad + B\mathcal{A}_{\omega V} W + C\mathcal{A}_{\omega V} W + \phi\mathcal{T}_W \omega V + \omega\mathcal{T}_W \omega V). 
\end{aligned}
\tag{3.60}
$$

From (3.59) and (3.60), we get

$$
\begin{aligned}
[V, W] &= B(\mathcal{T}_W \phi V - \mathcal{T}_V \phi W + \mathcal{A}_{\omega V} W - A_{\omega W} V) \\
&\quad + C(\mathcal{T}_W \phi V - \mathcal{T}_V \phi W + \mathcal{A}_{\omega V} W - A_{\omega W} V) \\
&\quad + \phi(\hat{\nabla}_W \phi V - \hat{\nabla}_V \phi W + \mathcal{T}_W \omega V - \mathcal{T}_V \omega W) \\
&\quad + \omega(\hat{\nabla}_W \phi V - \hat{\nabla}_V \phi W + \mathcal{T}_W \omega V - \mathcal{T}_V \omega W),
\end{aligned}
$$

for any $V, W \in \Gamma(\mathcal{D}_1) \subset \Gamma(ker\pi_*)$. As $(ker\pi_*)$ is integrable, therefore $[V, W] \in \Gamma(ker\pi_*)$, comparing the vertical part, we get

$$
\begin{aligned}
[V, W] &= B(\mathcal{T}_W \phi V - \mathcal{T}_V \phi W + \mathcal{A}_{\omega V} W - A_{\omega W} V) + \phi(\hat{\nabla}_W \phi V \\
&\quad -\hat{\nabla}_V \phi W + \mathcal{T}_W \omega V - \mathcal{T}_V \omega W). 
\end{aligned}
\tag{3.61}
$$

From (3.61) it follows that the distribution $\mathcal{D}_2$ is integrable if and only if

$$\hat{\nabla}_W \phi V - \hat{\nabla}_V \phi W + \mathcal{T}_W \omega V - \mathcal{T}_V \omega W \in \Gamma(\mathcal{D}_2)$$

for any $V$, $W \in \Gamma(\mathcal{D}_2)$.    $\square$

For the geometry of the leaves of the distributions $\mathcal{D}_1$ and $\mathcal{D}_2$ we have the following propositions.

**Proposition 28.** *[6] Let $\pi$ be a generic Riemannian submersion from a Kähler manifold $(M, g, J)$ onto a Riemannian manifold $(B, g_B)$. Then the distribution $\mathcal{D}_1$ defines a totally geodesic foliation if and only if*

$$\hat{\nabla}_{V_1} \phi W_1 \in \Gamma(\mathcal{D}_1) \ and \ \mathcal{T}_{V_1} \phi W_1 = 0,$$

*for any $V_1$, $W_1 \in \Gamma(\mathcal{D}_1)$.*

*Proof.* For $V_1$, $W_1 \in \Gamma(\mathcal{D}_1)$, using (3.5), (1.47), (3.51), and (3.53) we have

$$\begin{aligned}
\nabla_{V_1} W_1 &= -J \nabla_{V_1} J W_1 \\
&= -J(\nabla_{V_1} \phi W_1) \\
&= -J(\mathcal{T}_{V_1} \phi W_1 + \hat{\nabla}_{V_1} \phi W_1) \\
&= -(B\mathcal{T}_{V_1} \phi W_1 + C\mathcal{T}_{V_1} \phi W_1 + \phi \hat{\nabla}_{V_1} \phi W_1 + \omega \hat{\nabla}_{V_1} \phi W_1)
\end{aligned}$$

Hence $\nabla_{V_1} W_1 \in \Gamma(\mathcal{D}_1)$ if and only if $\hat{\nabla}_{V_1} \phi W_1 \in \Gamma(\mathcal{D}_1)$ and $\mathcal{T}_{V_1} \phi W_1 = 0$, which completes the proof.    $\square$

**Proposition 29.** *[6] Let $\pi$ be a generic Riemannian submersion from a Kähler manifold $(M, g, J)$ onto a Riemannian manifold $(B, g_B)$. Then the distribution $\mathcal{D}_2$ defines a totally geodesic foliation if and only if*

$$T_{V_2} \phi W_2 + \mathcal{A}_{\omega W_2} V_2 \in \Gamma(\omega \mathcal{D}_2) \ and \ \hat{\nabla}_{V_2} \phi W_2 + \mathcal{T}_{V_2} \omega W_2 = 0$$

*for any $V_2$, $W_2 \in \Gamma(\mathcal{D}_2)$.*

*Proof.* The proof follows from (3.59).    $\square$

From Proposition 28 and Proposition 29, we have the following theorem.

**Theorem 40.** *[6] Let $\pi$ be a generic Riemannian submersion from a Kähler manifold $(M, g, J)$ onto a Riemannian manifold $(B, g_B)$. Then the fibers of $\pi$ are the locally Riemannian product of leaves of $\mathcal{D}_1$ and $\mathcal{D}_2$ if and only if*

$$\hat{\nabla}_{V_1} \phi W_1 \in \Gamma(\mathcal{D}_1), \ T_{V_1} \phi W_1 = 0, \ and$$

$$\mathcal{T}_{V_2} \phi W_2 + \mathcal{A}_{\omega W_2} V_2 \in \Gamma(\omega \mathcal{D}_2), \ \hat{\nabla}_{V_2} \phi W_2 + \mathcal{T}_{V_2} \omega W_2 = 0,$$

*for any $V_1$, $W_1 \in \Gamma(\mathcal{D}_1)$ $V_2$, $W_2 \in \Gamma(\mathcal{D}_2)$.*

Now we discuss the geometry of the leaves of $(ker\pi_*)$ and $(ker\pi_*)^\perp$.

**Proposition 30.** *[6] Let $\pi : M \to B$ be generic Riemannian submersion from a Kähler manifold $(M, g, J)$ onto a Riemannian manifold $(B, g_B)$. Then the distribution $(ker\pi_*)^\perp$ defines a totally geodesic foliation if and only if*

$$\mathcal{A}_X BY + \mathcal{H}\nabla_X CY \in \Gamma(\mu) \text{ and } \mathcal{V}\nabla_X BY + \mathcal{A}_X CY = 0,$$

*for any $X$, $Y \in \Gamma(ker\pi_*)^\perp$.*

*Proof.* For any $X, Y \in \Gamma(ker\pi_*)^\perp$, from (3.2) we have

$$\nabla_X Y = -J\nabla_X JY$$

Then by using (3.53), (1.49), and (1.50), we get

$$\begin{aligned}
\nabla_X Y &= -J(\nabla_X BY + \nabla_X CY) \\
&= -J((\mathcal{A}_X BY + \mathcal{V}\nabla_X BY) + (\mathcal{H}\nabla_X CY + \mathcal{A}_X CY)) \\
&= -(B\mathcal{A}_X BY + C\mathcal{A}_X BY + \phi(\mathcal{V}\nabla_X BY) + \omega(\mathcal{V}\nabla_X BY) \\
&\quad + B(\mathcal{H}\nabla_X CY) + C(\mathcal{H}\nabla_X CY) + \phi(\mathcal{A}_X CY) + \omega(\mathcal{A}_X CY)) \quad (3.62)
\end{aligned}$$

From (3.62) it follows that $(ker\pi_*)^\perp$ defines a totally geodesic foliation if and only if

$$B(\mathcal{A}_X BY + \mathcal{H}\nabla_X CY) + \phi(\mathcal{V}\nabla_X BY + \mathcal{A}_X CY) = 0,$$

which then yields

$$B(\mathcal{A}_X BY + \mathcal{H}\nabla_X CY) = 0$$
$$\phi(\mathcal{V}\nabla_X BY + \mathcal{A}_X CY) = 0.$$

$\square$

For the distribution $(ker\pi_*)$, we have the following result.

**Proposition 31.** *[6] Let $\pi$ be a generic Riemannian submersion from a Kähler manifold $(M, g, J)$ onto a Riemannian manifold $(B, g_B)$. Then the distribution $(ker\pi_*)$ defines a totally geodesic foliation if and only if*

$$T_V \phi W + \mathcal{A}_{\omega W} V \in \Gamma(\omega \mathcal{D}_2) \text{ and } \hat{\nabla}_V \phi W + T_V \omega W \in \Gamma(\mathcal{D}_1),$$

*for any $V, W \in \Gamma(ker\pi_*)$.*

*Proof.* For any $V$, $W \in \Gamma(ker\pi_*)$ using (3.5), (1.47), (1.48), and (3.51), we get

$$\nabla_V W = -J\nabla_V JW$$
$$= -J(\nabla_V \phi\omega + \nabla_V \omega W)$$
$$= -J(\mathcal{T}_V \phi W + \hat{\nabla}_V \phi W + \mathcal{A}_{\omega W} V + \mathcal{T}_V \omega W)$$
$$= -(B\mathcal{T}_V \phi W + C\mathcal{T}_V \phi W + \phi\hat{\nabla}_V \phi W + \omega\hat{\nabla}_V \phi W$$
$$+ B\mathcal{A}_{\omega W} V + C\mathcal{A}_{\omega W} V + \phi\mathcal{T}_V \omega W + \omega\mathcal{T}_V \omega W)$$

or

$$\nabla_V W = -B(\mathcal{T}_V \phi W + \mathcal{A}_{\omega W} V) - \phi(\hat{\nabla}_V \phi W + \mathcal{V}\nabla_V \omega W)$$
$$- C(\mathcal{T}_V \phi W + \mathcal{A}_{\omega W} V) - \omega(\hat{\nabla}_V \phi W + \mathcal{V}\nabla_V \omega W). \tag{3.63}$$

From the above equation, it follows that $(ker\pi_*)$ defines a totally geodesic foliation if and only if

$$C(\mathcal{T}_V \phi W + \mathcal{A}_{\omega W} V) + \omega(\hat{\nabla}_V \phi W + \mathcal{V}\nabla_V \omega W) = 0.$$

which implies

$$\mathcal{T}_V \phi W + \mathcal{A}_{\omega W} V \in \Gamma(\omega\mathcal{D}_2) \text{ and } \hat{\nabla}_V \phi W + \mathcal{V}\nabla_V \omega W \in \Gamma(\mathcal{D}_1).$$

$\square$

From Theorem 40 and Proposition 30, we have the following decomposition for total space.

**Theorem 41.** *[6] Let $\pi$ be a generic Riemannian submersion from a Kähler manifold $(M, g, J)$ onto a Riemannian manifold $(B, g_B)$. Then the total space $M$ is a generic product manifold of the leaves of $\mathcal{D}_1$, $\mathcal{D}_2$ and $(ker\pi_*)^{\perp}$, i.e., $M = M_{\mathcal{D}_1} \times M_{\mathcal{D}_2} \times M_{(ker\pi_*)^{\perp}}$, if and only if*

$$\hat{\nabla}_{V_1}\phi W_1 \in \Gamma(\mathcal{D}_1), \; T_{V_1}\phi W_1 = 0,$$

$$\mathcal{T}_{V_2}\phi W_2 + \mathcal{A}_{\omega W_2} V_2 \in \Gamma(\omega\mathcal{D}_2), \; \hat{\nabla}_{V_2}\phi W_2 + \mathcal{T}_{V_2}\omega W_2 = 0 \text{ and}$$

$$\mathcal{A}_X BY + \mathcal{H}\nabla_X CY \in \Gamma(\mu), \; \mathcal{V}\nabla_X BY + \mathcal{A}_X CY = 0,$$

*for any $V_1$, $W_1 \in \Gamma(\mathcal{D}_1)$, $V_2$, $W_2 \in \Gamma(\mathcal{D}_2)$, and $X$, $Y \in \Gamma(ker\pi_*)^{\perp}$, where $M_{\mathcal{D}_1}$, $M_{\mathcal{D}_2}$ and $M_{(ker\pi_*)^{\perp}}$ are leaves of the distributions $\mathcal{D}_1$, $\mathcal{D}_2$ and $(ker\pi_*)^{\perp}$ respectively.*

From Proposition 30 and Proposition 31, we have the following result.

**Theorem 42.** *[6] Let $\pi$ be a generic Riemannian submersion from a Kähler manifold $(M, g, J)$ onto a Riemannian manifold $(B, g_B)$. Then $M$ is generic product manifold if and only if*

$$\mathcal{A}_X BY + \mathcal{H}\nabla_X CY \in \Gamma(\mu), \quad \mathcal{V}\nabla_X BY + \mathcal{A}_X CY = 0 \text{ and}$$

$$\mathcal{T}_V \phi W + \mathcal{A}_{\omega W} V \in \Gamma(\omega \mathcal{D}_2), \quad \hat{\nabla}_V \phi W + \mathcal{V}\nabla_V \omega W \in \Gamma(\mathcal{D}_1),$$

*for any $X, Y \in \Gamma(ker\pi_*)^\perp$ and $V, W \in \Gamma(ker\pi_*)$.*

Now we obtain the necessary and sufficient condition for generic Riemannian submersion to be totally geodesic.

**Theorem 43.** *Let $\pi$ be a generic Riemannian submersion from a Kähler manifold $(M, g, J)$ onto a Riemannian manifold $(B, g_B)$. Then $\pi$ is a totally geodesic map if and only if*

$$\hat{\nabla}_V \phi W + \mathcal{T}_V \omega W \in \Gamma(\mathcal{D}_1) \quad , \quad \mathcal{T}_V \phi W + \mathcal{A}_{\omega W} V \in \Gamma(\omega \mathcal{D}_2),$$
$$\hat{\nabla}_V BX + \mathcal{T}_V CX \in \Gamma(\mathcal{D}_1) \quad , \quad \mathcal{T}_V BX + \mathcal{A}_{CX} V \in \Gamma(\omega \mathcal{D}_2),$$

*for any $X \in \Gamma(ker\pi_*)^\perp$ and $V, W \in \Gamma(ker\pi_*)$.*

*Proof.* Since $\pi$ is a Riemannian submersion, we have

$$(\nabla \pi_*)(X, Y) = 0, \text{ for all } X, Y \in \Gamma(ker\pi_*)^\perp.$$

For any $V, W \in \Gamma(ker\pi_*)$ using (3.5), (1.47), (1.48), (1.71), and (3.51), we get

$$\begin{aligned}
(\nabla \pi_*)(V, W) &= -\pi_*(\nabla_V W) \\
&= -\pi_*(-J\nabla_V JW) \\
&= \pi_*(J\nabla_V(\phi W + \omega W)) \\
&= \pi_*(J(\mathcal{T}_V \phi W + \hat{\nabla}_V \phi W) + J(\mathcal{A}_{\omega W} V + \mathcal{T}_V \omega W)) \\
&= \pi_*((B\mathcal{T}_V \phi W + C\mathcal{T}_V \phi W) + (\phi \hat{\nabla}_V \phi W + \omega \hat{\nabla}_V \phi W) \\
&\quad + (B\mathcal{A}_{\omega W} V + C\mathcal{A}_{\omega W} V) + (\phi \mathcal{T}_V \omega W + \omega \mathcal{T}_V \omega W)).
\end{aligned}$$

Thus $(\nabla \pi_*)(V, W) = 0$ if and only if

$$\omega(\hat{\nabla}_V \phi W + \mathcal{T}_V \omega W) + C(\mathcal{T}_V \phi W + \mathcal{A}_{\omega W} V) = 0. \tag{3.64}$$

On the other hand using (3.5), (1.47), (1.48), and (3.53) for any $X \in \Gamma(ker\pi_*)^\perp$ and

$V \in \Gamma(ker\pi_*)$, we get

$$
\begin{aligned}
(\nabla\pi_*)(V, X) &= -\pi_*(\nabla_V X) \\
&= -\pi_*(-J\nabla_V JX) \\
&= \pi_*(J(\nabla_V BX + \nabla_V CX)) \\
&= \pi_*((B\mathcal{T}_V BX + C\mathcal{T}_V BX) + (\phi\hat{\nabla}_V BX + \omega\hat{\nabla}_V BX) \\
&\quad + (B\mathcal{A}_{CX}V + C\mathcal{A}_{CX}V) + (\phi\mathcal{T}_V CX + \omega\mathcal{T}_V CX)).
\end{aligned}
$$

Thus $(\nabla\pi_*)(V, X) = 0$ if and only if

$$
\omega(\hat{\nabla}_V BX + \mathcal{T}_V CX) + C(\mathcal{T}_V BX + \mathcal{A}_{CX}V) = 0. \tag{3.65}
$$

The result then follows from (3.64) and (3.65).    □

Finally, we investigate the geometry of generic submersions with totally umbilical fibers.

**Proposition 32.** *[6] Let $\pi$ be a generic Riemannian subersion with totally umbilical fibers from a Kähler manifold $(M, g, J)$ onto a Riemannian manifold $(B, g_B)$, then $H \in \Gamma(\omega\mathcal{D}_2)$.*

*Proof.* From (3.5), we have

$$
\nabla_V JW = J\nabla_V W,
$$

for any $V, W \in \Gamma(\mathcal{D}_1)$. Now, using (3.51) and (3.53), we obtain

$$
\begin{aligned}
\mathcal{T}_V JW + \hat{\nabla}_V JW &= J(\mathcal{T}_V W + \hat{\nabla}_V W) \\
&= B\mathcal{T}_V W + C\mathcal{T}_V W + \phi\hat{\nabla}_V W + \omega\hat{\nabla}_V W. \tag{3.66}
\end{aligned}
$$

Taking the inner product in (3.66) with $X \in \Gamma(\mu)$ and then using (1.65), we get

$$
g(\mathcal{T}_V JW, X) = g(C\mathcal{T}_V W, X).
$$

Since the fibers are totally umbilical, we get

$$
g(V, JW)g(H, X) = -g(V, W)g(H, JX). \tag{3.67}
$$

Interchanging $V$ and $W$ in (3.67), we get

$$
g(W, JV)g(H, X) = -g(V, W)g(H, JX) \tag{3.68}
$$

Combining (3.67) and (3.68), we get $g(H, JX) = 0$, which shows that $H \in \Gamma(\omega\mathcal{D}_2)$.

□

## 6. Slant submersions

In this section, as a generalization of almost Hermitian submersions and anti-invariant submersions, we introduce slant submersions from almost Hermitian manifolds onto Riemannian manifolds. We give examples, investigate the geometry of foliations which have arisen from the definition of a Riemannian submersion, and check the harmonicity of such submersions. We also find necessary and sufficient conditions for a slant submersion to be totally geodesic. Moreover, we obtain a decomposition theorem for the total manifold of such submersions.

We first recall the definition of the slant distribution. Given a submanifold $M$, isometrically immersed in an almost Hermitian manifold $(\bar{M}, g_{\bar{M}}, J)$, a differentiable distribution $D$ on $M$ is said to be a slant distribution if for any nonzero vector $X \in D_p$ ; $p \in M$, the angle between $JX$ and the vector space $D_p$ is constant, that is, it is independent of the choice of $p \in M$ and $X \in D_p$. This constant angle is called the slant angle of the slant distribution $D$ [47]. Following on from the above definition, we present the following notion.

**Definition 41.** [245] Let $F$ be a Riemannian submersion from an almost Hermitian manifold $(M_1, g_1, J_1)$ onto a Riemannian manifold $(M_2, g_2)$. If for any nonzero vector $X \in (ker F_{*p})$; $p \in M_1$, the angle $\theta(X)$ between $JX$ and the space $(ker F_{*p})$ is a constant, i.e., it is independent of the choice of the point $p \in M_1$ and choice of the tangent vector $X$ in $(ker F_{*p})$, then we say that $F$ is a *slant submersion*. In this case, the angle $\theta$ is called the *slant angle* of the slant submersion.

It is known that the distribution $(ker F_*)$ is integrable for a Riemannian submersion between Riemannian manifolds. In fact, its leaves are $F^{-1}(q)$, $q \in M_1$, i.e., fibers. Thus it follows from the above definition that the fibers of a slant submersion are slant submanifolds of $M_1$.

We first give some examples of slant submersions.

**Example 11.** Every almost Hermitian submersion from an almost Hermitian manifold onto an almost Hermitian manifold is a slant submersion with $\theta = 0$.

**Example 12.** Every anti-invariant Riemannian submersion from an almost Hermitian manifold to a Riemannian manifold is a slant submersion with $\theta = \frac{\pi}{2}$.

A slant submersion is said to be proper if it is neither Hermitian nor anti-invariant Riemannian submersion.

**Example 13.** [245] Consider the following Riemannian submersion given by

$$F: \quad \mathbb{R}^4 \quad \longrightarrow \quad \mathbb{R}^2$$
$$(x_1, x_2, x_3, x_4) \quad (x_1 \sin \alpha - x_3 \cos \alpha, x_4).$$

Then for any $0 < \alpha < \frac{\pi}{2}$, $F$ is a slant submersion with slant angle $\alpha$.

**Example 14.** [245] Consider the following Riemannian submersion given by

$$F: \quad \mathbb{R}^4 \quad \longrightarrow \quad \mathbb{R}^2$$
$$(x_1, x_2, x_3, x_4) \quad (\frac{x_1 - x_4}{\sqrt{2}}, x_2).$$

Then $F$ is a slant submersion with slant angle $\theta = \frac{\pi}{4}$.

Let $F$ be a Riemannian submersion from an almost Hermitian manifold $(M_1, g_1, J)$ onto a Riemannian manifold $(M_2, g_2)$. Then for $X \in \Gamma(ker F_*)$, we write

$$JX = \phi X + \omega X, \tag{3.69}$$

where $\phi X$ and $\omega X$ are vertical and horizontal parts of $JX$. Also for $Z \in \Gamma((ker F_*)^\perp)$, we have

$$JZ = \mathcal{B}Z + \mathcal{C}Z, \tag{3.70}$$

where $\mathcal{B}Z$ and $\mathcal{C}Z$ are vertical and horizontal components of $JZ$. Using (1.47), (1.48), (3.69), and (3.70), we obtain

$$(\nabla_X \omega)Y = \mathcal{C}\mathcal{T}_X Y - \mathcal{T}_X \phi Y \tag{3.71}$$
$$(\nabla_X \phi)Y = \mathcal{B}\mathcal{T}_X Y - \mathcal{T}_X \omega Y, \tag{3.72}$$

where $\nabla$ is the Levi-Civita connection on $M_1$ and

$$(\nabla_X \omega)Y = \mathcal{H}\nabla_X \omega Y - \omega \hat{\nabla}_X Y$$
$$(\nabla_X \phi)Y = \hat{\nabla}_X \phi Y - \phi \hat{\nabla}_X Y$$

for $X, Y \in \Gamma(ker F_*)$. Let $F$ be a proper slant submersion from an almost Hermitian manifold $(M_1, g_1, J_1)$ onto a Riemannian manifold $(M_2, g_2)$, then we say that $\omega$ is parallel with respect to the Levi-Civita connection $\nabla$ on $(ker F_*)$ if its covariant derivative with respect to $\nabla$ vanishes, i.e., we have

$$(\nabla_X \omega)Y = \nabla_X \omega Y - \omega(\nabla_X Y) = 0$$

for $X, Y \in \Gamma(ker F_*)$.

The proof of the following result is exactly the same as slant immersions (see [70] or [48] for the Sasakian case), therefore we omit its proof.

**Theorem 44.** *[245] Let $F$ be a Riemannian submersion from an almost Hermitian* manifold $(M_1, g_1, J)$ onto a Riemannian manifold $(M_2, g_2)$. Then $F$ is a proper slant submersion if and only if there exists a constant $\lambda \in [-1, 0]$ such that

$$\phi^2 X = \lambda X$$

for $X \in \Gamma(\ker F_*)$. If $F$ is a proper slant submersion, then $\lambda = -\cos^2 \theta$.

By using the above theorem, it is easy to see the following lemma.

**Lemma 19.** *[245] Let $F$ be a proper slant submersion from an almost Hermitian manifold $(M_1, g_1, J_1)$ onto a Riemannian manifold $(M_2, g_2)$ with slant angle $\theta$. Then, for any $X, Y \in \Gamma(\ker F_*)$, we have*

$$g_1(\phi X, \phi Y) = \cos^2 \theta g_1(X, Y), \tag{3.73}$$
$$g_1(\omega X, \omega Y) = \sin^2 \theta g_1(X, Y). \tag{3.74}$$

We now denote the orthogonal complementary distribution to $\omega(\ker F_*)$ in $(\ker F_*)^\perp$ by $\mu$. Then we have the following.

**Lemma 20.** *[245] Let $F$ be a proper slant submersion from an almost Hermitian manifold $(M_1, g_1, J_1)$ onto a Riemannian manifold $(M_2, g_2)$. Then $\mu$ is invariant with respect to $J_1$.*

*Proof.* For $V \in \Gamma(\mu)$, using (3.2), we have

$$g_1(J_1 V, \omega X) = g_1(J_1 V, J_1 X) - g_1(J_1 V, \phi X)$$
$$= -g_1(J_1 V, \phi X)$$

for $X \in \Gamma(\ker F_*)$. Then, using Theorem 44, we get

$$g_1(J_1 V, \omega X) = g_1(V, J_1 \phi X)$$
$$= g_1(V, \phi^2 X) + g_1(V, \omega \phi X)$$
$$= -\cos^2 \theta g_1(V, X) + g_1(V, \omega \phi X)$$
$$= g_1(V, \omega \phi X) = 0,$$

due to $\mu$ being orthogonal to $\omega(\ker F_*)$. In a similar way, we have $g_1(J_1 V, Y) = -g_1(V, J_1 Y) = 0$ due to $J_1 Y \in \Gamma((\ker F_*) \oplus \omega(\ker F_*))$ for $V \in \Gamma(\mu)$ and $Y \in \Gamma(\ker F_*)$. Thus the proof is complete. $\qquad \square$

**Corollary 12.** *[245] Let $F$ be a proper slant submersion from an almost Hermitian manifold $(M_1^m, g_1, J_1)$ onto a Riemannian manifold $(M_2^n, g_2)$. Let*

$$\{e_1, ..., e_{m-n}\}$$

*be a local orthonormal basis of $(ker F_*)$, then $\{\csc \omega e_1, ..., \csc \omega e_{m-n}\}$ is a local orthonormal basis of $\omega(ker F_*)$.*

*Proof.* It will be enough to show that $g_1(\csc \theta \omega e_i, \csc \theta \omega e_j) = \delta_{ij}$, $i, j \in \{1, ..., \frac{m-n}{2}\}$. By using (3.74), we have

$$g_1(\csc \theta \omega e_i, \csc \theta \omega e_j) = \csc^2 \theta \sin^2 \theta g_1(e_i, e_j) = \delta_{ij},$$

which proves the assertion.    □

We note that Lemma 20 tells that the distributions $\mu$ and $(ker F_*) \oplus \omega(ker F_*)$ are even-dimensional. In fact it implies that the distribution $(ker F_*)$ is even-dimensional. More precisely, we have the following result whose proof is similar to the above corollary.

**Lemma 21.** *[245] Let F be a proper slant submersion from an almost Hermitian manifold $(M_1^m, g_1, J_1)$ onto a Riemannian manifold $(M_2^n, g_2)$. If $e_1, ..., e_{\frac{m-n}{2}}$ are orthogonal unit vector fields in $(ker F_*)$, then*

$$\{e_1, \sec \theta \phi e_1, e_2, \sec \theta \phi e_2, ..., e_{\frac{m-n}{2}}, \sec \theta \phi e_{\frac{m-n}{2}}\}$$

*is a local orthonormal basis of $(ker F_*)$.*

Let $F$ be a proper slant submersion from an almost Hermitian manifold $(M_1^{2m}, J_1, g_1)$ onto a Riemannian manifold $(M_2^m, g_2)$. As in slant immersions, we call such an orthonormal frame

$$\{e_1, \sec \theta \phi e_1, e_2, \sec \theta \phi e_2, ..., e_n, \sec \theta \phi e_n, \csc \theta \omega e_1, \csc \theta \omega e_2, ..., \csc \theta \omega e_n\}$$

an *adapted slant frame* for slant submersions.

In the sequel, we show that the slant submersion puts some restrictions on the dimensions of the distributions and the base manifold.

**Lemma 22.** *[245] Let F be a proper slant submersion from an almost Hermitian manifold $(M_1^m, g_1, J_1)$ onto a Riemannian manifold $(M_2^n, g_2)$. Then $dim(\mu) = 2n - m$. If $\mu = \{0\}$, then $n = \frac{m}{2}$.*

*Proof.* First note that $dim(ker F_*) = m - n$. Thus, using Corollary 12, we have $dim((ker F_*) \oplus \omega(ker F_*)) = 2(m - n)$. Since $M_1$ is $m$–dimensional, we get $dim(\mu) = 2n - m$. Second assertion is clear.    □

We now obtain a new theorem for the harmonicity of slant submersions. However, we first give a preparatory lemma.

**Lemma 23.** *[245] Let F be a proper slant submersion from a Kähler manifold onto a Riemannian manifold. If $\omega$ is parallel with respect to $\nabla$ on $(\ker F_*)$, then we have*

$$\mathcal{T}_{\phi X}\phi X = -\cos^2\theta\,\mathcal{T}_X X \tag{3.75}$$

*for $X \in \Gamma(\ker F_*)$.*

*Proof.* If $\omega$ is parallel, then from (3.71) we have $C\mathcal{T}_X Y = \mathcal{T}_X \phi Y$ for $X, Y \in \Gamma(\ker F_*)$. Interchanging the role of $X$ and $Y$, we get $C\mathcal{T}_Y X = \mathcal{T}_Y \phi X$. Thus we have

$$C\mathcal{T}_X Y - C\mathcal{T}_Y X = \mathcal{T}_X \phi Y - \mathcal{T}_Y \phi X.$$

Using (1.45), we derive

$$\mathcal{T}_X \phi Y = \mathcal{T}_Y \phi X. \tag{3.76}$$

Then, substituting $Y$ by $\phi X$, we get $\mathcal{T}_X \phi^2 X = \mathcal{T}_{\phi X}\phi X$. Finally, using Theorem 44 we obtain (3.75). $\square$

We now give a sufficient condition for a proper slant submersion to be harmonic.

**Theorem 45.** *[245] Let F be a proper slant submersion from a Kähler manifold onto a Riemannian manifold. If $\omega$ is parallel with respect to $\nabla$ on $(\ker F_*)$, then F is a harmonic map.*

*Proof.* Since

$$(\nabla F_*)(Z_1, Z_2) = 0 \tag{3.77}$$

for $Z_1, Z_2 \in \Gamma((\ker F_*)^{\perp})$, a proper slant submersion $F$ is harmonic if and only if $\sum_{i=1}^{n}(\nabla F_*)(\tilde{e}_i, \tilde{e}_i) = -\sum_{i=1}^{n} F_*(\mathcal{T}_{\tilde{e}_i}\tilde{e}_i) = 0$, where $\{\tilde{e}_i\}_{i=1}^{m-n}$ is an orthonormal basis of $(\ker F_*)$. Thus, using Lemma 21, we can write

$$\tau = -\sum_{i=1}^{\frac{m-n}{2}} F_*(\mathcal{T}_{e_i}e_i + \mathcal{T}_{\sec\theta\phi e_i}\sec\theta\phi e_i).$$

Hence we have

$$\tau = -(\sum_{i=1}^{\frac{m-n}{2}} F_*(\mathcal{T}_{e_i}e_i + \sec^2\theta\,\mathcal{T}_{\phi e_i}\phi e_i)).$$

Then, using (3.75), we arrive at

$$\tau = -(\sum_{i=1}^{\frac{m-n}{2}} F_*(\mathcal{T}_{e_i} e_i - \mathcal{T}_{e_i} e_i)) = 0,$$

which shows us that $F$ is harmonic.                                          $\square$

We now investigate the geometry of the leaves of the distributions $(ker\, F_*)$ and $(ker\, F_*)^\perp$.

**Theorem 46.** *[245] Let $F$ be a proper slant submersion from a Kähler manifold $(M_1, g_1, J_1)$ onto a Riemannian manifold $(M_2, g_2)$. Then the distribution $(ker\, F_*)$ defines a totally geodesic foliation on $M_1$ if and only if*

$$g_1(\mathcal{H}\nabla_X \omega \phi Y, Z) = g_1(\mathcal{H}\nabla_X \omega Y, CZ) + g_1(\mathcal{T}_X \omega Y, \mathcal{B}Z)$$

*for $X, Y \in \Gamma(ker\, F_*)$ and $Z \in \Gamma((ker\, F_*)^\perp)$.*

*Proof.* For $X, Y \in \Gamma(ker\, F_*)$ and $Z \in \Gamma((ker\, F_*)^\perp)$, from (3.2) and (3.69) we have

$$g_1(\nabla_X Y, Z) = g_1(\nabla_X \phi Y, JZ) + g_1(\nabla_X \omega Y, JZ).$$

Using (3.2), (3.69), and (3.70) we get

$$\begin{aligned} g_1(\nabla_X Y, Z) &= -g_1(\nabla_X \phi^2 Y, Z) - g_1(\nabla_X \omega \phi Y, Z) \\ &+ g_1(\nabla_X \omega Y, \mathcal{B}Z) + g_1(\nabla_X \omega Y, CZ). \end{aligned}$$

Then from (1.48) and Theorem 44, we obtain

$$\begin{aligned} g_1(\nabla_X Y, Z) &= \cos^2\theta g_1(\nabla_X Y, Z) - g_1(\mathcal{H}\nabla_X \omega \phi Y, Z) \\ &+ g_1(\mathcal{T}_X \omega Y, \mathcal{B}Z) + g_1(\mathcal{H}\nabla_X \omega Y, CZ). \end{aligned}$$

Hence we have

$$\begin{aligned} \sin^2\theta g_1(\nabla_X Y, Z) &= -g_1(\mathcal{H}\nabla_X \omega \phi Y, Z) \\ &+ g_1(\mathcal{T}_X \omega Y, \mathcal{B}Z) + g_1(\mathcal{H}\nabla_X \omega Y, CZ), \end{aligned}$$

which proves the assertion.                                          $\square$

In a similar way, we have the following theorem for the distribution $(ker\, F_*)^\perp$.

**Theorem 47.** *[245] Let $F$ be a proper slant submersion from a Kähler manifold $(M_1, g_1, J_1)$ onto a Riemannian manifold $(M_2, g_2)$. Then the distribution $(ker F_*)^\perp$ de-*

*fines a totally geodesic foliation on $M_1$ if and only if*

$$g_1(\mathcal{H}\nabla_{Z_1}Z_2, \omega\phi X) = g_1(\mathcal{A}_{Z_1}\mathcal{B}Z_2 + \mathcal{H}\nabla_{Z_1}CZ_2, \omega X)$$

*for $X \in \Gamma(\ker F_*)$ and $Z_1, Z_2 \in \Gamma((\ker F_*)^\perp)$.*

From Theorem 46 and Theorem 47, we have the following result.

**Corollary 13.** *[245] Let $F$ be a proper slant submersion from a Kähler manifold $(M_1, g_1, J_1)$ onto a Riemannian manifold $(M_2, g_2)$. Then $M_1$ is a locally product Riemannian manifold if and only if*

$$g_1(\mathcal{H}\nabla_{Z_1}Z_2, \omega\phi X) = g_1(\mathcal{A}_{Z_1}\mathcal{B}Z_2 + \mathcal{H}\nabla_{Z_1}CZ_2, \omega X)$$

*and*

$$g_1(\mathcal{H}\nabla_X\omega\phi Y, Z_1) = g_1(\mathcal{H}\nabla_X\omega Y, CZ_1) + g_1(\mathcal{T}_X\omega Y, \mathcal{B}Z_1)$$

*for $X, Y, \in \Gamma(\ker F_*)$ and $Z_1, Z_2 \in \Gamma((\ker F_*)^\perp)$.*

Finally, we give necessary and sufficient conditions for a proper slant submersion to be totally geodesic.

**Theorem 48.** *[245] Let $F$ be a proper slant submersion from a Kähler manifold $(M_1, g_1, J_1)$ onto a Riemannian manifold $(M_2, g_2)$. Then $F$ is totally geodesic if and only if*

$$g_1(\mathcal{T}_X\omega Y, \mathcal{B}Z_1) + g_1(\mathcal{H}\nabla_X\omega Y, CZ_1) = g_1(\mathcal{H}\nabla_X\omega\phi Y, Z_1)$$

*and*

$$g_1(\mathcal{A}_{Z_1}\mathcal{B}Z_2 + \mathcal{H}\nabla_{Z_1}CZ_2, \omega X) = -g_1(\mathcal{H}\nabla_{Z_1}\omega\phi X, Z_2)$$

*for $Z_1, Z_2 \in \Gamma((\ker F_*)^\perp)$ and $X, Y \in \Gamma(\ker F_*)$.*

*Proof.* First of all, since $F$ is a Riemannian submersion, we have

$$(\nabla F_*)(Z_1, Z_2) = 0$$

for $Z_1, Z_2 \in \Gamma((\ker F_*)^\perp)$. Thus it is enough to show that $(\nabla F_*)(X, Y) = 0$ and $(\nabla F_*)(X, Z) = 0$ for $X, Y \in \Gamma(\ker F_*)$ and $Z \in \Gamma((\ker F_*)^\perp)$. For $X, Y \in \Gamma(\ker F_*)$ and $Z_1 \in \Gamma((\ker F_*)^\perp)$, since $F$ is a Riemannian submersion, from (3.2), (3.69), and (3.70) we have

$$g_2((\nabla F_*)(X, Y), F_*Z_1) = g_1(\nabla_X J\phi Y, Z) - g_1(\nabla_X\omega Y, JZ).$$

Using again (3.69) and (3.70), we get

$$
\begin{aligned}
g_2((\nabla F_*)(X, Y), F_* Z_1) &= g_1(\nabla_X \phi^2 Y, Z) + g_1(\nabla_X \omega \phi Y, Z) \\
&\quad - g_1(\nabla_X \omega Y, \mathcal{B}Z) - g_1(\nabla_X \omega Y, \mathcal{C}Z).
\end{aligned}
$$

Then (1.47) and (1.48) imply that

$$
\begin{aligned}
g_2((\nabla F_*)(X, Y), F_* Z_1) &= -\cos^2 \theta g_1(\nabla_X Y, Z) + g_1(\nabla_X \omega \phi Y, Z) \\
&\quad - g_1(\mathcal{T}_X \omega Y, \mathcal{B}Z) - g_1(\mathcal{H} \nabla_X \omega Y, \mathcal{C}Z).
\end{aligned}
$$

Hence we obtain

$$
\begin{aligned}
(1 + \cos^2 \theta) g_2((\nabla F_*)(X, Y), F_* Z_1) &= g_1(\nabla_X \omega \phi Y, Z) - g_1(\mathcal{T}_X \omega Y, \mathcal{B}Z) \\
&\quad - g_1(\mathcal{H} \nabla_X \omega Y, \mathcal{C}Z). \tag{3.78}
\end{aligned}
$$

In a similar way, we get

$$
\begin{aligned}
(1 + \cos^2 \theta) g_2((\nabla F_*)(X, Z_1), F_*(Z_2)) &= g_1(\mathcal{A}_{Z_1} \mathcal{B}Z_2 + \mathcal{H} \nabla_{Z_1} \mathcal{C}Z_2, \omega X) \\
&\quad + g_1(\mathcal{H} \nabla_{Z_1} \omega \phi X, Z_2). \tag{3.79}
\end{aligned}
$$

Then the proof follows from (3.78) and (3.79).    □

## 7. Semi-slant submersions

In this section, we introduce semi-slant submersions from almost Hermitian manifolds onto Riemannian manifolds as a generalization of slant submersions, semi-invariant submersions, and anti-invariant submersions, and give examples. We also obtain characterizations, investigate the integrability of distributions and the geometry of foliations and find a condition for such submersions to be harmonic.

**Definition 42.** [221] Let $(M, g_M, J)$ be an almost Hermitian manifold and $(N, g_N)$ a Riemannian manifold. A Riemannian submersion $F : (M, g_M, J) \mapsto (N, g_N)$ is called a *semi-slant submersion* if there is a distribution $\mathcal{D}_1 \subset (\ker F_*)$ such that

$$
(\ker F_*) = \mathcal{D}_1 \oplus \mathcal{D}_2, \quad J(\mathcal{D}_1) = \mathcal{D}_1,
$$

and the angle $\theta = \theta(X)$ between $JX$ and the space $(\mathcal{D}_2)_q$ is constant for nonzero $X \in (\mathcal{D}_2)_q$ and $q \in M$, where $\mathcal{D}_2$ is the orthogonal complement of $\mathcal{D}_1$ in $(\ker F_*)$.

We call the angle $\theta$ a *semi-slant angle*.

**Example 15.** Let $F$ be a slant submersion from an almost Hermitian manifold $(M, g_M, J)$ onto a Riemannian manifold $(N, g_N)$. Then the map $F$ is a semi-slant submersion with $\mathcal{D}_2 = (\ker F_*)$.

**Example 16.** Let $F$ be a semi-invariant submersion from an almost Hermitian manifold $(M, g_M, J)$ onto a Riemannian manifold $(N, g_N)$. Then the map $F$ is a semi-slant submersion with the semi-slant angle $\theta = \frac{\pi}{2}$.

**Example 17.** [221] Define a map $F : \mathbb{R}^6 \mapsto \mathbb{R}^2$ by

$$F(x_1, x_2, \cdots, x_6) = (x_3 \sin \alpha - x_5 \cos \alpha, x_6),$$

where $\alpha \in (0, \frac{\pi}{2})$. Then the map $F$ is a semi-slant submersion such that

$$\mathcal{D}_1 = \text{Span}\{\frac{\partial}{\partial x_1}, \frac{\partial}{\partial x_2}\} \text{ and } \mathcal{D}_2 = span\{\frac{\partial}{\partial x_4}, \cos \alpha \frac{\partial}{\partial x_3} + \sin \alpha \frac{\partial}{\partial x_5}\}$$

with the semi-slant angle $\theta = \alpha$.

**Example 18.** [221] Define a map $F : \mathbb{R}^8 \mapsto \mathbb{R}^2$ by

$$F(x_1, x_2, \cdots, x_8) = (\frac{x_5 - x_8}{\sqrt{2}}, x_6).$$

Then the map $F$ is a semi-slant submersion such that

$$\mathcal{D}_1 = \text{Span}\{\frac{\partial}{\partial x_1}, \frac{\partial}{\partial x_2}, \frac{\partial}{\partial x_3}, \frac{\partial}{\partial x_4}\} \text{ and } \mathcal{D}_2 = \text{Span}\{\frac{\partial}{\partial x_7}, \frac{\partial}{\partial x_5} + \frac{\partial}{\partial x_8}\}$$

with the semi-slant angle $\theta = \frac{\pi}{4}$.

**Example 19.** [221] Define a map $F : \mathbb{R}^{10} \mapsto \mathbb{R}^5$ by

$$F(x_1, x_2, \cdots, x_{10}) = (x_2, x_1, \frac{x_5 + x_6}{\sqrt{2}}, \frac{x_7 + x_9}{\sqrt{2}}, \frac{x_8 + x_{10}}{\sqrt{2}}).$$

Then the map $F$ is a semi-slant submersion such that

$$\mathcal{D}_1 = \text{Span}\{\frac{\partial}{\partial x_3}, \frac{\partial}{\partial x_4}, -\frac{\partial}{\partial x_7} + \frac{\partial}{\partial x_9}, -\frac{\partial}{\partial x_8} + \frac{\partial}{\partial x_{10}}\} \text{ and } \mathcal{D}_2 = \text{Span}\{-\frac{\partial}{\partial x_5} + \frac{\partial}{\partial x_6}\}$$

with the semi-slant angle $\theta = \frac{\pi}{2}$.

**Example 20.** [221] Define a map $F : \mathbb{R}^{10} \mapsto \mathbb{R}^4$ by

$$F(x_1, x_2, \cdots, x_{10}) = (\frac{x_3 - x_5}{\sqrt{2}}, x_6, \frac{x_7 - x_9}{\sqrt{2}}, x_8).$$

Then the map $F$ is a semi-slant submersion such that

$$\mathcal{D}_1 = \text{Span}\{\frac{\partial}{\partial x_1}, \frac{\partial}{\partial x_2}\} \text{ and } \mathcal{D}_2 = \text{Span}\{\frac{\partial}{\partial x_3} + \frac{\partial}{\partial x_5}, \frac{\partial}{\partial x_7} + \frac{\partial}{\partial x_9}, \frac{\partial}{\partial x_4}, \frac{\partial}{\partial x_{10}}\}$$

with the semi-slant angle $\theta = \frac{\pi}{4}$.

**Example 21.** [221] Define a map $F : \mathbb{R}^8 \mapsto \mathbb{R}^4$ by

$$F(x_1, x_2, \cdots, x_8) = (x_1, x_2, x_3 \cos \alpha - x_5 \sin \alpha, x_4 \sin \beta - x_6 \cos \beta),$$

where $\alpha$ and $\beta$ are constant. Then the map $F$ is a semi-slant submersion such that

$$\mathcal{D}_1 = \text{Span}\{\frac{\partial}{\partial x_7}, \frac{\partial}{\partial x_8}\} \text{ and } \mathcal{D}_2 = \text{Span}\{\sin \alpha \frac{\partial}{\partial x_3} + \cos \alpha \frac{\partial}{\partial x_5}, \cos \beta \frac{\partial}{\partial x_4} + \sin \beta \frac{\partial}{\partial x_6}\}$$

with the semi-slant angle $\theta$ with $\cos \theta = |\sin(\alpha + \beta)|$.

**Example 22.** [221] Let G be a slant submersion from an almost Hermitian manifold $(M_1, g_{M_1}, J_1)$ onto a Riemannian manifold $(N, g_N)$ with the slant angle $\theta$ and $(M_2, g_{M_2}, J_2)$ an almost Hermitian manifold. Denote by $(M, g, J)$ the warped product of $(M_1, g_{M_1}, J_1)$ and $(M_2, g_{M_2}, J_2)$ by a positive function $f$ on $M_1$, where $J = J_1 \times J_2$. Define a map $F : (M, g, J) \mapsto (N, g_N)$ by

$$F(x, y) = G(x) \quad \text{for } x \in M_1 \text{ and } y \in M_2.$$

Then the map $F$ is a semi-slant submersion such that

$$\mathcal{D}_1 = TM_2 \text{ and } \mathcal{D}_2 = \ker G_*,$$

with the semi-slant angle $\theta$.

Let $F : (M, g_M, J) \mapsto (N, g_N)$ be a semi-slant submersion. Then for $X \in \Gamma((\ker F_*))$, we write

$$X = PX + QX,$$

where $PX \in \Gamma(\mathcal{D}_1)$ and $QX \in \Gamma(\mathcal{D}_2)$. For $X \in \Gamma(\ker F_*)$, we get

$$JX = \phi X + \omega X,$$

where $\phi X \in \Gamma(\ker F_*)$ and $\omega X \in \Gamma((\ker F_*)^{\perp})$. For $Z \in \Gamma((\ker F_*)^{\perp})$, we obtain

$$JZ = BZ + CZ,$$

where $BZ \in \Gamma(\ker F_*)$ and $CZ \in \Gamma((\ker F_*)^{\perp})$. For $U \in \Gamma(TM)$, we write

$$U = \mathcal{V}U + \mathcal{H}U,$$

where $\mathcal{V}U \in \Gamma(\ker F_*)$ and $\mathcal{H}U \in \Gamma((\ker F_*)^{\perp})$. Then we have

$$(\ker F_*)^{\perp} = \omega \mathcal{D}_2 \oplus \mu,$$

where $\mu$ is the orthogonal complement of $\omega \mathcal{D}_2$ in $(\ker F_*)^{\perp}$ and is invariant under $J$.

Furthermore, we have the following relations:

$$\phi \mathcal{D}_1 = \mathcal{D}_1, \omega \mathcal{D}_1 = 0, \phi \mathcal{D}_2 \subset \mathcal{D}_2, B((\ker F_*)^\perp) = \mathcal{D}_2$$
$$\phi^2 + B\omega = -id, C^2 + \omega B = -id, \omega\phi + C\omega = 0, BC + \phi B = 0.$$

Define

$$(\nabla_X \phi)Y := \widehat{\nabla}_X \phi Y - \phi \widehat{\nabla}_X Y$$

and

$$(\nabla_X \omega)Y := \mathcal{H}\nabla_X \omega Y - \omega \widehat{\nabla}_X Y$$

for $X, Y \in \Gamma(\ker F_*)$, where $\widehat{\nabla}_X Y := \mathcal{V}\nabla_X Y$. Then we easily have the following expressions.

**Lemma 24.** *[221] Let $(M, g_M, J)$ be a Kähler manifold and $(N, g_N)$ a Riemannian manifold. Let $F : (M, g_M, J) \mapsto (N, g_N)$ be a semi-slant submersion. Then:*
**1.** *for $X, Y \in \Gamma(\ker F_*)$, we get*

$$\widehat{\nabla}_X \phi Y + \mathcal{T}_X \omega Y = \phi \widehat{\nabla}_X Y + B\mathcal{T}_X Y$$
$$\mathcal{T}_X \phi Y + \mathcal{H}\nabla_X \omega Y = \omega \widehat{\nabla}_X Y + C\mathcal{T}_X Y,$$

**2.** *for $Z, W \in \Gamma((\ker F_*)^\perp)$, we have*

$$\mathcal{V}\nabla_Z BW + \mathcal{A}_Z CW = \phi \mathcal{A}_Z W + B\mathcal{H}\nabla_Z W$$
$$\mathcal{A}_Z BW + \mathcal{H}\nabla_Z CW = \omega \mathcal{A}_Z W + C\mathcal{H}\nabla_Z W,$$

**3.** *for $X \in \Gamma(\ker F_*)$ and $Z \in \Gamma((\ker F_*)^\perp)$, we get*

$$\widehat{\nabla}_X BZ + \mathcal{T}_X CZ = \phi \mathcal{T}_X Z + B\mathcal{H}\nabla_X Z$$
$$\mathcal{T}_X BZ + \mathcal{H}\nabla_X CZ = \omega \mathcal{T}_X Z + C\mathcal{H}\nabla_X Z.$$

We now investigate the integrability of distributions.

**Theorem 49.** *[221] Let $F$ be a semi-slant submersion from an almost Hermitian manifold $(M, g_M, J)$ onto a Riemannian manifold $(N, g_N)$. Then the complex distribution $\mathcal{D}_1$ is integrable if and only if we have*

$$\omega(\widehat{\nabla}_X Y - \widehat{\nabla}_Y X) = 0 \quad \text{for } X, Y \in \Gamma(\mathcal{D}_1).$$

*Proof.* For $X, Y \in \Gamma(\mathcal{D}_1)$ and $Z \in \Gamma((\ker F_*)^\perp)$, since $[X, Y] \in \Gamma(\ker F_*)$, we obtain

$$\begin{aligned}
g_M(J[X, Y], Z) &= g_M(J(\nabla_X Y - \nabla_Y X), Z) \\
&= g_M(\phi \widehat{\nabla}_X Y + \omega \widehat{\nabla}_X Y + B\mathcal{T}_X Y + C\mathcal{T}_X Y - \phi \widehat{\nabla}_Y X - \omega \widehat{\nabla}_Y X \\
&\quad - B\mathcal{T}_Y X - C\mathcal{T}_Y X, Z) \\
&= g_M(\omega \widehat{\nabla}_X Y + C\mathcal{T}_X Y - \omega \widehat{\nabla}_Y X - C\mathcal{T}_Y X, Z) \\
&= g_M(\omega(\widehat{\nabla}_X Y - \widehat{\nabla}_Y X), Z).
\end{aligned}$$

Therefore, we have the result. □

In a similar way, we have the following result.

**Theorem 50.** *[221] Let F be a semi-slant submersion from an almost Hermitian manifold $(M, g_M, J)$ onto a Riemannian manifold $(N, g_N)$. Then the slant distribution $\mathcal{D}_2$ is integrable if and only if we obtain*

$$P(\phi(\widehat{\nabla}_X Y - \widehat{\nabla}_Y X)) = 0 \quad \text{for } X, Y \in \Gamma(\mathcal{D}_2).$$

**Lemma 25.** *[221] Let $(M, g_M, J)$ be a Kähler manifold and $(N, g_N)$ a Riemannian manifold. Let $F : (M, g_M, J) \mapsto (N, g_N)$ be a semi-slant submersion. Then the slant distribution $\mathcal{D}_2$ is integrable if and only if we obtain*

$$P(\widehat{\nabla}_X \phi Y - \widehat{\nabla}_Y \phi X + \mathcal{T}_X \omega Y - \mathcal{T}_Y \omega X) = 0 \quad \text{for } X, Y \in \Gamma(\mathcal{D}_2).$$

*Proof.* For $X, Y \in \Gamma(\mathcal{D}_2)$ and $Z \in \Gamma(\mathcal{D}_1)$, since $[X, Y] \in \Gamma(\ker F_*)$, we have

$$\begin{aligned}
g_M(J[X, Y], Z) &= g_M(\nabla_X JY - \nabla_Y JX, Z) \\
&= g_M(\widehat{\nabla}_X \phi Y + \mathcal{T}_X \phi Y + \mathcal{T}_X \omega Y + \mathcal{H}\nabla_X \omega Y - \widehat{\nabla}_Y \phi X - \mathcal{T}_Y \phi X \\
&\quad - \mathcal{T}_Y \omega X - \mathcal{H}\nabla_Y \omega X, Z) \\
&= g_M(\widehat{\nabla}_X \phi Y + \mathcal{T}_X \omega Y - \widehat{\nabla}_Y \phi X - \mathcal{T}_Y \omega X, Z).
\end{aligned}$$

Therefore, the result follows. □

We also obtain the following lemma.

**Lemma 26.** *[221] Let $(M, g_M, J)$ be a Kähler manifold and $(N, g_N)$ a Riemannian manifold. Let $F : (M, g_M, J) \mapsto (N, g_N)$ be a semi-slant submersion. Then the complex distribution $\mathcal{D}_1$ is integrable if and only if we get*

$$Q(\widehat{\nabla}_X \phi Y - \widehat{\nabla}_Y \phi X) = 0 \text{ and } \mathcal{T}_X \phi Y = \mathcal{T}_Y \phi X \quad \text{for } X, Y \in \Gamma(\mathcal{D}_1).$$

Define an endomorphism $\widehat{F}$ of (ker $F_*$) by

$$\widehat{F} := JP + \phi Q,$$

where $(\nabla_X \widehat{F})Y := \widehat{\nabla}_X \widehat{F}Y - \widehat{F}\widehat{\nabla}_X Y$ for $X, Y \in \Gamma(\ker F_*)$. Then it is not difficult to find the following expression.

**Lemma 27.** *[221] Let F be a semi-slant submersion from a Kähler manifold $(M, g_M, J)$ onto a Riemannian manifold $(N, g_N)$. Then we have*

$$(\nabla_X \widehat{F})Y = \phi(\widehat{\nabla}_X PY - \widehat{\nabla}_X Y) + B\mathcal{T}_X PY + \widehat{\nabla}_X \phi QY \quad \text{for } X, Y \in \Gamma(\ker F_*).$$

**Proposition 33.** *[221] Let F be a semi-slant submersion from an almost Hermitian manifold $(M, g_M, J)$ onto a Riemannian manifold $(N, g_N)$. Then we obtain*

$$\phi^2 X = -\cos^2\theta X \quad \text{for } X \in \Gamma(\mathcal{D}_2),$$

*where $\theta$ denotes the semi-slant angle of $\mathcal{D}_2$.*

*Proof.* Since

$$\cos\theta = \frac{g_M(JX, \phi X)}{|JX| \cdot |\phi X|} = \frac{-g_M(X, \phi^2 X)}{|X| \cdot |\phi X|}$$

and $\cos\theta = \dfrac{|\phi X|}{|JX|}$, we have

$$\cos^2\theta = -\frac{g_M(X, \phi^2 X)}{|X|^2} \quad \text{for } X \in \Gamma(\mathcal{D}_2).$$

Hence,

$$\phi^2 X = -\cos^2\theta X \quad \text{for } X \in \Gamma(\mathcal{D}_2),$$

which gives us the assertion □

Assume that the semi-slant angle $\theta$ is not equal to $\dfrac{\pi}{2}$ and define an endomorphism $\widehat{J}$ of (ker $F_*$) by

$$\widehat{J} := JP + \frac{1}{\cos\theta}\phi Q.$$

Then,

$$\widehat{J}^2 = -id \quad \text{on(ker } F_*). \tag{3.80}$$

**Remark 7.** Let $F$ be a semi-slant submersion from an almost Hermitian manifold

$(M, g_M, J)$ onto a Riemannian manifold $(N, g_N)$. Assume that $\dim M = 2m$, $\dim N = n$, and $\theta \in [0, \frac{\pi}{2})$. From (3.80), we have

$$\dim(\ker(F_*)_p) = 2k \text{ and } \dim((\ker(F_*)_p)^\perp) = 2m - 2k \quad \text{for } p \in M,$$

where $k$ is a non-negative integer. Therefore, $n$ must be even.

**Theorem 51.** *[221] Let $F$ be a semi-slant submersion from an almost Hermitian manifold $(M, g_M, J)$ onto a Riemannian manifold $(N, g_N)$ with the semi-slant angle $\theta \in [0, \frac{\pi}{2})$. Then $N$ is an even-dimensional manifold.*

We now find necessary and sufficient conditions for distributions to define totally geodesic foliations on total space.

**Proposition 34.** *[221] Let $F$ be a semi-slant submersion from a Kähler manifold $(M, g_M, J)$ onto a Riemannian manifold $(N, g_N)$. Then the distribution $(\ker F_*)$ defines a totally geodesic foliation if and only if*

$$\omega(\widehat{\nabla}_X \phi Y + \mathcal{T}_X \omega Y) + C(\mathcal{T}_X \phi Y + \mathcal{H} \nabla_X \omega Y) = 0 \quad \text{for } X, Y \in \Gamma(\ker F_*).$$

*Proof.* For $X, Y \in \Gamma(\ker F_*)$,

$$\nabla_X Y = -J \nabla_X J Y = -J(\widehat{\nabla}_X \phi Y + \mathcal{T}_X \phi Y + \mathcal{T}_X \omega Y + \mathcal{H} \nabla_X \omega Y)$$
$$= -(\phi \widehat{\nabla}_X \phi Y + \omega \widehat{\nabla}_X \phi Y + B\mathcal{T}_X \phi Y + C\mathcal{T}_X \phi Y + \phi \mathcal{T}_X \omega Y + \omega \mathcal{T}_X \omega Y$$
$$+ B\mathcal{H} \nabla_X \omega Y + C\mathcal{H} \nabla_X \omega Y).$$

Thus, we obtain

$$\nabla_X Y \in \Gamma(\ker F_*) \Leftrightarrow \omega(\widehat{\nabla}_X \phi Y + \mathcal{T}_X \omega Y) + C(\mathcal{T}_X \phi Y + \mathcal{H} \nabla_X \omega Y) = 0.$$

$\square$

Similarly, we have the following result for horizontal distribution.

**Proposition 35.** *[221] Let $F$ be a semi-slant submersion from a Kähler manifold $(M, g_M, J)$ onto a Riemannian manifold $(N, g_N)$. Then the distribution $(\ker F_*)^\perp$ defines a totally geodesic foliation if and only if*

$$\phi(\mathcal{V} \nabla_X BY + \mathcal{A}_X CY) + B(\mathcal{A}_X BY + \mathcal{H} \nabla_X CY) = 0 \quad \text{for } X, Y \in \Gamma((\ker F_*)^\perp).$$

**Proposition 36.** *[221] Let $F$ be a semi-slant submersion from a Kähler manifold $(M, g_M, J)$ onto a Riemannian manifold $(N, g_N)$. Then the distribution $\mathcal{D}_1$ defines a* *totally geodesic foliation if and only if*

$$Q(\phi\widehat{\nabla}_X\phi Y + B\mathcal{T}_X\phi Y) = 0 \text{ and } \omega\widehat{\nabla}_X\phi Y + C\mathcal{T}_X\phi Y = 0$$

*for $X, Y \in \Gamma(\mathcal{D}_1)$.*

*Proof.* For $X, Y \in \Gamma(\mathcal{D}_1)$, we get

$$\nabla_X Y = -J\nabla_X JY = -J(\widehat{\nabla}_X\phi Y + \mathcal{T}_X\phi Y)$$
$$= -(\phi\widehat{\nabla}_X\phi Y + \omega\widehat{\nabla}_X\phi Y + B\mathcal{T}_X\phi Y + C\mathcal{T}_X\phi Y).$$

Hence, we get

$$\nabla_X Y \in \Gamma(\mathcal{D}_1) \Leftrightarrow Q(\phi\widehat{\nabla}_X\phi Y + B\mathcal{T}_X\phi Y) = 0 \text{ and } \omega\widehat{\nabla}_X\phi Y + C\mathcal{T}_X\phi Y = 0.$$

$\square$

In a similar way, we obtain the following result.

**Proposition 37.** *[221] Let $F$ be a semi-slant submersion from a Kähler manifold $(M, g_M, J)$ onto a Riemannian manifold $(N, g_N)$. Then the distribution $\mathcal{D}_2$ defines a totally geodesic foliation if and only if*

$$P(\phi(\widehat{\nabla}_X\phi Y + \mathcal{T}_X\omega Y) + B(\mathcal{T}_X\phi Y + \mathcal{H}\nabla_X\omega Y)) = 0$$
$$\omega(\widehat{\nabla}_X\phi Y + \mathcal{T}_X\omega Y) + C(\mathcal{T}_X\phi Y + \mathcal{H}\nabla_X\omega Y) = 0$$

*for $X, Y \in \Gamma(\mathcal{D}_2)$.*

The following theorem shows that the notion of semi-slant submersions produce new conditions for a Riemannian submersion to be totally geodesic.

**Theorem 52.** *[221] Let $F$ be a semi-slant submersion from a Kähler manifold $(M, g_M, J)$ onto a Riemannian manifold $(N, g_N)$. Then $F$ is a totally geodesic map if and only if*

$$\omega(\widehat{\nabla}_X\phi Y + \mathcal{T}_X\omega Y) + C(\mathcal{T}_X\phi Y + \mathcal{H}\nabla_X\omega Y) = 0$$
$$\omega(\widehat{\nabla}_X BZ + \mathcal{T}_X CZ) + C(\mathcal{T}_X BZ + \mathcal{H}\nabla_X CZ) = 0$$

*for $X, Y \in \Gamma(\ker F_*)$ and $Z \in \Gamma((\ker F_*)^\perp)$.*

*Proof.* Since $F$ is a Riemannian submersion, we have

$$(\nabla F_*)(Z_1, Z_2) = 0 \quad \text{for } Z_1, Z_2 \in \Gamma((\ker F_*)^\perp).$$

For $X, Y \in \Gamma(\ker F_*)$, we obtain

$$(\nabla F_*)(X, Y) = -F_*(\nabla_X Y) = F_*(J\nabla_X(\phi Y + \omega Y))$$
$$= F_*(\phi\widehat{\nabla}_X\phi Y + \omega\widehat{\nabla}_X\phi Y + B\mathcal{T}_X\phi Y + C\mathcal{T}_X\phi Y + \phi\mathcal{T}_X\omega Y + \omega\mathcal{T}_X\omega Y$$
$$+ B\mathcal{H}\nabla_X\omega Y + C\mathcal{H}\nabla_X\omega Y).$$

Thus, we have

$$(\nabla F_*)(X, Y) = 0 \Leftrightarrow \omega(\widehat{\nabla}_X\phi Y + \mathcal{T}_X\omega Y) + C(\mathcal{T}_X\phi Y + \mathcal{H}\nabla_X\omega Y) = 0.$$

For $X \in \Gamma(\ker F_*)$ and $Z \in \Gamma((\ker F_*)^\perp)$, we get

$$(\nabla F_*)(X, Z) = -F_*(\nabla_X Z) = F_*(J\nabla_X(BZ + CZ))$$
$$= F_*(\phi\widehat{\nabla}_X BZ + \omega\widehat{\nabla}_X BZ + B\mathcal{T}_X BZ + C\mathcal{T}_X BZ + \phi\mathcal{T}_X CZ + \omega\mathcal{T}_X CZ$$
$$+ B\mathcal{H}\nabla_X CZ + C\mathcal{H}\nabla_X CZ).$$

Hence, we obtain

$$(\nabla F_*)(X, Z) = 0 \Leftrightarrow \omega(\widehat{\nabla}_X BZ + \mathcal{T}_X CZ) + C(\mathcal{T}_X BZ + \mathcal{H}\nabla_X CZ) = 0.$$

Since $(\nabla F_*)(X, Z) = (\nabla F_*)(Z, X)$, we get the result. $\qquad\square$

Let $F$ be a semi-slant submersion from a Kähler manifold $(M, g_M, J)$ onto a Riemannian manifold $(N, g_N)$. Assume that $\mathcal{D}_1$ is integrable. Choose a local orthonormal frame $\{v_1, \cdots, v_l\}$ of $\mathcal{D}_2$ and a local orthonormal frame $\{e_1, \cdots, e_{2k}\}$ of $\mathcal{D}_1$ such that $e_{2i} = Je_{2i-1}$ for $1 \le i \le k$. Since

$$F_*(\nabla_{Je_{2i-1}}Je_{2i-1}) = -F_*(\nabla_{e_{2i-1}}e_{2i-1})$$

for $1 \le i \le k$, we have

$$trace(\nabla F_*) = 0 \Leftrightarrow \sum_{j=1}^{l} F_*(\nabla_{v_j}v_j) = 0.$$

**Theorem 53.** *[221] Let F be a semi-slant submersion from a Kähler manifold $(M, g_M, J)$ onto a Riemannian manifold $(N, g_N)$ such that $\mathcal{D}_1$ is integrable. Then F is a harmonic map if and only if*

$$trace(\nabla F_*) = 0 \quad on \ \mathcal{D}_2.$$

Now, we investigate semi-slant submersion with totally umbilical fibers.

**Lemma 28.** *[221] Let F be a semi-slant submersion with totally umbilical fibers from* a Kähler manifold $(M, g_M, J)$ onto a Riemannian manifold $(N, g_N)$. Then we have

$$H \in \Gamma(\omega \mathcal{D}_2).$$

*Proof.* For $X, Y \in \Gamma(\mathcal{D}_1)$ and $W \in \Gamma(\mu)$, we get

$$\mathcal{T}_X JY + \widehat{\nabla}_X JY = \nabla_X JY = J\nabla_X Y = B\mathcal{T}_X Y + C\mathcal{T}_X Y + \phi\widehat{\nabla}_X Y + \omega\widehat{\nabla}_X Y$$

so that

$$g_M(\mathcal{T}_X JY, W) = g_M(C\mathcal{T}_X Y, W).$$

By (1.65), with a simple calculation we obtain

$$g_M(X, JY)g_M(H, W) = -g_M(X, Y)g_M(H, JW).$$

Interchanging the role of $X$ and $Y$, we get

$$g_M(Y, JX)g_M(H, W) = -g_M(Y, X)g_M(H, JW)$$

so that combining the above two equations, we have

$$g_M(X, Y)g_M(H, JW) = 0$$

which means $H \in \Gamma(\omega \mathcal{D}_2)$, since $J\mu = \mu$. Therefore, we obtain the result. $\quad\square$

**Remark 8.** We note that here are some similarities between slant submanifolds and semi-slant submanifolds. Indeed, for totally geodesic conditions, two cases have the same condition.

## 8. Hemi-slant submersions

In this section, we introduce the notion of hemi-slant submersions as natural generalizations of both the notions of semi-invariant and slant submersions from almost Hermitian manifolds onto Riemannian manifolds. We focus mainly on hemi-slant submersions from a Kähler manifold onto a Riemannian manifold.

**Definition 43.** [286] Let $M$ be a $2m$-dimensional almost Hermitian manifold with Hermitian metric $g$ and almost complex structure $J$, and $N$ be a Riemannian manifold with Riemannian metric $g_N$. A Riemannian submersion $\pi : (M, g, J) \to (N, g_N)$ is called a *hemi-slant submersion* if the vertical distribution $ker\pi_*$ of $\pi$ admits two orthogonal complementary distributions $\mathcal{D}^\theta$ and $\mathcal{D}^\perp$ such that $\mathcal{D}^\theta$ is slant and $\mathcal{D}^\perp$ is anti-invariant, i.e., we have

$$ker\pi_* = \mathcal{D}^\theta \oplus \mathcal{D}^\perp. \tag{3.81}$$

In this case, the angle $\theta$ is called the hemi-slant angle of the submersion.

We can easily observe that the notion of hemi-slant submersions is a natural generalization of the notions of anti-invariant and semi-invariant submersions and slant submersions. More precisely, if we denote the dimension of $\mathcal{D}^{\perp}$ and $\mathcal{D}^{\theta}$ by $m_1$ and $m_2$, respectively, then we have the following:

**(a)** If $m_2 = 0$, then $M$ is an anti-invariant submersion.
**(b)** If $m_1 = 0$ and $\theta = 0$, then $M$ is an invariant submersion.
**(c)** If $m_1 = 0$ and $\theta \neq 0, \frac{\pi}{2}$, then $M$ is a proper slant submersion with slant angle $\theta$.
**(d)** If $\theta = \frac{\pi}{2}$, then $M$ is an anti-invariant submersion.

We say that the hemi-slant submersion $\pi : (M, g, J) \to (N, g_N)$ is *proper* if $\mathcal{D}^{\perp} \neq \{0\}$ and $\theta \neq 0, \frac{\pi}{2}$. As we have seen from the above argument, anti-invariant submersions, semi-invariant submersions, and slant submersions are all examples of hemi-slant submersions. We now present an example of proper hemi-slant submersions.

**Example 23.** [286] Define a map $\pi : \mathbb{R}^6 \to \mathbb{R}^3$ by $\pi(x_1, ..., x_6) = \left( \frac{x_1 - x_4}{\sqrt{2}}, x_2, x_5 \right)$.

Then the map $\pi$ is a proper hemi-slant submersion such that

$$ker\pi_* = \mathcal{D}^{\theta} \oplus \mathcal{D}^{\perp},$$

where

$$\mathcal{D}^{\theta} = span\{\frac{1}{\sqrt{2}}(\partial_1 + \partial_4), \partial_3\}$$

with the slant angle $\theta = \frac{\pi}{4}$ and

$$\mathcal{D}^{\perp} = span\{\partial_6\}.$$

Moreover,

$$(ker\pi_*)^{\perp} = span\{\frac{1}{\sqrt{2}}(\partial_1 - \partial_4), \partial_2, \partial_5\}, \ \partial_i = \frac{\partial}{\partial x_i}.$$

For any $V \in \Gamma(ker\pi_*)$, we put

$$V = \mathcal{P}V + QV, \tag{3.82}$$

where $\mathcal{P}V \in \Gamma(\mathcal{D}^{\theta})$ and $QV \in \Gamma(\mathcal{D}^{\perp})$ and put

$$JV = \phi V + \omega V, \tag{3.83}$$

where $\phi V \in \Gamma(ker\pi_*)$ and $\omega V \in \Gamma((ker\pi_*)^{\perp})$. Also for any $\xi \in \Gamma((ker\pi_*)^{\perp})$, we have

$$J\xi = \mathcal{B}\xi + C\xi, \tag{3.84}$$

where $\mathcal{B}\xi \in \Gamma(ker\pi_*)$ and $C\xi \in \Gamma((ker\pi_*)^{\perp})$.

The horizontal distribution $(ker\pi_*)^\perp$ is then decomposed as

$$(ker\pi_*)^\perp = \omega\mathcal{D}^\theta \oplus J\mathcal{D}^\perp \oplus \mu, \tag{3.85}$$

where $\mu$ is the orthogonal complementary distribution of $\omega\mathcal{D}^\theta \oplus J\mathcal{D}^\perp$ and it is an invariant distribution of $(ker\pi_*)^\perp$ with respect to $J$.

Thus, using (3.83), (3.84), and (3.85), we get the following:

**Lemma 29.** *[286] Let $\pi$ be a hemi-slant submersion from an almost Hermitian manifold $(M, g, J)$ onto a Riemannian manifold $(N, g_N)$. Then, we have*

$$\begin{array}{llll} (a) & \phi\mathcal{D}^\theta = \mathcal{D}^\theta, & (b) & \phi\mathcal{D}^\perp = \{0\}, \\ (c) & \mathcal{B}\omega\mathcal{D}^\theta = \mathcal{D}^\theta, & (d) & \mathcal{B}J\mathcal{D}^\perp = \mathcal{D}^\perp, \end{array} \tag{3.86}$$

On the other hand, using (3.83), (3.84) and the fact that $J^2 = -I$, we obtain the following:

**Lemma 30.** *[286] Let $\pi$ be a hemi-slant submersion from an almost Hermitian manifold $(M, g, J)$ onto a Riemannian manifold $(N, g_N)$. Then, we have*

$$\begin{array}{llll} (a) & \phi^2 + \mathcal{B}\omega = -I, & (b) & C^2 + \omega\mathcal{B} = -I, \\ (c) & \phi\mathcal{B} + \mathcal{B}C = 0, & (d) & \omega\phi + C\omega = 0, \end{array} \tag{3.87}$$

*where $I$ is the identity operator on the total space of $\pi$.*

The proof of the following theorem is exactly the same as hemi-slant submanifolds; see Theorem 3.2. of [238]. we therefore omit it.

**Theorem 54.** *[286] Let $\pi$ be a Riemannian submersion from an almost Hermitian manifold $(M, g, J)$ onto a Riemannian manifold. Then $\pi$ is a hemi-slant submersion if and only if there exists a constant $\lambda \in [-1, 0]$ and a distribution $\mathcal{D}$ on $ker\pi_*$ such that:*
**(a)** $\mathcal{D} = \{V \in \Gamma(ker\pi_*) \mid \phi^2 V = \lambda V\}$, *and*
**(b)** *for any $V \in \Gamma(ker\pi_*)$ orthogonal to $\mathcal{D}$, we have $\phi V = 0$.*
*Moreover, in this case $\lambda = -\cos^2\theta$, where $\theta$ is the slant angle of $\pi$.*

Thus, for any $Z \in \Gamma(\mathcal{D}^\theta)$, we conclude that

$$\phi^2 Z = -\cos^2\theta Z \tag{3.88}$$

from Theorem 3.7. On the other hand, for any $Z, W \in \Gamma(\mathcal{D}^\theta)$, using (3.2), (3.83), and (3.88), we get

$$g(\phi Z, \phi W) = \cos^2\theta g(Z, W). \tag{3.89}$$

Also, using (3.2), (3.83), (3.87)-(a), and (3.88), we find

$$g(\omega Z, \omega W) = \sin^2\theta g(Z, W). \tag{3.90}$$

Let $(M, g, J)$ be a Kähler manifold and $(N, g_N)$ be a Riemannian manifold. We now examine how the Kähler structure on $M$ effects the tensor fields $\mathcal{T}$ and $\mathcal{A}$ of a hemi-slant submersion $\pi : (M, g, J) \to (N, g_N)$.

**Lemma 31.** *[286] Let $\pi$ be a hemi-slant submersion from a Kähler manifold $(M, g, J)$ onto a Riemannian manifold $(N, g_N)$. Then*

$$\hat{\nabla}_U \phi V + \mathcal{T}_U \omega V = \phi \hat{\nabla}_U V + \mathcal{B}\mathcal{T}_U V, \tag{3.91}$$

$$\mathcal{T}_U \phi V + \mathcal{H}\nabla_U \omega V = \omega \hat{\nabla}_U V + C\mathcal{T}_U V, \tag{3.92}$$

$$\mathcal{V}\nabla_\xi \eta + \mathcal{A}_\xi C\eta = \phi \mathcal{A}_\xi \eta + \mathcal{B}\mathcal{H}\nabla_\xi \eta, \tag{3.93}$$

$$\mathcal{A}_\xi \mathcal{B}\eta + \mathcal{H}\nabla_\xi C\eta = \omega \mathcal{A}_\xi \eta + C\mathcal{H}\nabla_\xi \eta, \tag{3.94}$$

$$\hat{\nabla}_U \mathcal{B}\xi + \mathcal{T}_U C\xi = \phi \mathcal{T}_U \xi + \mathcal{B}\mathcal{H}\nabla_U \xi, \tag{3.95}$$

$$\mathcal{T}_U \mathcal{B}\xi + \mathcal{H}\nabla_U C\xi = \omega \mathcal{T}_U \xi + C\mathcal{H}\nabla_U \xi, \tag{3.96}$$

*where $U, V \in \Gamma(ker\pi_*)$ and $\xi, \eta \in \Gamma((ker\pi_*)^\perp)$.*

*Proof.* Using (1.47)–(1.50), (3.83), (3.84), and (3.2), we can easily obtain all assertions. □

Now, we define

$$(\nabla_U \phi)V = \hat{\nabla}_U \phi V - \phi \hat{\nabla}_U V \tag{3.97}$$

and

$$(\nabla_U \omega)V = \mathcal{H}\nabla_U \omega V - \omega \hat{\nabla}_U V \tag{3.98}$$

for $U, V \in \Gamma(ker\pi_*)$. Then from (3.91) and (3.92), we have the following result.

**Corollary 14.** *Let $\pi$ be a hemi-slant submersion from a Kähler manifold $(M, g, J)$ onto a Riemannian manifold $(N, g_N)$. Then*

$$\phi \quad is \quad parallel, \quad i.e., \nabla\phi \equiv 0 \Leftrightarrow \mathcal{T}_U \omega V = \mathcal{B}\mathcal{T}_U V \tag{3.99}$$

*and*

$$\omega \quad is \quad parallel, \quad i.e., \nabla\omega \equiv 0 \Leftrightarrow \mathcal{T}_U \phi V = C\mathcal{T}_U V, \tag{3.100}$$

*where $U, V \in \Gamma(ker\pi_*)$.*

The above corollary will be useful when we deal with the harmonicity of hemi-slant submersions. We now examine the integrability conditions for slant distributions and anti-invariant distributions.

**Theorem 55.** *[286] Let $\pi$ be a proper hemi-slant submersion from a Kähler manifold $(M, g, J)$ onto a Riemannian manifold $(N, g_N)$. Then, the slant distribution $\mathcal{D}^\theta$ is integrable if and only if*

$$g(\mathcal{H}\nabla_Z \omega W - \mathcal{H}\nabla_W \omega Z, JX) = g(\mathcal{T}_Z \omega \phi W - \mathcal{T}_W \omega \phi Z, X), \qquad (3.101)$$

*where $Z, W \in \mathcal{D}^\theta$ and $X \in \mathcal{D}^\perp$.*

*Proof.* For $Z, W \in \Gamma(\mathcal{D}^\theta)$ and $X \in \Gamma(\mathcal{D}^\perp)$, using (3.2) and (3.83), we have

$$\begin{aligned}
g([Z, W], X) &= g(\nabla_Z \omega W, JX) - g(\nabla_Z J\phi W, X) - g(\nabla_W \omega Z, JX) \\
&\quad + g(\nabla_W J\phi Z, X).
\end{aligned}$$

Then from (1.48), (3.83), and (3.89), we obtain

$$\begin{aligned}
g([Z, W], X) &= g(\mathcal{H}\nabla_Z \omega W - \mathcal{H}\nabla_W \omega Z, JX) + cos^2\theta g([Z, W], X) \\
&\quad - g(\mathcal{T}_Z \omega \phi W - \mathcal{T}_W \omega \phi Z, X)
\end{aligned}$$

or

$$\begin{aligned}
sin^2\theta g([Z, W], X) &= g(\mathcal{H}\nabla_Z \omega W - \mathcal{H}\nabla_W \omega Z, JX) - g(\mathcal{T}_Z \omega \phi W \\
&\quad - \mathcal{T}_W \omega \phi Z, X),
\end{aligned}$$

which completes the proof. □

From Theorem 29, we have the following sufficient conditions for the slant distribution to be integrable.

**Corollary 15.** *Let $\pi$ be a hemi-slant submersion from a Kähler manifold $(M, g, J)$ onto a Riemannian manifold $(N, g_N)$. If*

$$\mathcal{H}\nabla_Z \omega W - \mathcal{H}\nabla_W \omega Z \in \mu \oplus \omega \mathcal{D}^\theta$$

*and*

$$\mathcal{T}_Z \omega \phi W - \mathcal{T}_W \omega \phi Z \in \Gamma(\mathcal{D}^\theta),$$

*then, the slant distribution $\mathcal{D}^\theta$ is integrable.*

*Proof.* The proof can be seen from the above theorem.                                    □

**Theorem 56.** *[286] Let $\pi$ be a hemi-slant submersion from a Kähler manifold $(M, g, J)$ onto a Riemannian manifold $(N, g_N)$. Then, the anti-invariant distribution $\mathcal{D}^\perp$ is always integrable.*

*Proof.* Since $M$ is a Kähler manifold, then its Kähler form $\Omega$ defined by $\Omega(E, F) = g(E, JF)$ for $E, F \in TM$, is closed, i.e., $d\Omega \equiv 0$. Next, for any $X, Y \in \Gamma(\mathcal{D}^\perp)$ and $U \in \Gamma(ker\pi_*)$, using the integrability of $ker\pi_*$, we have

$$0 = 3 \, d\Omega(U, X, Y) = U \, \Omega(X, Y) + X \, \Omega(Y, U) + Y \, \Omega(U, X)$$

$$-\Omega([U, X], Y) - \Omega([X, Y], U) - \Omega([Y, U], X)$$

$$= g(\phi[X, Y], U) \, .$$

Since $g$ is Riemannian, it follows that $\phi[X, Y] = 0$. Because of (3.164)-(b), we can deduce that $[X, Y] \in \Gamma(\mathcal{D}^\perp)$.                                    □

We now investigate the geometry of leaves of distributions. We first give the following preparatory result.

**Corollary 16.** *Let $\pi$ be a hemi-slant submersion from a Kähler manifold $(M, g, J)$ onto a Riemannian manifold $(N, g_N)$. Then, for any $X, Y \in \Gamma(\mathcal{D}^\perp)$, we have*

$$\mathcal{T}_X JY = \mathcal{T}_Y JX. \tag{3.102}$$

*Proof.* Using (3.164)-(b), from (3.91), we have

$$\mathcal{T}_X \omega Y = \phi(\hat{\nabla}_X Y) + \mathcal{B} \mathcal{T}_Y X \tag{3.103}$$

for all $X, Y \in \Gamma(\mathcal{D}^\perp)$. By interchanging the role of $X$ and $Y$ in (3.103), then subtracting it from (3.103), we obtain

$$\mathcal{T}_X \omega Y - \mathcal{T}_Y \omega X = \phi[X, Y]. \tag{3.104}$$

By Theorem 56 and (3.164)-(b), we get $\phi[X, Y] = 0$ from (3.104). This gives us (3.102), since $\omega X = JX$ for any $X \in \Gamma(\mathcal{D}^\perp)$.                                    □

We now obtain necessary and sufficient conditions for the horizontal distribution and the vertical distribution to be totally geodesic. We first give the following result about the horizontal distribution.

**Theorem 57.** *[286] Let $\pi$ be a proper hemi-slant submersion from a Kähler manifold*

$(M, g, J)$ onto a Riemannian manifold $(N, g_N)$. Then the horizontal distribution defines a totally geodesic foliation on $M$ if and only if

$$g(\mathcal{A}_\xi \eta, \mathcal{P}V) = \sec^2\theta\{g(\mathcal{H}\nabla_\xi \eta, \omega\phi\mathcal{P}V) - g(\mathcal{A}_\xi \mathcal{B}\eta + \mathcal{H}\nabla_\xi C\eta, \omega V)\} \qquad (3.105)$$

for $\xi, \eta \in (ker\pi_*)^\perp$ and $V \in ker\pi_*$.

*Proof.* From (3.82), we have

$$g(\nabla_\xi \eta, V) = g(\nabla_\xi \eta, \mathcal{P}V) + g(\nabla_\xi \eta, \mathcal{Q}V).$$

Then Kähler $M$ gives us

$$g(\nabla_\xi \eta, V) = g(\nabla_\xi J\eta, J\mathcal{P}V) + g(\nabla_\xi J\eta, J\mathcal{Q}V).$$

Using (3.83), we get

$$g(\nabla_\xi \eta, V) = g(\nabla_\xi J\eta, \phi\mathcal{P}V) + g(\nabla_\xi J\eta, \omega\mathcal{P}V) + g(\nabla_\xi J\eta, J\mathcal{Q}V).$$

Again using the Kähler character of $M$ and (3.83), we obtain

$$g(\nabla_\xi \eta, V) = -g(\nabla_\xi \eta, \phi^2\mathcal{P}V) - g(\nabla_\xi \eta, \omega\phi\mathcal{P}V) + g(\nabla_\xi J\eta, \omega\mathcal{P}V) + g(\nabla_\xi J\eta, J\mathcal{Q}V).$$

Thus from Theorem 54 and (3.84), (1.50), and (1.49) we get

$$\begin{aligned} g(\nabla_\xi \eta, V) &= \cos^2\theta g(\mathcal{A}_\xi \eta, \mathcal{P}V) - g(\mathcal{H}\nabla_\xi \eta, \omega\phi\mathcal{P}V) + g(\mathcal{A}_\xi \mathcal{B}\eta, \omega\mathcal{P}V) \\ &\quad + g(\mathcal{H}\nabla_\xi C\eta, \omega\mathcal{P}V) + g(\mathcal{A}_\xi \mathcal{B}\eta, J\mathcal{Q}V) + g(\mathcal{H}\nabla_\xi C\eta, J\mathcal{Q}V). \end{aligned}$$

Since $\omega V = \omega\mathcal{P}V + J\mathcal{Q}V$, we find

$$\begin{aligned} g(\nabla_\xi \eta, V) &= \cos^2\theta g(\mathcal{A}_\xi \eta, \mathcal{P}V) - g(\mathcal{H}\nabla_\xi \eta, \omega\phi\mathcal{P}V) + g(\mathcal{A}_\xi \mathcal{B}\eta, \omega V) \\ &\quad + g(\mathcal{H}\nabla_\xi C\eta, \omega V), \end{aligned}$$

which gives us the proof.                                                        $\square$

In a similar way, we have the following result.

**Theorem 58.** *[286] Let $\pi$ be a proper hemi-slant submersion from a Kähler manifold $(M, g, J)$ onto a Riemannian manifold $(N, g_N)$. Then the vertical distribution defines a totally geodesic foliation on $M$ if and only if*

$$g(\mathcal{T}_U\mathcal{P}V, \xi) = \sec^2\theta\{g(\mathcal{H}\nabla_U \omega\phi\mathcal{P}V, \xi) - g(\mathcal{T}_U\omega V, \mathcal{B}\xi) - g(\mathcal{H}\nabla_U \omega V, C\xi)\} \qquad (3.106)$$

*for $U, V \in ker\pi_*$ and $\xi \in (ker\pi_*)^\perp$.*

Theorem 57 and Theorem 58 show that these conditions are different from those conditions found in section 7. From Theorem 57 and Theorem 58, we also have the following decomposition result.

**Corollary 17.** *Let $\pi$ be a proper hemi-slant submersion from a Kähler manifold $(M, g, J)$ onto a Riemannian manifold $(N, g_N)$. Then the total space is a locally product manifold if and only if (3.105) and (3.106) are satisfied.*

We now investigate the geometry of leaves of anti-invariant distribution and slant distribution.

**Theorem 59.** *[286] Let $\pi$ be a proper hemi-slant submersion from a Kähler manifold $(M, g, J)$ onto a Riemannian manifold $(N, g_N)$. Then, the anti-invariant distribution $\mathcal{D}^\perp$ defines a totally geodesic foliation if and only if*

$$g(\mathcal{H}\nabla_X \omega Y, \omega Z) = g(\mathcal{T}_X Y, \omega \phi Z) \tag{3.107}$$

*and*

$$g(\omega \mathcal{T}_X JY, \xi) = g(\mathcal{H}\nabla_X JY, C\xi), \tag{3.108}$$

*where $X, Y \in \Gamma(\mathcal{D}^\perp)$, $Z \in \Gamma(\mathcal{D}^\theta)$ and $\xi \in \Gamma((ker\pi_*)^\perp)$.*

*Proof.* For $X, Y \in \Gamma(\mathcal{D}^\perp)$ and $Z \in \Gamma(\mathcal{D}^\theta)$, using (3.2) and (3.83) we have

$$g(\nabla_X Y, Z) \;\; = g(\nabla_X JY, \phi Z) + g(\nabla_X JY, \omega Z).$$

Then (1.48) and (3.89) imply that

$$\begin{aligned} g(\nabla_X Y, Z) \;\; &= \;\; cos^2\theta g(\nabla_X Y, Z) - g(\mathcal{T}_X Y, \omega \phi Z) \\ &\quad + g(\mathcal{H}\nabla_X \omega Y, \omega Z) \end{aligned}$$

*or*

$$sin^2\theta g(\nabla_X Y, Z) \;\; = \;\; g(\mathcal{H}\nabla_X \omega Y, \omega Z) - g(\mathcal{T}_X Y, \omega \phi Z) \tag{3.109}$$

which gives us (3.107). Now, for $X, Y \in \Gamma(\mathcal{D}^\perp)$ and $\xi \in \Gamma((ker\pi_*)^\perp)$, from (3.2) we have

$$g(\nabla_X Y, \xi) \;\; = \;\; g(\nabla_X JY, J\xi)$$

Using (1.48), (3.83) and (3.84), we get

$$g(\nabla_X Y, \xi) \;\; = \;\; g(\mathcal{T}_X JY, \mathcal{B}\xi) + g(\mathcal{H}\nabla_X JY, C\xi). \tag{3.110}$$

The proof follows from (3.109) and (3.110).    □

**Theorem 60.** *[286] Let $\pi$ be a proper hemi-slant submersion from a Kähler manifold $(M, g, J)$ onto a Riemannian manifold $(N, g_N)$. Then, the slant distribution $\mathcal{D}^\theta$ defines* *a totally geodesic foliation if and only if*

$$g(\mathcal{H}\nabla_Z\omega W, \omega X) = g(\mathcal{T}_Z\omega\phi W, X) \tag{3.111}$$

*and*

$$g(\mathcal{H}\nabla_Z\omega\phi W, \xi) = g(\mathcal{T}_Z\omega W, \mathcal{B}\xi) + g(\mathcal{H}\nabla_Z\omega W, \mathcal{C}\xi), \tag{3.112}$$

*where* $Z, W \in \Gamma(\mathcal{D}^\theta)$, $X \in \Gamma(\mathcal{D}^\perp)$ *and* $\xi \in (ker\pi_*)^\perp$.

*Proof.* For $Z, W \in \Gamma(\mathcal{D}^\theta)$ and $X \in \Gamma(\mathcal{D}^\perp)$, (3.2), and (3.83) we have

$$g(\nabla_Z W, X) \;=\; g(\nabla_Z\phi W, JX) + g(\nabla_Z\omega W, JX).$$

Then (3.2) and (1.48) imply that

$$g(\nabla_Z W, X) \;=\; -g(\nabla_Z J\phi W, X) + g(\mathcal{T}_Z\omega W + \mathcal{H}\nabla_Z\omega W, JX)$$

Thus using (3.83) and Theorem 54,

$$sin^2\theta g(\nabla_Z W, X) \;=\; g(\mathcal{H}\nabla_Z\omega W, JX) - g(\mathcal{T}_Z\omega\phi W, X). \tag{3.113}$$

Similarly, we get

$$sin^2\theta g(\nabla_Z W, \xi) = g(\mathcal{T}_Z\omega W, \mathcal{B}\xi) + g(\mathcal{H}\nabla_Z\omega W, \mathcal{C}\xi) - g(\mathcal{H}\nabla_Z\omega\phi W, \xi). \tag{3.114}$$

Thus, from (3.113) and (3.114), we have the assertion.    □

We are now going to obtain new conditions for hemi-slant submersions to be harmonic and totally geodesic. We also introduce the notion of mixed geodesic hemi-slant submersions and obtain necessary and sufficient conditions for such submersions to be mixed geodesic.

**Theorem 61.** *[286] Let $\pi$ be a proper hemi-slant submersion from a Kähler manifold $(M, g, J)$ onto a Riemannian manifold. If $\omega$ is parallel with respect to $\nabla$ on $ker\pi_*$, then $\pi$ is a harmonic map.*

*Proof.* From Corollary 3, we know that a Riemannian submersion is harmonic if and only if it has minimal fibers. Thus the submersion $\pi$ is harmonic if and only if $\sum_{k=1}^{2p+q} \mathcal{T}_{v_k} v_k = 0$, where $\{v_1, ..., v_{2p+q}\}$ is a local orthonormal frame of $ker\pi_*$ such that $\{v_1, ..., v_{2p}\}$ is a local orthonormal frame of $\mathcal{D}^\theta$ and $\{v_{2p+1}, ..., v_{2p+q}\}$ is a local orthonormal frame of $\mathcal{D}^\perp$. Since $\omega$ is parallel, for any $X \in \Gamma(\mathcal{D}^\perp)$, we have $\mathcal{T}_X X = 0$ from

(3.92). So, we get

$$\sum_{k=1}^{q} \mathcal{T}_{v_{2p+k}} v_{2p+k} = 0. \tag{3.115}$$

On the other hand, by the same way as in the proof of Theorem 45, we find

$$\sum_{k=1}^{2p} \mathcal{T}_{v_k} v_k = 0. \tag{3.116}$$

Thus, from (3.194) and (3.195), we get $\sum_{k=1}^{2p+q} \mathcal{T}_{v_k} v_k = 0.$    □

We now obtain a necessary and sufficient conditions for a hemi-slant submersion to be totally geodesic.

**Theorem 62.** *[286] Let $\pi$ be a hemi-slant submersion from a Kähler manifold $(M, g, J)$ onto a Riemannian manifold $(N, g_N)$. Then, $\pi$ is a totally geodesic map if and only if*

$$g(\mathcal{H}\nabla_U \omega \phi PV \quad - \quad cos^2\theta \mathcal{T}_U PV, \xi) = g(\mathcal{T}_U \omega PV + JQ\xi, \mathcal{B}\xi)$$
$$+ g(\mathcal{H}\nabla_U PV + \mathcal{H}\nabla_U JQ\xi, C\xi) \tag{3.117}$$

*and*

$$(cos^2\theta \mathcal{A}_\eta PV \quad - \quad \mathcal{H}\nabla_\eta \omega \phi PV, \xi) = g(\mathcal{A}_\eta \omega PV + \mathcal{A}_\eta JQV, \mathcal{B}\xi)$$
$$- g(\mathcal{H}\nabla_\eta \omega PV + \mathcal{H}\nabla_\eta J\phi V, C\xi) \tag{3.118}$$

*for $U, V \in \Gamma(ker\pi_*)$ and $\xi, \eta \in \Gamma((ker\pi_*)^\perp)$.*

*Proof.* For $U, V \in \Gamma(ker\pi_*)$ and $\xi \in \Gamma((ker\pi_*)^\perp)$, from (3.2) and (3.83) we have

$$g(\nabla_U V, \xi) \quad = \quad -g(\nabla_U \phi PV, J\xi) - g(\nabla_U \omega PV, J\xi) - g(\nabla_U JQV, J\xi)$$

Again using (1.48), (3.2), (3.83), and (3.84), we get

$$g(\nabla_U V, \xi) \quad = \quad g(\nabla_U J\phi PV, \xi) - g(\mathcal{T}_U \omega PV + \mathcal{H}\nabla_U PV, \mathcal{B}\xi + C\xi)$$
$$- \quad g(\mathcal{T}_U JQV + \mathcal{H}\nabla_U JQV, J\xi).$$

Then Theorem 54 and (1.48) imply

$$g(\nabla_U V, \xi) \quad = \quad -cos^2\theta \, g(\nabla_U PV, \xi) + g(\nabla_U \omega \phi PV, X)$$
$$- \quad g(\mathcal{T}_U \omega PV + JQV, \mathcal{B}\xi) - g(\mathcal{H}\nabla_U PV + \mathcal{H}\nabla_U JQV, C\xi).$$

Thus using (1.47) and (1.48) we have

$$
\begin{aligned}
g(\nabla_U V, \xi) ={} & g(\mathcal{H}\nabla_U \omega \phi \mathcal{P}V - \cos^2\theta \mathcal{T}_U \mathcal{P}V, \xi) \\
& - g(\mathcal{T}_U \omega \mathcal{P}V + JQV, \mathcal{B}\xi) - g(\mathcal{H}\nabla_U \mathcal{P}V + \mathcal{H}\nabla_U JQV, \mathcal{C}\xi),
\end{aligned}
$$

which gives us the first assertion. The other assertion can be found in a similar way.
□

Let $\pi$ be a proper hemi-slant submersion from a Kähler manifold $(M, g, J)$ onto a Riemannian manifold. Then we say that the fibers of $\pi$ are *mixed geodesic*, if $\mathcal{T}_Z X = 0$, where $Z \in \Gamma(\mathcal{D}^\theta)$ and $X \in \mathcal{D}^\perp$.

**Theorem 63.** *[286] Let $\pi$ be a proper hemi-slant submersion from a Kähler manifold $(M, g, J)$ onto a Riemannian manifold and $\omega$ be parallel with respect to $\nabla$ on $\ker\pi_*$, then*

$$
\mathcal{T}_X Z = 0 \quad and \quad \mathcal{T}_W Z = 0, \quad if \quad C \equiv 0 \tag{3.119}
$$

*and*

$$
\mathcal{T}_Y X = 0 \quad and \quad \mathcal{T}_Z X = 0, \quad if \quad C \neq 0, \tag{3.120}
$$

*where $Z, W \in \Gamma(\mathcal{D}^\theta)$ and $X, Y \in \mathcal{D}^\perp$. In any cases, the fibers of $\pi$ are mixed geodesic.*

*Proof.* If $C \equiv 0$, then for $V \in \Gamma(\ker\pi_*)$ and $Z \in \Gamma(\mathcal{D}^\theta)$, we have

$$
\mathcal{T}_V \phi Z = 0. \tag{3.121}
$$

By putting $Z = \phi Z$ in (3.121) and using (3.88), we obtain

$$
\cos^2\theta \mathcal{T}_V Z = 0. \tag{3.122}
$$

Since $\theta \neq \frac{\pi}{2}$, we get

$$
\mathcal{T}_V Z = 0. \tag{3.123}
$$

This equation gives us (3.119).

Next, if $C \neq 0$, then for $V \in \Gamma(\ker\pi_*)$ and $X \in \Gamma(\mathcal{D}^\perp)$, we have

$$
C\mathcal{T}_V X = \mathcal{T}_V \phi X = 0 \tag{3.124}
$$

from (3.100), since $\phi X = 0$. From (3.124), we deduce that

$$
\mathcal{T}_V X = 0. \tag{3.125}
$$

This equation gives us (3.120). Moreover, from (3.119) and (3.120), we see easily that the fibers of $\pi$ are mixed geodesic.
□

As a result of this theorem, we have the following:

**Corollary 18.** *Let $\pi$ be a proper hemi-slant submersion from a Kähler manifold $(M, g, J)$ onto a Riemannian manifold and $\omega$ be parallel with respect to $\nabla$ on $\ker\pi_*$, then*

$$\mathcal{T} \equiv 0 \Leftrightarrow \mathcal{T}_X Y = 0, \quad if \quad C \equiv 0 \tag{3.126}$$

*and*

$$\mathcal{T} \equiv 0 \Leftrightarrow \mathcal{T}_Z W = 0, \quad if \quad C \neq 0, \tag{3.127}$$

*where $Z, W \in \Gamma(\mathcal{D}^\theta)$ and $X, Y \in \mathcal{D}^\perp$.*

In the rest of this section, we study the geometry hemi-slant submersions with umbilical fibers by using the previous results.

We first recall that a fiber of a Riemannian submersion $\pi$ is called *minimal*, if $H = 0$, identically. We now give a characterization theorem for the proper hemi-slant submersion with totally umbilical fibers.

**Theorem 64.** *[286] Let $\pi$ be a proper hemi-slant submersion with totally umbilical fibers from a Kähler manifold $(M, g, J)$ onto a Riemannian manifold. Then, either the anti-invariant distribution $\mathcal{D}^\perp$ is one-dimensional or the mean curvature vector field $H$ of any fiber $\pi^{-1}(y)$, $y \in N$ is perpendicular to $J\mathcal{D}^\perp$. Moreover, if $\phi$ is parallel, then $H \in \mu$. Furthermore, if $\omega$ is parallel, then $\mathcal{T} \equiv 0$.*

*Proof.* Since $\pi$ is a proper hemi-slant submersion, then either $Dim(\mathcal{D}^\perp) = 1$ or $Dim(\mathcal{D}^\perp) > 1$. If $Dim(\mathcal{D}^\perp) = 1$, it is obvious. If $Dim(\mathcal{D}^\perp) > 1$, then we can choose $X, Y \in \mathcal{D}^\perp$ such that $\{X, Y\}$ is orthonormal. By using (1.45), (1.46), (3.83), (3.84), and (3.2), we get

$$\mathcal{T}_X JY + \mathcal{H}\nabla_X JY = \nabla_X JY = J\nabla_X Y = \phi\hat{\nabla}_X Y + \omega\hat{\nabla}_X Y + \mathcal{B}\mathcal{T}_X Y + C\mathcal{T}_X Y,$$

so that $g(\mathcal{T}_X JY, X) = g(\phi\hat{\nabla}_X Y, X) + g(\mathcal{B}\mathcal{T}_X Y, X)$. Hence, using (3.2), we obtain

$$g(\mathcal{T}_X JY, X) = -g(\mathcal{T}_X Y, JX). \tag{3.128}$$

Thus, using (1.65) and (3.128), we have

$$g(H, JY) = g(\mathcal{T}_X X, JY) = -g(\mathcal{T}_X JY, X) = g((\mathcal{T}_X Y, JX) = g(X, Y)g(H, JX) = 0,$$

since $g(X, Y) = 0$. We therefore deduce that

$$H \perp \mathcal{D}^\perp. \tag{3.129}$$

Moreover, if $\phi$ is parallel, then using (3.99) and (3.2), for $Z \in \Gamma(\mathcal{D}^\theta)$, we get

$$g(H, \omega Z) = g(\mathcal{T}_X X, \omega Z) = -g(\mathcal{T}_X \omega Z, X) = -g(\mathcal{B}\mathcal{T}_X \omega Z, X) = -g(J\mathcal{T}_X \omega Z, X)$$
$$= g(\mathcal{T}_X \omega Z, JX) = 0.$$

Hence, it follows that

$$H \perp \omega \mathcal{D}^\theta. \tag{3.130}$$

By (3.129) and (3.130), we conclude that $H \in \mu$ from (3.85). Furthermore, if $\omega$ is parallel, then using (3.100) and (3.2), for the unit vector field $X \in \Gamma(\mathcal{D}^\perp)$ and $\xi \in \mu$, we get

$$g(H, \xi) = g(\mathcal{T}_X X, \xi) = g(J\mathcal{T}_X X, J\xi) = g((C\mathcal{T}_X X, J\xi) = g(\mathcal{T}_X \phi X, J\xi) = 0,$$

since $\phi X = 0$. Thus, we deduce that $H = 0$- that is, the fibers are minimal. Since the fibers are also totally umbilical, we get $\mathcal{T} \equiv 0$ from (1.65).  □

## 9. Pointwise slant submersions

In this section, we present recent results on the geometry of pointwise slant submersions from almost Hermitian manifolds.

**Definition 44.** [172] Let $F$ be a Riemannian submersion from an almost Hermitian manifold $(M, g_M, J)$ onto a Riemannian manifold $(N, g_N)$. If, at each given point $p \in M$, the Wirtinger angle $\theta(X)$ between $JX$ and the space $(ker\, F_*)_p$ is independent of the choice of the nonzero vector $X \in (ker\, F_*)$, then we say that $F$ is a *pointwise slant submersion*. In this case, the angle $\theta$ can be regarded as a function $M$, which is called the *slant function* of the pointwise slant submersion.

**Definition 45.** [172] A point $p$ in a pointwise slant submersion is called *totally real* if its slant function $\theta = \frac{\pi}{2}$ at $p$. Similarly, a point $p$ is called a *complex point* if its slant function $\theta = 0$ at $p$. A pointwise slant submersion is said to be *proper* if it is neither a totally real nor a complex Riemannian submersion.

**Remark 9.** A pointwise slant submersion is called *slant* if its slant function $\theta$ is globally constant, i.e., $\theta$ is also independent of the choice of the point on $M$. In this case, the constant $\theta$ is called the *slant angle* of the slant submersion. A pointwise slant submersion $F$ is called totally real if every point of $M$ is a totally real point.

It is obvious that invariant, anti-invariant, and slant submersions are examples of pointwise slant submersions. We now give an example for proper pointwise slant submersions.

**Example 24.** [172] Let $(\mathbb{R}^4, g_0)$ be the standard Euclidean space with the standard metric $g_0$. Consider $\{J_0, J_1\}$, a pair of almost complex structures on $\mathbb{R}^4$ satisfying $J_0 J_1 = -J_1 J_0$, where

$$
\begin{aligned}
J_0(a,b,c,d) &= (-c,-d,a,b) \\
J_1(a,b,c,d) &= (-b,a,-d,c).
\end{aligned}
$$

For any real-valued function $f : \mathbb{R}^4 \longrightarrow \mathbb{R}$, we define a new almost complex structure $J_f$ on $\mathbb{R}^4$ by

$$
J_f = (\cos f)J_0 + (\sin f)J_1.
$$

Then $\mathbb{R}^4_f = (\mathbb{R}^4, J_f, g_0)$ is an almost Hermitian manifold. Consider a Riemannian submersion $F : \mathbb{R}^4_f \longrightarrow \mathbb{R}^2$ by $F(x_1, x_2, x_3, x_4) = (\frac{x_1 - x_2}{\sqrt{2}}, \frac{x_3 - x_4}{\sqrt{2}})$. Then $F$ is a pointwise slant submersion with the slant function $f$.

The above example shows us that if a Riemannian submersion is invariant or holomorphic with respect to $J_1$, then it will always produce a pointwise slant submersion with the slant function $f$. This shows us that there are many pointwise slant submersions.

Now, we assume that $F$ is a pointwise slant submersion from an almost Hermitian manifold $(M, g_M, J)$ onto Riemannian manifold $(N, g_N)$. Then for $V \in \Gamma(\ker F_*)$, we have

$$
JV = \varphi V + \omega V, \tag{3.131}
$$

where $\varphi V(\omega V$, resp.) is vertical( horizontal, resp.) of $JV$. Also for $X \in \Gamma((\ker F_*)^\perp)$, we have

$$
JX = \mathcal{B}X + \mathcal{C}X, \tag{3.132}
$$

where $\mathcal{B}X$ ($\mathcal{C}X$, resp.) is vertical( horizontal, resp.) of $\varphi X$. We denote the orthogonal complementary distribution to $\omega(\ker F_*)$ in $(\ker F_*)^\perp$ by $\mu$. That is,

$$
(\ker F_*)^\perp = \omega(\ker F_*)\perp\mu.
$$

Using (1.47), (1.48), (3.131), and (3.132), we obtain

$$
\begin{aligned}
(\nabla_V \omega) W &= \mathcal{C} \mathcal{T}_V W - \mathcal{T}_V t W \tag{3.133}\\
(\nabla_V t) W &= \mathcal{B} \mathcal{T}_V W - \mathcal{T}_V \omega W + (\nabla_V \varphi)W, \tag{3.134}
\end{aligned}
$$

where $\nabla$ is the Levi-Civita connection on $M$ and

$$
\begin{aligned}
(\nabla_V \omega) W &= \mathcal{H}\nabla_V \omega W - \omega \hat{\nabla}_V W \\
(\nabla_V t) W &= \hat{\nabla}_V t W - t \hat{\nabla}_V W,
\end{aligned}
$$

for $V, W \in \Gamma(\ker F_*)$. We say that $\omega$ is parallel with respect to the Levi-Civita connection $\nabla$ on $\ker F_*$ if its covariant derivative with respect to $\nabla$ vanishes, i.e., $(\nabla_V \omega)W = 0$, for $V, W \in \Gamma(\ker F_*)$.

The proof of the following result is the same as slant immersions (see [81]); therefore we omit its proof.

**Theorem 65.** *[172] Let F be a Riemannian submersion from an almost Hermitian manifold $(M, g_M, J)$ onto a Riemannian manifold $(N, g_N)$. Then F is a pointwise slant submersion if and only if there exists a real-valued functioin $\theta$ defined on $\ker F_*$ such that*

$$\varphi^2 = -\left(\cos^2 \theta\right) I.$$

**Theorem 66.** *[172] Let F be a pointwise proper slant submersion from an almost Hermitian manifold $(M, g_M, J)$ onto a Riemannian manifold $(N, g_N)$. Then $\mu$ is invariant with respect to $\varphi$.*

*Proof.* For $Z \in \Gamma(\mu_p)$ and $V \in \Gamma(\ker F_{*p})$, we have

$$
\begin{aligned}
g_M(JZ, \omega V) &= g_M(JZ, JV) - g_M(JZ, \varphi V) \\
&= -g_M(JZ, \varphi V) \\
&= g_M(Z, \varphi^2 V) \\
&= g_M(Z, -\cos^2 \theta(p)V) \\
&= -\cos^2 \theta(p) g_M(Z, V) = 0.
\end{aligned}
$$

On the other hand, for $Z \in \Gamma(\mu_p)$ and $U \in \Gamma(\ker F_{*p})$, we have

$$g_M(JZ, U) = -g_M(Z, JU) = -g_M(Z, \omega U) = 0.$$

Therefore, $JZ \in \Gamma(\mu_p)$. $\qquad\square$

**Corollary 19.** *[172] Let F be a pointwise proper slant submersion from an almost Hermitian manifold $(M^m, g_M, J)$ onto a Riemannian manifold $(N^n, g_N)$ with the slant function $\theta$. If $e_1, e_2, \cdots, e_{m-n}$ are locally orthonormal basis for $\ker F_*$, then*

$$\{\csc \theta \omega e_1, \csc \theta \omega e_2, \cdots, \csc \theta \omega e_{m-n}\}$$

*is a local orthonormal basis of $\omega(\ker F_*)$.*

The following result is a consequence of Theorem 66 and Corollary 19.

**Corollary 20.** *[172] Let F be a pointwise proper slant submersion from an almost Hermitian manifold $(M, g_M, J)$ onto a Riemannian manifold $(N, g_N)$. Then the distributions $\mu$ and $\ker F_* \oplus \omega(\ker F_*)$ are even-dimensional.*

The following proposition gives a characterization of pointwise slant submersions

**Proposition 38.** *[172] Let F be a pointwise slant submersion from an almost Hermitian manifold $(M, g_M, J)$ onto a Riemannian manifold $(N, g_N)$ if and only if $\varphi : \ker F_* \longrightarrow \ker F_*$ preserves orthogonality, $\varphi$ carries each pair of orthogonal vectors into orthogonal vectors.*

*Proof.* Denote by $S(\ker F_*)$ the unit kernel bundle of $\ker F_*$. Then consider the function $\theta : S(\ker F_*) \longrightarrow R$ on the unit kernel bundle $S(\ker F_*)$. With respect to the induced metric, $S_p(\ker F_*)$ is the unit sphere in $\ker F_{*p}$ centered at 0. At a given point $p \in M$, we have

$$g_M(\varphi X, \varphi X) = \cos^2 \theta(X)$$

for any unit vector $X \in \ker F_{*p}$. For each unit vector $Y$ tagent to $\ker F_{*p}$ at $X \in \ker F_{*p}$, we have

$$2g_M(\varphi X, \varphi Y) = Y g_M(\varphi X, \varphi X) = -(Y\theta) \sin 2\theta(X).$$

In the long run, $\varphi$ carries each pair of orthogonal vector in $\ker F_{*p}$ into a pair of orthogonal vectors in $\ker F_{*p}$ if and only if the slant function $\theta$ is independent of $X \in \ker F_{*p}$.  □

**Proposition 39.** *[172] Let F be a pointwise slant submersion from a Kähler manifold $(M, g_M, J)$ onto a Riemannian manifold $(N, g_N)$. F is slant submersion if and only if*

$$\mathcal{T}_X \omega \varphi X = \mathcal{T}_{\varphi X} \omega X$$

*for $X \in \Gamma(\ker F_*)$.*

*Proof.* Let $F$ be a pointwise slant submersion from a Kähler manifold $(M, g_M, J)$ onto a Riemannian manifold $(N, g_N)$ with the slant function $\theta$. For any unit vector field $X \in \Gamma(\ker F_*)$, we may put

$$\varphi X = (\cos \theta) X^*,$$

where $X^*$ is a unit vector field orthogonal ot $X$. Then for $Y \in \Gamma(ker\, F_*)$, we have

$$
\begin{aligned}
\nabla_Y(JX) &= \nabla_Y((\cos\theta)\, X^*) + \nabla_Y(\omega X) \\
&= -(\sin\theta)(Y\theta)\, X^* + (\cos\theta)\left(\mathcal{T}_Y X^* + \hat{\nabla}_Y X^*\right) \qquad (3.135) \\
&\quad + \mathcal{H}\nabla_Y(\omega X) + \mathcal{T}_Y(\omega X).
\end{aligned}
$$

On the other hand, we also have

$$
\begin{aligned}
\nabla_Y(JX) &= J(\nabla_Y X) \\
&= \mathcal{B}\mathcal{T}_Y X + C\mathcal{T}_Y X + \varphi\hat{\nabla}_Y X + \omega\hat{\nabla}_Y X. \qquad (3.136)
\end{aligned}
$$

Comparing the vertical components of (3.135) and (3.136),

$$
-(\sin\theta)(Y\theta)\, X^* + (\cos\theta)\,\hat{\nabla}_Y X^* + \mathcal{T}_Y(\omega X) = \mathcal{B}\mathcal{T}_Y X + \varphi\hat{\nabla}_Y X \qquad (3.137)
$$

Therefore, by taking the inner product of (3.137) with $X^*$, we have

$$
-(\sin\theta)(Y\theta) + g_M(\mathcal{T}_{X^*}\omega X, Y) = g_M(Y, \mathcal{T}_X\omega X^*).
$$

Consequently, the pointwise slant submersion is slant if and only if

$$
\mathcal{T}_{X^*}\omega X = \mathcal{T}_X\omega X^*.
$$

$\square$

**Corollary 21.** *[172] Let $F$ be a pointwise slant submersion from a Kähler manifold $(M, g_M, J)$ onto a Riemannian manifold $(N, g_N)$. If $F$ has totally geodesic fibers, it is slant.*

*Proof.* If $F$ has totally geodesic fibers, $\mathcal{T}$ vanishes in the same way. Thus the proof follows from Proposition 39. $\square$

We now consider totally umbilical pointwise proper slant submanifolds of a Kähler manifold.

**Corollary 22.** *[172] Let $F$ be a pointwise proper slant submersion with totally umbilical fibers from a Kähler manifold $(M^{2n}, g_M, J)$ onto a Riemannian manifold $(N^n, g_N)$. If $F$ has no totally geodesic fibers, then it is always non-slant.*

*Proof.* Assume that $F$ is a pointwise proper slant submersion with totally umbilical fibers. Then we have

$$
\mathcal{T}_X Y = g_M(X, Y)H,
$$

for $X, Y \in \Gamma(ker\, F_*)$, where $H$ is the mean curvature vector field of the fiber. Therefore, we have

$$g_M(\mathcal{T}_X \omega \varphi X, Y) = -g_M(X, Y)g_M(\omega \varphi X, H) \qquad (3.138)$$
$$g_M(\mathcal{T}_{\varphi X} \omega X, Y) = -g_M(\varphi X, Y)g_M(\omega X, H). \qquad (3.139)$$

If the submersion is slant, then Proposition 39 and (3.138)-(3.139) imply that

$$g_M(\omega \varphi X, H)X = g_M(\omega X, H)\varphi X.$$

Since $g_M(X, \varphi X) = 0$,

$$g_M(\omega \varphi X, H) = g_M(\omega X, H) = 0,$$

which implies $H \in \Gamma(\mu)$. On the other hand, we note that $dim(ker F_*) = 2n - n = n$. Thus using, we have $dim((ker F_*) \oplus \omega(ker F_*)) = 2(2n - n) = 2n$. Therefore, we get $dim(\mu) = 2n - 2n = 0$, which contradicts that $H$ vanishes in the same way. $\qquad \square$

We now investigate certain properties of pointwise slant submersions. The following theorem proposes a new condition for Riemannian submersions to be harmonic. We recall again that if $F : (M_1, g_1) \longrightarrow (M_2, g_2)$ is a map between Riemannian manifolds $(M_1, g_1)$ and $(M_2, g_2)$. Then the adjoint map $^*F_*$ of $F_*$ is characterized by $g_1(x, {}^*F_{*p_1}y) = g_2(F_{*p_1}x, y)$ for $x \in T_{p_1}M_1$, $y \in T_{F(p_1)}M_2$ and $p_1 \in M_1$. Considering $F_*^h$ at each $p_1 \in M_1$ as a linear transformation

$$F_{*p_1}^h : ((ker F_*)^\perp(p_1), g_{1p_1((kerF_*)^\perp(p_1))}) \longrightarrow (rangeF_*(p_2), g_{2p_2(rangeF_*)(p_2))}),$$

we will denote the adjoint of $F_*^h$ by $^*F^h_{*p_1}$. Let $^*F_{*p_1}$ be the adjoint of $F_{*p_1} : (T_{p_1}M_1, g_{1p_1}) \longrightarrow (T_{p_2}M_2, g_{2p_2})$. Then the linear transformation

$$({}^*F_{*p_1})^h : rangeF_*(p_2) \longrightarrow (ker F_*)^\perp(p_1),$$

defined by $({}^*F_{*p_1})^h y = {}^*F_{*p_1}y$, where $y \in \Gamma(rangeF_{*p_1})$, $p_2 = F(p_1)$, is an isomorphism and $(F_{*p_1}^h)^{-1} = ({}^*F_{*p_1})^h = {}^*(F_{*p_1}^h)$.

**Theorem 67.** *[172] Let $F$ be a pointwise slant submersion from Kähler manifold $(M_1, g_1, J)$ onto a Riemannian manifold $(M_2, g_2)$. Then $F$ is harmonic if and only if*

$$trace^*F_*((\nabla F_*)((.), \omega \varphi(.))) - trace\omega \mathcal{T}_{(.)}\omega(.) + traceC^*F_*(\nabla F_*)((.), \omega(.)) = 0.$$

*Proof.* From (1.47), (3.3), (3.5), (3.131), and (3.132), we have

$$g_1(\mathcal{T}_U U, X) = -g_1(\nabla_U J\varphi U, X) + g_1(\nabla_U \omega U, JX)$$

for $U \in \Gamma(\mathcal{V})$ and $X \in \Gamma(\mathcal{H})$. Then Theorem 3.1, (3.131) and (1.48) imply

$$
\begin{aligned}
g_1(\mathcal{T}_U U, X) \; = \; & -g_1(\sin 2\theta U(\theta), X) + g_1(\cos^2 \theta \nabla_U U, X) \\
& -g_1(\nabla_U \omega \varphi U, X) + g_1(\mathcal{T}_U \omega U, \mathcal{B} X) \\
& +g_1(\nabla_U \omega U, CX).
\end{aligned}
$$

Using (1.71), we arrive at

$$
\begin{aligned}
\sin^2 \theta g_1(\mathcal{T}_U U, X) \; = \; & g_2((\nabla F_*)(U, \omega \varphi U), F_*(X)) + g_1(\mathcal{T}_U \omega U, \mathcal{B} X) \\
& -g_2((\nabla F_*)(U, \omega U), F_*(CX)).
\end{aligned}
$$

Then from (3.132), we get

$$
\begin{aligned}
\sin^2 \theta g_1(\mathcal{T}_U U, X) \; = \; & g_1(^*F_*((\nabla F_*)(U, \omega \varphi U)), X) - g_1(\omega \mathcal{T}_U \omega U, X) \\
& +g_1(C^* F_*((\nabla F_*)(U, \omega U)), X).
\end{aligned}
$$

Thus the proof follows from a fact that a Riemannian submersion is harmonic if and only if its fibers are minimal submanifolds in $M_1$.  □

The next result gives us a necessary and sufficient condition for fibers of a pointwise slant submersion to be totally geodesic submanifolds.

**Theorem 68.** *[172] Let $F$ be a pointwise slant submersion from Kähler manifold $(M_1, g_1, J)$ onto a Riemannian manifold $(M_2, g_2)$. Then the fibers are totally geodesic submanifolds in $M_1$ if and only if*

$$
\begin{aligned}
g_2(\nabla^2_{X'} F_*(\omega U), F_*(\omega V)) \; = \; & -g_1([U, X], V) + 2 \cot \theta X(\theta) g_1(U, V) \\
& + \sec^2 \theta g_1(\mathcal{A}_X \omega \varphi U, V) - \sec^2 \theta g_1(\mathcal{A}_X \omega U, \varphi V),
\end{aligned}
$$

*where $X$ and $X'$ are $F$-related vector fields and $\nabla^2$ is the Levi-Civita connection on $M_2$.*

*Proof.* Since $\mathcal{T}$ is skew-symmetric with respect to $g_1$ and $\nabla$ is a torsion-free connection on $M_1$, using (1.48) we have

$$
g_1(\mathcal{T}_U V, X) = -g_1([U, X], V) - g_1(\nabla_X U, V)
$$

for $U, V \in \Gamma(\mathcal{V})$ and $X \in \Gamma(\mathcal{H})$. Then from (3.131), (3.132), (3.3), and (3.5), we obtain

$$
\begin{aligned}
g_1(\mathcal{T}_U V, X) \; = \; & -g_1([U, X], V) + g_1(\nabla_X \varphi^2 U + \omega \varphi U, V) \\
& -g_1(\nabla_X \omega U, \varphi V) - g_1(\nabla_X \omega U, \omega V).
\end{aligned}
$$

Thus Theorem 65, (1.50), and Lemma 1, imply

$$
\begin{aligned}
g_1(\mathcal{T}_U V, X) &= -g_1([U, X], V) + \sin 2\theta\, X(\theta) g_1(U, V) r - \cos^2 \theta g_1(\nabla_X U, V) \\
&\quad + g_1(\mathcal{A}_X \omega\varphi U, V) - g_1(\mathcal{A}_X \omega U, \varphi V) \\
&\quad - g_2(\nabla^2_{X'} F_*(\omega U), F_*(\omega V)).
\end{aligned}
$$

Since $\nabla$ is a torsion-free connection, using the skew-symmetry property of $\mathcal{T}$ and (1.47) we get

$$
\begin{aligned}
\sin^2 \theta g_1(\mathcal{T}_U V, X) &= -\sin^2 \theta g_1([U, X], V) + \sin 2\theta\, X(\theta) g_1(U, V) \\
&\quad + g_1(\mathcal{A}_X \omega\varphi U, V) - g_1(\mathcal{A}_X \omega U, \varphi V) \\
&\quad - g_2(\nabla^2_{X'} F_*(\omega U), F_*(\omega V)),
\end{aligned}
$$

which gives us the assertion.                                    $\square$

Finally, we give a result for pointwise slant submersions to be totally geodesic maps.

**Theorem 69.** *[172] Let $F$ be a pointwise slant submersion from Kähler manifold $(M_1, g_1, J)$ onto a Riemannian manifold $(M_2, g_2)$. Then $F$ is a totally geodesic map if and only if*

$$
\begin{aligned}
g_2(\nabla^2_{X'} F_*(\omega U), F_*(\omega V)) &= -g_1([U, X], V) + 2\cot\theta\, X(\theta) g_1(U, V) \\
&\quad + \sec^2 \theta g_1(\mathcal{A}_X \omega\varphi U, V) - \sec^2 \theta g_1(\mathcal{A}_X \omega U, \varphi V),
\end{aligned} \tag{3.140}
$$

*and*

$$
g_1(\mathcal{A}_X \omega U, \mathcal{B}Y) = -\left[ g_2(\nabla^F_X F_*(\omega\varphi U), F_*(Y)) + g_2(\nabla^F_X F_*(\omega U), F_*(CY)) \right], \tag{3.141}
$$

*where $X$ and $X'$ are $F$−related vector fields and $\nabla^F$ is the pullback connection on along $F$.*

*Proof.* By definition, it follows that $F$ is totally geodesic if and only if $(\nabla F_*)(X, Y) = 0$, for $X, Y \in \Gamma(\mathcal{H})$, $(\nabla F_*)(X, V) = 0$ for $V \in \Gamma(\mathcal{V})$ and $(\nabla F_*)(U, V) = 0$ for $U \in \Gamma(\mathcal{V})$. From Lemma 1 of section 4 of Chapter 1, it follows that $(\nabla F_*)(X, Y) = 0$. Also from Theorem 3.10, it follows that $(\nabla F_*)(X, V) = 0$ if and only if (3.140) is satisfied. On the other hand, since $F$ is a Riemannian submersion, using (1.71) we have

$$
g_2((\nabla F_*)(X, U), F_*(Y)) = -g_1(\nabla_X U, Y).
$$

Then from Theorem 65, (3.3), (3.5), (3.131), (3.132), and (1.50), we obtain

$$
\begin{aligned}
g_2((\nabla F_*)(X, U), F_*(Y)) &= -\sin 2\theta X(\theta) g_1(U, Y) - \cos^2 \theta g_1(\nabla_X U, Y) \\
&\quad + g_1(\nabla_X \omega\varphi U, Y) - g_1(\mathcal{A}_X \omega U, \mathcal{B}Y) \\
&\quad - g_1(\nabla_X \omega U, CY).
\end{aligned}
$$

Thus using (1.71) and (3.26), we get

$$
\begin{aligned}
\sin^2 \theta g_2((\nabla F_*)(X, U), F_*(Y)) &= -\left[ g_2(\overset{F}{\nabla}_X F_*(\omega\varphi U), F_*(Y)) + g_2(\overset{F}{\nabla}_X F_*(\omega U), F_*(CY)) \right] \\
&\quad - g_1(\mathcal{A}_X \omega U, \mathcal{B}Y)
\end{aligned}
$$

which completes the proof. $\qquad\qquad\square$

## 10. Einstein metrics on the total space of an anti-invariant submersion

In this section, we assume that total space is an Einstein manifold and the necessary and sufficient condition for an anti-invariant Riemannian submersion is obtained. To achieve this aim, we give curvature relations for an anti-invariant Riemannian submersion from a Kähler manifold onto a Riemannian manifold. First of all, however, we give the following preparatory lemma.

**Lemma 32.** *[169] Let $F$ be an anti-invariant Riemannian submersion from a Kähler manifold $(M, J, g_M)$ onto a Riemannian manifold $(N, g_N)$. Then for $X \in \Gamma(\mathcal{H})$ and $U, V \in \Gamma(\mathcal{V})$, we have the following relations.*

$$
\begin{aligned}
B\mathcal{T}_U V &= \mathcal{T}_U JV & (3.142) \\
B\mathcal{A}_X U &= \mathcal{A}_X JU. & (3.143)
\end{aligned}
$$

*Proof.* From (3.5) and (1.48), we have

$$
J\nabla_U V = \mathcal{H}\nabla_U JV + \mathcal{T}_U JV.
$$

Using (1.47), we get

$$
J\mathcal{T}_U V + J\hat{\nabla}_U V = \mathcal{H}\nabla_U JV + \mathcal{T}_U JV.
$$

Then (3.18) implies that

$$
B\mathcal{T}_U V + C\mathcal{T}_U V + J\hat{\nabla}_U V = \mathcal{H}\nabla_U JV + \mathcal{T}_U JV.
$$

Taking the vertical and horizontal parts of this equation, we obtain (3.142). The other assertions can be obtained in a similar way. $\qquad\square$

**Lemma 33.** *[169] Let F be an anti-invariant Riemannian submersion from a Kähler manifold* $(M, g_1, J)$ *onto a Riemannian manifold* $(N, g)$. *Then we have the following curvature relations:*

$$
\begin{aligned}
R(U, V, W, W') &= R^*(JU, JV, JW, JW') + 2g_1(B\mathcal{A}_{JU}V, B\mathcal{A}_{JW}W') \\
&\quad - g_1(B\mathcal{A}_{JV}W, B\mathcal{A}_{JU}W') + g_1(B\mathcal{A}_{JU}W, B\mathcal{A}_{JV}W'), \quad (3.144)
\end{aligned}
$$

$$
\begin{aligned}
R(X, U, V, W) &= R^*(CX, JU, JV, JW) + 2g_1(B\mathcal{A}_{CX}U, B\mathcal{A}_{JV}W) \\
&\quad - g_1(B\mathcal{A}_{JU}V, B\mathcal{A}_{CX}W) + g_1(B\mathcal{A}_{CX}V, B\mathcal{A}_{JU}W) \\
&\quad + g_1((\nabla_{JU}\mathcal{A})(JV, JW), BX) + g_1(B\mathcal{A}_{JV}W, B\mathcal{T}_{BX}U) \\
&\quad - g_1(B\mathcal{A}_{JW}U, B\mathcal{T}_{BX}V) + g_1(B\mathcal{A}_{JU}V, B\mathcal{T}_{BX}W), \quad (3.145)
\end{aligned}
$$

$$
\begin{aligned}
R(X, U, Y, W) &= R^*(CX, JU, CY, JW) + 2g_1(B\mathcal{A}_{CX}U, B\mathcal{A}_{CY}W) \\
&\quad - g_1((\nabla_{JU}\mathcal{T})(BX, BY), JW) - g_1(\mathcal{A}_{JU}CY, B\mathcal{A}_{CX}W) \\
&\quad - g_1((\nabla_{BX}\mathcal{A})(JU, JW), BY) + g_1(B\mathcal{A}_{CX}CY, B\mathcal{A}_{JU}W) \\
&\quad - g_1(\mathcal{A}_{JU}BX, \mathcal{A}_{JW}BY) + g_1(B\mathcal{T}_{BX}U, B\mathcal{T}_{BY}W) \\
&\quad + g_1((\nabla_{JU}\mathcal{A})(CY, JW), BX) + g_1(B\mathcal{A}_{CY}W, B\mathcal{T}_{BX}U) \\
&\quad - g_1(B\mathcal{A}_{JW}U, \mathcal{T}_{BX}CY) + g_1(B\mathcal{A}_{CY}U, B\mathcal{T}_{BX}W) \\
&\quad + g_1((\nabla_{JW}\mathcal{A})(CX, JU), BY) + g_1(B\mathcal{A}_{CX}U, B\mathcal{T}_{BY}W) \\
&\quad - g_1(B\mathcal{A}_{JU}W, \mathcal{T}_{BY}CX) + g_1(B\mathcal{A}_{CX}W, B\mathcal{T}_{BY}U). \quad (3.146)
\end{aligned}
$$

*for any* $U, V, W, W' \in \Gamma(\mathcal{V})$ *and* $X, Y \in \Gamma(\mathcal{H})$, *where* $R^*$ *is Riemannian curvature of N.*

*Proof.* From (3.13) we have $R(U, V, W, W') = R(JU, JV, W, W')$. Also from (3.2), we get $R(U, V, W, W') = g(JR(JU, JV)W, JW')$. Again using (3.13), we obtain $R(U, V, W, W') = R(JU, JV, JW, JW')$. Now using (1.61) and (3.143) in this equation, we derive (3.144). By using (3.142) and (3.143), the other curvature relations can be obtained in a similar way. □

**Lemma 34.** *[169] Let F be an anti-invariant Riemannian submersion from a Kähler manifold* $(M, g_1, J)$ *onto a Riemannian manifold* $(N, g)$. *Then the Ricci tensor* $\rho$ *is* *given by*

$$
\rho(U, V) = -3 \sum_{i}^{r+s} g_1(B\mathcal{A}_{E_i}U, B\mathcal{A}_{E_i}V) + \sum_{i}^{r} g_1((\nabla_{JU}\mathcal{T})(u_i, u_i), JV)
$$

$$
+ \sum_{i}^{r} g_1((\nabla_{u_i}\mathcal{A})(JU, JV), u_i) - \sum_{i}^{r} g_1(B\mathcal{T}_{u_i}U, B\mathcal{T}_{u_i}V)
$$

$$
+ \sum_{i}^{r} g_1(\mathcal{A}_{JU}u_i, \mathcal{A}_{JV}u_i) + \rho^*(JU, JV), \tag{3.147}
$$

$$
\rho(U, X) = -3 \sum_{i}^{r+s} g_1(\mathcal{A}_{E_i}JU, \mathcal{A}_{E_i}CX) + \sum_{i}^{r+s} g_1((\nabla_{E_i}\mathcal{A})(E_i, JU), BX)
$$

$$
-2 \sum_{i}^{r+s} g_1(\mathcal{A}_{JU}E_i, \mathcal{T}_{BX}E_i) + \sum_{i}^{r} g_1((\nabla_{BX}\mathcal{T})(u_i, u_i), JU)
$$

$$
+ \sum_{i}^{r} g_1((\nabla_{JU}\mathcal{T})(u_i, u_i), CX) + \sum_{i}^{r} g_1((\nabla_{u_i}\mathcal{A})(JU, CX), u_i)
$$

$$
+ \sum_{i}^{r} g_1(\mathcal{A}_{JU}u_i, \mathcal{A}_{CX}u_i) - \sum_{i}^{r} g_1((\nabla_{u_i}\mathcal{T})(BX, u_i), JU)
$$

$$
- \sum_{i}^{r} g_1(\mathcal{T}_{u_i}JU, \mathcal{T}_{u_i}CX) + \rho^*(JU, CX), \tag{3.148}
$$

$$
\rho(X, Y) = \sum_{i}^{r+s} g_1((\nabla_{E_i}\mathcal{A})(E_i, CX), BY) + \sum_{i}^{r+s} g_1((\nabla_{E_i}\mathcal{T})(BX, BY), E_i)
$$

$$
-2 \sum_{i}^{r+s} g_1(\mathcal{A}_{CX}E_i, \mathcal{T}_{BY}E_i) - 3 \sum_{i}^{r+s} g_1(\mathcal{A}_{E_i}CX, \mathcal{A}_{E_i}CY)
$$

$$
- \sum_{i}^{r+s} g_1(\mathcal{T}_{BX}E_i, \mathcal{T}_{BY}E_i) + \sum_{i}^{r+s} g_1(\mathcal{A}_{E_i}BX, \mathcal{A}_{E_i}BY)
$$

$$
- \sum_{i}^{r+s} g_1((\nabla_{E_i}\mathcal{A})(CY, E_i), BX) + \sum_{i}^{r} g_1((\nabla_{BY}\mathcal{T})(u_i, u_i), CX)
$$

$$
- \sum_{i}^{r} g_1((\nabla_{u_i}\mathcal{T})(BY, u_i), CX) + \sum_{i}^{r} g_1((\nabla_{u_i}\mathcal{A})(CX, CY), u_i)
$$

$$
- \sum_{i}^{r} g_1(\mathcal{T}_{u_i}CX, \mathcal{T}_{u_i}CY) + \sum_{i}^{r} g_1(\mathcal{A}_{CX}u_i, \mathcal{A}_{CY}u_i)
$$

$$- \sum_{i}^{r} g_1(\mathcal{T}_{u_i} u_i, \mathcal{T}_{BX} BY) + \sum_{i}^{r} g_1(\mathcal{T}_{BX} u_i, \mathcal{T}_{BY} u_i)$$

$$- \sum_{i}^{r} g_1((\nabla_{u_i} \mathcal{T})(BX, u_i), CY) + \sum_{i}^{r} g_1((\nabla_{BX} \mathcal{T})(u_i, u_i), CY)$$

$$+ \sum_{i}^{r} g_1((\nabla_{CX} \mathcal{T})(u_i, u_i), CY) + \rho^*(CX, CY) + \hat{\rho}(BX, BY)$$

$$-2 \sum_{i}^{r+s} g_1(\mathcal{A}_{CY} E_i, \mathcal{T}_{BX} E_i), \tag{3.149}$$

*for $X, Y \in \Gamma(\mathcal{H})$, $U, V \in \Gamma(\mathcal{V})$, where $\{u_1, ..., u_r\}$, $\{E_1, ..., E_{r+s}\}$ and $\{\mu_1, ..., \mu_s\}$ are orthonormal frames of $(ker F_*)$, $J(ker F_*) \oplus \mu$ and $\mu$, respectively, $\rho^*$ is the Ricci tensor of $N$ and $\hat{\rho}$ is the Ricci tensor of any fiber.*

*Proof.* We see that for $X \in Jker F_*$, $CX$ is zero and for $X \in \mu$, $BX$ is zero. Thus Lemma 34 comes from (3.144)-(3.146).    □

From (3.147)-(3.149), the scalar curvature $\tau$ is given by

$$\tau = \tau^* + \hat{\tau} - 3 \sum_{i,j}^{r+s} g_1(\mathcal{A}_{E_i} E_j, \mathcal{A}_{E_i} E_j) + 2 \sum_{j}^{r+s} \sum_{i}^{r} g_1((\nabla_{E_j} \mathcal{T})(u_i, u_i), E_j)$$

$$- 2 \sum_{j}^{r+s} \sum_{i}^{r} g_1(\mathcal{T}_{u_i} E_j, \mathcal{T}_{u_i} E_j) + 2 \sum_{j}^{r+s} \sum_{i}^{r} g_1(\mathcal{A}_{E_j} u_i, \mathcal{A}_{E_j} u_i) \tag{3.150}$$

$$- \sum_{i,j}^{r} g_1(\mathcal{T}_{u_i} u_i, \mathcal{T}_{u_j} u_j) + \sum_{i,j}^{r} g_1(\mathcal{T}_{u_i} u_j, \mathcal{T}_{u_i} u_j),$$

where $\tau^*$ is the scalar curvature of $N$ and $\hat{\tau}$ is the scalar curvature of any fiber.

We now investigate necessary and sufficient conditions for the total space of anti-invariant Riemannian submersion to be an Einstein manifold. We have the following proposition.

**Proposition 40.** *[169] Let F be an anti-invariant Riemannian submersion with totally geodesic fibers from a Kähler manifold $(M, g_1, J)$ onto a Riemannian manifold $(N, g)$.*

*Then, $(M, g_1)$ is Einstein if and only if the following relations hold:*

$(i)$ $-\dfrac{\tau}{m} g_1(U, V) + \rho^*(JU, JV) - 3 \displaystyle\sum_{i}^{r+s} g_1(B\mathcal{A}_{E_i} U, B\mathcal{A}_{E_i} V)$

$+ \displaystyle\sum_{i}^{r} g_1((\nabla_{u_i}\mathcal{A})(JU, JV), u_i) + \sum_{i}^{r} g_1(\mathcal{A}_{JU} u_i, \mathcal{A}_{JV} u_i) = 0,$

$(ii)$ $-\dfrac{\tau}{m} g_1(X, Y) + \displaystyle\sum_{i}^{r+s} g_1((\nabla_{E_i}\mathcal{A})(E_i, CX), BY) - 3 \sum_{i}^{r+s} g_1(\mathcal{A}_{E_i} CX, \mathcal{A}_{E_i} CY)$

$+ \displaystyle\sum_{i}^{r+s} g_1(\mathcal{A}_{E_i} BX, \mathcal{A}_{E_i} BY) - \sum_{i}^{r+s} g_1((\nabla_{E_i}\mathcal{A})(CY, E_i), BX) + \hat{\rho}(BX, BY)$    (3.151)

$+ \displaystyle\sum_{i}^{r} g_1((\nabla_{u_i}\mathcal{A})(CX, CY), u_i) + \sum_{i}^{r} g_1(\mathcal{A}_{CX} u_i, \mathcal{A}_{CY} u_i) + \rho^*(CX, CY) = 0,$

$(iii)$ $\rho^*(JU, CX) - 3 \displaystyle\sum_{i}^{r+s} g_1(\mathcal{A}_{E_i} JU, \mathcal{A}_{E_i} CX) + \sum_{i}^{r+s} g_1((\nabla_{E_i}\mathcal{A})(E_i, JU), BX)$

$+ \displaystyle\sum_{i}^{r} g_1((\nabla_{u_i}\mathcal{A})(JU, CX), u_i) + \sum_{i}^{r} g_1(\mathcal{A}_{JU} u_i, \mathcal{A}_{CX} u_i) = 0,$

*where $\rho^*$ and $\hat{\rho}$ are the Ricci tensor of N and the Ricci tensor of the fiber, respectively, and $m = dim(M)$, $r = dim(ker\ F_*)$, $s = dim(\mu)$.*

From the above proposition, we have the following theorem.

**Theorem 70.** *[169] Let $F : M \longrightarrow N$ be a Lagrangian submersion with totally geodesic fibers from a Kähler manifold $(M, g_1, J)$ to a Riemannian manifold $(N, g)$. Then, $(M, g_1, J)$ is Einstein if and only if the fibers and the base space $(N, g)$ are Einstein.*

*Proof.* From Proposition 52 and (3.150), the following relations hold:

$$-\frac{\tau^* + \hat{\tau}}{m} g_1(U, V) + \rho^*(JU, JV) = 0,$$
$$-\frac{\tau^* + \hat{\tau}}{m} g_1(X, Y) + \hat{\rho}(BX, BY) = 0.$$    (3.152)

From (3.152), for $U, V \in \Gamma(\mathcal{V})$ we have

$$\rho^*(JU, JV) = \hat{\rho}(U, V).$$    (3.153)

This completes the proof.    □

## 11. Clairaut submersions from almost Hermitian manifolds

In this section, we give new characterizations for several Riemannian submersions to be Clairaut submersions. We also check the effect of Clairaut's condition on the geometry of the fibers. In classical differential geometry, an important tool analyzing geodesics on ordinary surfaces of revolution is Clairaut's relation. More precisely we have the following theorem.

**Theorem 71.** *(Clairaut's Theorem) see [229, page:228] Let $\alpha$ be a geodesic on a surface of revolution $S$, let $\sigma$ be the distance of a point of $S$ from the axis of rotation, and let $\psi$ be the angle between $\dot{\alpha}$ and the meridians of $S$. Then, $\sigma \sin \psi$ is constant along $\alpha$. Conversely, if $r \sin \psi$ is constant along some curve $\alpha$ in the surface of revolution $S$, and if no part of $\alpha$ is part of some parallel of $S$, then $\alpha$ is geodesic.*

As we have seen, the converse part of above theorem gives a method how to find the geodesics on the surface of a revolution. Clairaut's relation also has a mechanical interpretation. For this, we quote the following notes from [209]. *Consider a surface M and suppose a particle is constrained to move on M, but is otherwise free of external forces. D'Alembert's principle in Mechanics (see [292]) states that the constraint force F is normal to the surface, so Newton's law becomes $|F|U = m\alpha''$, where $|F|$ denotes the magnitude of F, U is M′ s unit normal and $\alpha(t)$ is the motion of curve of the particle. Taking $m = 1$, Newton's law tells us that $x_u$ and $x_v$ components of acceleration vanish and this leads to the geodesic equations of the surface. Hence, a freely moving particle constrained to move on a surface moves along geodesics. Now suppose M is a surface of a revolution parametrized by $x(u, v) = (h(u) \cos v, h(u) \sin v, g(u))$. Along the surface, the radial vector r from the origin is simply $r = (h(u) \cos v, h(u) \sin v, g(u))$ and the momentum vector of the particle's trajectory is given by $\mathbf{p} = \dot{\alpha} = u' x_u + v' x_v$ (since $m = 1$). Note that, since $\alpha$ is a geodesic, we may assume its speed is constant. The angular momentum of the particle about the origin may be calculated to be*

$$\mathbf{L} = r \times \mathbf{p} = ((g'h - gh')u' \sin v - ghv' \cos v, -(g'h - gh')u' \cos v - ghv' \sin v, h^2v').$$

*Because no external forces (or, more appropriately, torques) act on the particle, angular momentum, and in particular its z−component , is conserved. Therefore, we see that $h^2v' = C$, where C is constant. Further if $\phi$ is the angle from $x_u$ to $\alpha'$, then*

$$| \alpha' | h \sin \phi = \alpha' x_v = (u' x_u + v' x_v) x_v = h^2 v' = C,$$

*since $\cos(\frac{\pi}{2} - \phi) = \sin \phi$ and $x_v.x_v = h^2$. Because $| \alpha' |$ is constant as well, we have $h \sin \phi = constant$. But this is precisely Clairaut's relation since $h = \sqrt{G} = \sqrt{x_v.x_v}$. In this way Clairaut's relation is a physical phenomenon as well as a mathematical one.*

In the submersion theory, this notion was defined by Bishop. According to his definition, a submersion $F : M \to N$ to be a Clairaut submersion if there is a function $r : M \to \mathbb{R}^+$ such that for every geodesic, making angles $\theta$ with the horizontal subspaces, $r \sin \theta$ is constant. Bishop found the following characterization.

**Theorem 72.** *[35] Let $F : (M, g_1) \to (B, g)$ be a Riemannian submersion with connected fibers. Then $F$ is a Clairaut submersion with $r = e^f$ if and only if each fiber is totally umbilical and has the mean curvature vector field $H = -\operatorname{grad} f$.*

Clairaut submersions have been studied in Lorentizan spaces and timelike, spacelike, and null geodesics of Lorentzian Clairaut submersion with one-dimensional fibers have been investigated in detail [10]. Such submersions have been further generalized [17, 153].

As we have seen above, the origin of the notion of Clairaut submersion comes from geodesic curves on a surface. Therefore we are going to find necessary conditions for a curve on the total space to be geodesic.

**Lemma 35.** *[169] Let $F$ be an anti-invariant Riemannian submersion from a Kähler manifold $(M, g_1, J)$ onto a Riemannian manifold $(N, g)$. If $\alpha : I \to M$ is a regular curve and $X(t)$, $V(t)$ denote the horizontal and vertical components of its tangent vector field, then $\alpha$ is a geodesic on $M$ if and only if*

$$\bar{\nabla}_{\dot{\alpha}} CX \ + \ \bar{\nabla}_{\dot{\alpha}} JV + \mathcal{A}_X BX + \mathcal{T}_V BX = 0 \qquad (3.154)$$

$$\bar{\nabla}_{\dot{\alpha}} BX \ + \ \mathcal{V}\nabla_{\dot{\alpha}} JV + \mathcal{A}_X CX + \mathcal{T}_V CX = 0, \qquad (3.155)$$

*where $\bar{\nabla}$ is the Schouten connection associated with the mutually orthogonal distributions $\mathcal{V}$ and $\mathcal{H}$.*

*Proof.* Since $\nabla_{\dot{\alpha}}\dot{\alpha} = -J\nabla_{\dot{\alpha}} J\dot{\alpha}$, using (3.18), we get

$$\begin{aligned}\nabla_{\dot{\alpha}}\dot{\alpha} \ = \ &-J(\mathcal{V}\nabla_{\dot{\alpha}} BX(t) + \mathcal{H}\nabla_{\dot{\alpha}} BX(t) \\ &+ \ \mathcal{V}\nabla_{\dot{\alpha}} CX(t) + \mathcal{H}\nabla_{\dot{\alpha}} CX(t) + \nabla_{\dot{\alpha}} JV(t)).\end{aligned}$$

Then the nonsingular $J$ implies that $\alpha$ is geodesic if and only if

$$\mathcal{V}\nabla_{\dot{\alpha}} BX(t) + \mathcal{H}\nabla_{\dot{\alpha}} BX(t) + \mathcal{V}\nabla_{\dot{\alpha}} CX(t) + \mathcal{H}\nabla_{\dot{\alpha}} CX(t) + \nabla_{\dot{\alpha}} JV(t) = 0.$$

Thus, using (1.47)–(1.50) and the Schouten connection $\bar{\nabla}_X W' = \mathcal{V}\nabla_X \mathcal{V} W' + \mathcal{H}\nabla_X \mathcal{H} W'$ for $X, W' \in \Gamma(TM)$, we obtain (3.154) and (3.155). □

**Theorem 73.** *[169] Let $F$ be an anti-invariant Riemannian submersion from a Kähler manifold $(M, J, g_1)$ onto a Riemannian manifold $(N, g)$. Then $F$ is a Clairaut anti-*

*invariant submersion with $r = e^f$ if and only if*

$$g_1(V, V)g_1(X, grad f) - g_1(\mathcal{A}_Z BZ + \mathcal{T}_V BZ + \bar{\nabla}_{\dot{\alpha}(t)} CZ, JV) = 0,$$

*where $Z(t)$ and $V(t)$ denote the horizontal and vertical components of $\dot{\alpha}(t) = X$. Moreover, if $F$ is a Clairaut anti-invariant submersion with $e^f$, then at least one of the following statements is true (a) $f$ is constant on $J(\ker F_*)$ (b) the fibers are one-dimensional (c)*

$$\mathcal{A}_{JW} JY = Y(f)W$$

*for $Y \in \Gamma(\mu)$ and $W \in \Gamma(\ker F_*)$.*

*Proof.* For a geodesic $\alpha(t)$ on $M$, putting $a = \| \dot{\alpha}(t) \|^2$ which is constant, we obtain

$$g_{1\alpha(t)}(Z(t), Z(t)) = a \cos^2 \theta, \; g_{1\alpha(t)}(V(t), V(t)) = a \sin^2 \theta. \tag{3.156}$$

Differentiating the second expression, we have

$$\frac{d}{dt} g_{1\alpha(t)}(V(t), V(t)) = 2g_{1\alpha(t)}(\nabla_{\dot{\alpha}(t)} V(t), V(t)) = 2a \sin \theta \cos \theta \frac{d\theta}{dt}.$$

Hence we get

$$g_{1\alpha(t)}(\mathcal{H}\nabla_{\dot{\alpha}(t)} JV(t), JV(t)) = a \sin \theta \cos \theta \frac{d\theta}{dt}.$$

Using (3.154), we derive

$$-g_{1\alpha(t)}(\bar{\nabla}_{\dot{\alpha}} CZ(t) + \mathcal{A}_{Z(t)} BZ(t) + \mathcal{T}_{V(t)} BZ(t), JV(t)) = a \sin \theta \cos \theta \frac{d\theta}{dt}. \tag{3.157}$$

On the other hand, $F$ is a Clairaut submersion with $e^f$ if and only if

$$\frac{d}{dt}(e^f \sin \theta) = 0.$$

Multiplying with the nonzero factor $a \sin \theta(t)$ and using (3.156) and (3.157), it follows that $F$ is a Clairaut submersion if and only if

$$g_1(V, V)\frac{df(\alpha(t))}{dt} - g_1(\mathcal{A}_Z BZ + \mathcal{T}_V BZ + \bar{\nabla}_{\dot{\alpha}(t)} CZ, JV) = 0,$$

which gives us the first part of the theorem. Now suppose that $F$ is a Clairaut anti-invariant submersion with $r = e^f$, then from Bishop's theorem, the fibers of $F$ are totally umbilical with mean curvature vector field $H = -grad f$. Thus we have

$$\mathcal{T}_U V = -g_1(U, V)grad f,$$

for $U, V \in \Gamma(\ker F_*)$. Multiplying this expression with $JW$ for $W \in \Gamma(\ker F_*)$ and using (3.5) and (1.47), we get

$$g_1(\nabla_U JV, W) = g_1(U, V)g_1(grad f, JW).$$

Since $\nabla$ is a metric connection, using again (1.47), we derive

$$g_1(U, W)g_1(grad f, JV) = g_1(U, V)g_1(grad f, JW). \tag{3.158}$$

Taking $U = W$, interchanging the role of $U$ and $V$ in (3.158), we obtain

$$g_1(V, V)g_1(grad f, JU) = g_1(V, U)g_1(grad f, JV). \tag{3.159}$$

Using (3.158) with $W = U$ and (3.159), we have

$$g_1(grad f, JU) = \frac{(g_1(U, V))^2}{\| U \|^2 \| V \|^2} g_1(grad f, JU). \tag{3.160}$$

On the other hand, from (3.5) and (1.47), we find

$$g_1(\nabla_V JW, JY) = -g_1(V, W)g_1(grad f, Y)$$

for $V, W \in \Gamma(\mathcal{V})$ and $Y \in \Gamma(\mu)$. Since $\mu$ is invariant with respect to $J$ and $[V, JW]$ belongs to $\mathcal{V}$ for basic $JW$, by using (1.49), we get

$$g_1(\mathcal{A}_{JW}V, JY) = -g_1(V, W)g_1(grad f, Y).$$

Since $\mathcal{A}_{JW}$ is skew-symmetric with respect to $g_1$ and since $\mathcal{A}_{JW}JY$, $V$ and $W$ are vertical and $grad f$ is horizontal, and the above equation holds for all vertical $V$ and $W$, hence we derive

$$\mathcal{A}_{JW}JY = Y(f)W. \tag{3.161}$$

Now if $grad f \in \Gamma(J(\mathcal{V}))$, then (3.160) and the equality case of the Schwarz inequality implies that either $f$ is constant on $J(\mathcal{V})$ or the fibers are one-dimensional. If $grad f \in \Gamma(\mu)$, then (3.161) implies (c). This completes the proof. $\qquad \square$

From Theorem 73, we have the following results.

**Corollary 23.** *[169] Let $F$ be a Clairaut anti-invariant submersion from a Kähler manifold $(M, g_1, J)$ onto a Riemannian manifold $(N, g)$ with $r = e^f$ and $dim(\mathcal{V}) > 1$. Then the fibers of $F$ are totally geodesic if and only if $\mathcal{A}_{JW}JX = 0$ for $W \in \Gamma(\mathcal{V})$ and $X \in \Gamma(\mu)$.*

For a Lagrangian submersion, we have the following.

**Corollary 24.** *[169] Let $F$ be a Clairaut Lagrangian submersion from a Kähler manifold $(M, g_1, J)$ onto a Riemannian manifold $(N, g)$ with $r = e^f$. Then either the fibers* *of F are totally geodesic or they are one-dimensional.*

In fact, this case is valid for general case of Lagrangian submersions with totally umbilical fibers. We omit its proof, which is similar to the proof of Theorem 73.

**Proposition 41.** *[169], [284] Let F be a Lagrangian submersion from Kähler manifold $(M, g_1, J)$ onto a Riemannian manifold $(N, g)$ with totally umbilical fibers with $dim(\mathcal{V}) > 1$. Then the fibers are totally geodesic.*

We note that Proposition 41 was first given in [284] as Proposition 5.4.

**Lemma 36.** *[169] For an anti-invariant Riemannian submersion, we have*

$$g_1(\mathcal{T}_V U, X) = -g_1(\mathcal{T}_V BX, JU) + g_1(\mathcal{A}_{CX} JU, V),$$

*for $U, V \in \Gamma(\mathcal{V})$ and $X \in \Gamma(\mathcal{H})$. Observe that if F is a Clairaut anti-invariant submersion, we have*

$$\mathcal{A}_{JU} CX = X(f)U + JU(f)BX,$$

*and if $dim(\mathcal{V}) > 1$, we get*

$$\mathcal{A}_{JU} CX = X(f)U,$$

*for basic CX.*

From Lemma 36, we also have the following expressions.

**Lemma 37.** *[169] Let F be a Clairaut anti-invariant submersion from a Kähler manifold $(M, g_1, J)$ onto a Riemannian manifold $(N, g)$ with $r = e^f$ and $dim(\mathcal{V}) > 1$. Then we have*

$$\sum_{i=1}^{r+s} g_1(\mathcal{A}_{CY} E_i, \mathcal{T}_{BX} E_i) \;\; = \;\; \sum_{j=1}^{s} g_1(\mathcal{A}_{CY} \mu_j, \mathcal{T}_{BX} \mu_j), \tag{3.162}$$

$$\sum_{i=1}^{r+s} g_1(\mathcal{A}_{E_i} CX, \mathcal{A}_{E_i} CY) \;\; = \;\; rX(f)Y(f) + \sum_{j=1}^{s} g_1(\mathcal{A}_{\mu_j} CX, \mathcal{A}_{\mu_j} CY) \tag{3.163}$$

*for $X, Y \in \Gamma(\mathcal{H})$ such that CX and CY are basic.*

*Proof.* For an anti-invariant Riemannian submersion, we can write

$$\sum_{i=1}^{r+s} g_1(\mathcal{A}_{CY}E_i, \mathcal{T}_{BX}E_i) = \sum_{i=1}^{r} g_1(\mathcal{A}_{CY}Ju_i, \mathcal{T}_{BX}Ju_i) + \sum_{j=1}^{s} g_1(\mathcal{A}_{CY}\mu_j, \mathcal{T}_{BX}\mu_j).$$

Since $F$ is a Clairaut submersion, $\mathcal{A}$ is anti-symmetric on $\mathcal{H}$ and $dim(\mathcal{V}) > 1$, from Lemma 36, we have

$$\sum_{i=1}^{r+s} g_1(\mathcal{A}_{CY}E_i, \mathcal{T}_{BX}E_i) = -\sum_{i=1}^{r} Y(f)g_1(u_i, \mathcal{T}_{BX}Ju_i) + \sum_{j=1}^{s} g_1(\mathcal{A}_{CY}\mu_j, \mathcal{T}_{BX}\mu_j).$$

Also, since $\mathcal{T}$ is anti-symmetric with respect to $g_1$, using Theorem 72, we obtain

$$\sum_{i=1}^{r+s} g_1(\mathcal{A}_{CY}E_i, \mathcal{T}_{BX}E_i) = -\sum_{i=1}^{r} Y(f)g_1(BX, u_i)g_1(grad f, Ju_i)$$

$$+ \sum_{j=1}^{s} g_1(\mathcal{A}_{CY}\mu_j, \mathcal{T}_{BX}\mu_j).$$

Then we get (3.162) since $f$ is constant on $J\mathcal{V}$ from Theorem 73. In a similar way, we obtain (3.163). $\qquad\square$

From (3.163), (3.162), and Lemma 34, we have the following result, which characterizes the fibers.

**Proposition 42.** *[169] Let F be a Clairaut anti-invariant submersion from a Kähler manifold $(M, g_1, J)$ onto a Riemannian manifold $(N, g)$ with $r = e^f$ and $dim(\mathcal{V}) > 1$. Then we have*

$$\rho(X, X) \leq \sum_{i}^{r+s} \{2g_1((\nabla_{E_i}\mathcal{A})(E_i, CX), BX) + g_1((\nabla_{E_i}\mathcal{T})(BX, BX), E_i)\}$$

$$+ \parallel trace\mathcal{A}_{(.)}BX \parallel^2_{J\mathcal{V}\oplus\mu} - 4\sum_{i}^{s} g_1(\mathcal{A}_{\mu_i}CX, \mathcal{T}_{BX}\mu_i)$$

$$+ \sum_{i}^{r} \{2g_1((\nabla_{BX}\mathcal{T})(u_i, u_i), CX) + g_1((\nabla_{CX}\mathcal{T})(u_i, u_i), CX)$$

$$- 2g_1((\nabla_{u_i}\mathcal{T})(BX, u_i), CX) + g_1((\nabla_{u_i}\mathcal{A})(CX, CX), u_i)$$

$$- g_1(\mathcal{T}_{u_i}u_i, \mathcal{T}_{BX}BX)\} + \parallel trace\mathcal{A}_{CX}(.) \parallel^2_{\mathcal{V}} + \parallel trace\mathcal{T}_{BX}(.) \parallel^2_{\mathcal{V}}$$

$$- 3 \parallel trace\mathcal{A}_{(.)}CX \parallel^2_{\mu} + \rho^*(CX, CX) + \hat{\rho}(BX, BX),$$

*for $X \in \Gamma(\mathcal{H})$ such that $CX$ and $CY$ are basic. The equality case is satisfied if and only if $f$ is constant on the horizontal distribution, that is, F has totally geodesic fibers. In* *the equality case, it takes the following form,*

$$\rho(X,X) = \sum_{i}^{r+s} \{2g_1((\nabla_{E_i}\mathcal{A})(E_i, CX), BX) + \| \, trace\mathcal{A}_{CX}(.) \, \|_{\mathcal{V}}^2$$

$$+ g_1((\nabla_{u_i}\mathcal{A})(CX, CX), u_i) + \| \, trace\mathcal{A}_{(.)}BX \, \|_{J\mathcal{V}\oplus\mu}^2 + \rho^*(CX, CX)$$

$$- 3 \| \, trace\mathcal{A}_{(.)}CX \, \|_{\mu}^2 + \hat{\rho}(BX, BX).$$

We also have the following result.

**Corollary 25.** *[169] Let F be a Clairaut Lagrangian submersion from a Kähler manifold $(M, g_1, J)$ onto a Riemannian manifold $(N, g)$ with $r = e^f$ and $dimM > 2$. Then $(M, g_1, J)$ is Einstein if and only if the fibers and the base space $(N, g)$ are Einstein.*

We now investigate Clairaut conditions for other Riemannian submersions given in this chapter. We first recall the following definition.

**Definition 46.** *([283]) Let $\pi : (M, g, J) \to (N, g_N)$ be a semi-invariant submersion. Then, we call $\pi$ an anti-holomorphic semi-invariant submersion if $(ker\pi_*)^{\perp} = J\mathcal{D}^{\perp}$, i.e., $\mu = \{0\}$.*

We can observe that an anti-holomorphic semi-invariant submersion is a natural generalization of a Lagrangian submersion. Now suppose that the dimension of distribution $\mathcal{D}$ (resp. $\mathcal{D}^{\perp}$) is $2p$ (resp. $q$). Then, we have $dim(M) = 2p + 2q$ and $dim(N) = q$. An anti-holomorphic semi-invariant submersion is called a *proper anti-holomorphic semi-invariant submersion* if $p \neq 0$ and $q \neq 0$.

Thus, using (3.34), (3.35), and (3.33), we get the following:

**Lemma 38.** *[261] Let $\pi$ be a semi-invariant submersion from an almost Hermitian manifold $(M, g, J)$ onto a Riemannian manifold $(N, g_N)$. Then, we have*

$$\begin{array}{llll} (a) & \phi\mathcal{D} = \mathcal{D}, & (b) & \phi\mathcal{D}^{\perp} = \{0\}, \\ (c) & \omega\mathcal{D} = \{0\}, & (d) & \mathcal{B}J\mathcal{D}^{\perp} = \mathcal{D}^{\perp}. \end{array} \tag{3.164}$$

We now examine how the Kähler structure on $M$ affects the tensor fields $\mathcal{T}$ and $\mathcal{A}$ of a semi-invariant submersion $\pi : (M, g, J) \to (N, g_N)$.

**Lemma 39.** *[261] Let $\pi$ be a semi-invariant submersion from a Kähler manifold*

$(M, g, J)$ *onto a Riemannian manifold* $(N, g_N)$. *Then*

$$\hat{\nabla}_U \phi V + \mathcal{T}_U \omega V = \phi \hat{\nabla}_U V + \mathcal{B}\mathcal{T}_U V, \tag{3.165}$$

$$\mathcal{T}_U \phi V + \mathcal{H}\nabla_U \omega V = \omega \hat{\nabla}_U V + C\mathcal{T}_U V, \tag{3.166}$$

$$\mathcal{V}\nabla_\xi \eta + \mathcal{A}_\xi C\eta = \phi \mathcal{A}_\xi \eta + \mathcal{B}\mathcal{H}\nabla_\xi \eta, \tag{3.167}$$

$$\mathcal{A}_\xi \mathcal{B}\eta + \mathcal{H}\nabla_\xi C\eta = \omega \mathcal{A}_\xi \eta + C\mathcal{H}\nabla_\xi \eta, \tag{3.168}$$

$$\hat{\nabla}_U \mathcal{B}\xi + \mathcal{T}_U C\xi = \phi \mathcal{T}_U \xi + \mathcal{B}\mathcal{H}\nabla_U \xi, \tag{3.169}$$

$$\mathcal{T}_U \mathcal{B}\xi + \mathcal{H}\nabla_U C\xi = \omega \mathcal{T}_U \xi + C\mathcal{H}\nabla_U \xi, \tag{3.170}$$

*where* $U, V \in \Gamma(ker\pi_*)$ *and* $\xi, \eta \in \Gamma((ker\pi_*)^\perp)$.

*Proof.* Using (1.47)–(1.50), (3.34), (3.35), and (3.5), we can easily obtain all the assertions.  □

**Corollary 26.** *[261] Let* $\pi$ *be an anti-holomorphic semi-invariant submersion from a Kähler manifold* $(M, g, J)$ *onto a Riemannian manifold* $(N, g_N)$. *Then*

$$\hat{\nabla}_U \phi V + \mathcal{T}_U \omega V = \phi \hat{\nabla}_U V + J\mathcal{T}_U V, \tag{3.171}$$

$$\mathcal{T}_U \phi V + \mathcal{H}\nabla_U \omega V = \omega \hat{\nabla}_U V, \tag{3.172}$$

$$\mathcal{V}\nabla_\xi \eta = \phi \mathcal{A}_\xi \eta + J\mathcal{H}\nabla_\xi \eta, \tag{3.173}$$

$$\mathcal{A}_\xi J\eta = \omega \mathcal{A}_\xi \eta, \tag{3.174}$$

$$\hat{\nabla}_U J\xi = \phi \mathcal{T}_U \xi + J\mathcal{H}\nabla_U \xi, \tag{3.175}$$

$$\mathcal{T}_U J\xi = \omega \mathcal{T}_U \xi, \tag{3.176}$$

*where* $U, V \in \Gamma(ker\pi_*)$ *and* $\xi, \eta \in \Gamma((ker\pi_*)^\perp)$.

**Corollary 27.** *[261] Let* $\pi$ *be an anti-holomorphic semi-invariant submersion from a Kähler manifold* $(M, g, J)$ *onto a Riemannian manifold* $(N, g_N)$. *Then we get*

$$\mathcal{T}_U JX = \phi \hat{\nabla}_U X + J\mathcal{T}_U X, \tag{3.177}$$

$$\mathcal{H}\nabla_U JX = \omega \hat{\nabla}_U X, \tag{3.178}$$

$$\hat{\nabla}_U JW = \phi \hat{\nabla}_U W + J \mathcal{T}_U W, \tag{3.179}$$

$$\mathcal{T}_U JW = \omega \hat{\nabla}_U W, \tag{3.180}$$

where $U \in \Gamma(ker\pi_*)$, $X \in \Gamma(\mathcal{D}^\perp)$ and $W \in \mathcal{D}$.

**Corollary 28.** *[261] Let $\pi$ be an anti-holomorphic semi-invariant submersion from a Kähler manifold $(M, g, J)$ onto a Riemannian manifold $(N, g_N)$. Then we have*

$$\mathcal{A}_\xi J\eta = -\mathcal{A}_\eta J\xi, \tag{3.181}$$

where $\xi, \eta \in \Gamma((ker\pi_*)^\perp)$.

*Proof.* For $\xi, \eta \in \Gamma((ker\pi_*)^\perp)$, using (1.46) and (3.174), we obtain

$$\mathcal{A}_\xi J\eta = \omega \mathcal{A}_\eta \xi = -\omega \mathcal{A}_\eta \xi = -\mathcal{A}_\eta J\xi.$$

$\square$

**Lemma 40.** *[261] Let $\pi$ be an anti-holomorphic semi-invariant submersion from a Kähler manifold $(M, g, J)$ onto a Riemannian manifold $(N, g_N)$. Then,*

$$\mathcal{A}_\xi \mathcal{D}^\perp = 0, \tag{3.182}$$

where $\xi \in \Gamma((ker\pi_*)^\perp)$.

*Proof.* For $\xi, \eta$ and $\zeta \in \Gamma((ker\pi_*)^\perp)$, using (3.5), (3.34), and (3.174), we obtain

$$g(\mathcal{A}_\xi J\eta, \zeta) = g(\omega \mathcal{A}_\xi \eta, \zeta) = g(J\mathcal{A}_\xi \eta, \zeta) = -g(\mathcal{A}_\xi \eta, J\zeta) = 0.$$

Hence, we conclude that

$$\mathcal{A}_\xi J(ker\pi_*)^\perp = 0. \tag{3.183}$$

However, by (3.33), we have $J(ker\pi_*)^\perp = \mathcal{D}^\perp$ since $\mu = \{0\}$. Our assertion follows from (3.183).    $\square$

We now investigate a necessary and sufficient condition for a curve on the total space to be geodesic.

**Lemma 41.** *[261] Let $\pi$ be a semi-invariant submersion from a Kähler manifold $(M, g, J)$ onto a Riemannian manifold $(N, g_N)$. If $\alpha : I \subset \mathbb{R} \to M$ is a regular curve and $V(t)$ and $\xi(t)$ are the vertical and horizontal components of the tangent vector field $\dot{\alpha}(t)$, then $\alpha$ is geodesic if and only if along $\alpha$ the following equations*

$$\overline{\nabla}_{\dot{\alpha}} \phi V + \overline{\nabla}_{\dot{\alpha}} \mathcal{B}\xi + \mathcal{T}_V \omega V + \mathcal{T}_V C\xi + \mathcal{A}_\xi \omega V + \mathcal{A}_\xi C\xi = 0, \tag{3.184}$$

$$\overline{\nabla}_{\dot{\alpha}}\omega V + \overline{\nabla}_{\dot{\alpha}}C\xi + \mathcal{T}_V\phi V + \mathcal{T}_V\mathcal{B}\xi + \mathcal{A}_\xi\phi V + \mathcal{A}_\xi\mathcal{B}\xi = 0 \ , \tag{3.185}$$

hold, where $\overline{\nabla}$ is the Schouten connection associated with the mutually orthogonal distributions $\mathcal{V}$ and $\mathcal{H}$.

*Proof.* Using (3.5), we have

$$\nabla_{\dot{\alpha}}\dot{\alpha} = -J\nabla_{\dot{\alpha}}J\dot{\alpha} \ .$$

Hence, using (3.34) and (3.35), we get

$$\begin{aligned}
\nabla_{\dot{\alpha}}\dot{\alpha} &= -J\nabla_{\dot{\alpha}}J(V(t) + \xi(t)) = -J\{\nabla_{\dot{\alpha}}JV(t) + \nabla_{\dot{\alpha}}J\xi(t)\} \\
&= -J\{\mathcal{V}\nabla_{\dot{\alpha}}\phi V + \mathcal{H}\nabla_{\dot{\alpha}}\phi V + \mathcal{V}\nabla_{\dot{\alpha}}\omega V + \mathcal{H}\nabla_{\dot{\alpha}}\omega V\} \\
&\quad -J\{\mathcal{V}\nabla_{\dot{\alpha}}\mathcal{B}\xi + \mathcal{H}\nabla_{\dot{\alpha}}\mathcal{B}\xi + \mathcal{V}\nabla_{\dot{\alpha}}C\xi + \mathcal{H}\nabla_{\dot{\alpha}}C\xi\}.
\end{aligned}$$

Since $J$ is nonsingular, $\alpha$ is geodesic if and only if

$$\begin{aligned}
&\mathcal{V}\nabla_{\dot{\alpha}}\phi V + \mathcal{H}\nabla_{\dot{\alpha}}\phi V + \mathcal{V}\nabla_{\dot{\alpha}}\omega V + \mathcal{H}\nabla_{\dot{\alpha}}\omega V \\
&+\mathcal{V}\nabla_{\dot{\alpha}}\mathcal{B}\xi + \mathcal{H}\nabla_{\dot{\alpha}}\mathcal{B}\xi + \mathcal{V}\nabla_{\dot{\alpha}}C\xi + \mathcal{H}\nabla_{\dot{\alpha}}C\xi = 0.
\end{aligned}$$

Thus, using (1.47)–(1.50) and the Schouten connection $\overline{\nabla}_E F = \mathcal{V}\nabla_E\mathcal{V}F + \mathcal{H}\nabla_E\mathcal{H}F$ for $E, F \in TM$, we find (3.184) and (3.185). $\qquad\square$

For a semi-invariant submersion to be the Clairaut type, we have the following necessary and sufficient conditions.

**Theorem 74.** *[261] Let $\pi$ be a semi-invariant submersion with connected fibers from a Kähler manifold $(M, g, J)$ onto a Riemannian manifold $(N, g_N)$. Then $\pi$ is a Clairaut submersion with $r = e^f$ if and only if along $\alpha$*

$$g(V, V)g(\text{grad } f, \xi) = g(\overline{\nabla}_{\dot{\alpha}}J\xi + \mathcal{T}_V JV + \mathcal{T}_V J\xi + \mathcal{A}_\xi JV + \mathcal{A}_\xi J\xi, JV) \ , \tag{3.186}$$

*holds, where $V(t)$ and $\xi(t)$ are the vertical and horizontal components of the tangent vector field $\dot{\alpha}(t) = E$ of the geodesic $\alpha(t)$ on $M$.*

*Proof.* Let $\alpha(t)$ be a geodesic on $M$, then we have

$$c = \|\dot{\alpha}(t)\|^2,$$

where $c$ is a constant. From this equality, we deduce that

$$g(V(t), V(t)) = c\sin^2\gamma(t) \quad \textit{and} \quad g(\xi(t), \xi(t)) = c\cos^2\gamma(t), \tag{3.187}$$

where $\gamma(t)$ is the angle between $\dot{\alpha}(t)$ and the horizontal space at $\alpha(t)$. Differentiating the first expression, we obtain

$$\frac{d}{dt}g(V(t), V(t)) = 2g(\nabla_{\dot{\alpha}(t)}V(t), V(t)) = 2c\cos\gamma\sin\gamma\frac{d\gamma}{dt}(t).$$

Hence, using (3.5), we get

$$g(\nabla_{\dot{\alpha}(t)}JV(t), JV(t)) = c\cos\gamma\sin\gamma\frac{d\gamma}{dt}(t).$$

By using (3.184) and (3.185), we find along $\alpha$

$$-g(\overline{\nabla}_{\dot{\alpha}}J\xi + \mathcal{T}_V JV + \mathcal{T}_V J\xi + \mathcal{A}_\xi JV + \mathcal{A}_\xi J\xi, JV) = c\cos\gamma\sin\gamma\frac{d\gamma}{dt}. \qquad (3.188)$$

On the other hand, $\pi$ is a Clairaut submersion with $r = e^f$ if and only if

$$\frac{d}{dt}(e^f \sin\gamma) = 0.$$

Multiplying last equation with the nonzero factor $c\sin\gamma$ and using (3.187) and (3.188), it follows that $\pi$ is a Clairaut submersion with $r = e^f$ if and only if (3.186) holds, since

$$\frac{df}{dt}(\alpha(t)) = \dot{\alpha}(t)[f] = E[f] = g(\text{grad } f, E) = g(\text{grad } f, \xi).$$

$\square$

Let $\pi$ be a Clairaut semi-invariant submersion with $r = e^f$ from a Kähler manifold $(M, g, J)$ onto a Riemannian manifold. Then, from Theorem 72, we can write

$$\mathcal{T}_U V = -g(U, V)\text{grad } f, \qquad (3.189)$$

where $U, V \in \Gamma(ker\pi_*)$. Thus, from (3.189), we have the following.

**Corollary 29.** *[261] Let $\pi$ be a Clairaut semi-invariant submersion with $r = e^f$ from a Kähler manifold $(M, g, J)$ onto a Riemannian manifold $(N, g_N)$. Then, the fibers of $\pi$ are always mixed totally geodesic.*

From Lemma 15, if $\pi$ is a Clairaut semi-invariant submersion with $r = e^f$ from a Kähler manifold $(M, g, J)$ onto a Riemannian manifold $(N, g_N)$. Then, we have grad $f \in J\mathcal{D}^\perp$, since the fibers are totally umbilical in this case.

We now give some applications of Lemma 15.

**Theorem 75.** *[261] Let $\pi$ be a Clairaut proper semi-invariant submersion with $r = e^f$ from a Kähler manifold $(M, g, J)$ onto a Riemannian manifold $(N, g_N)$. If $\mathcal{T}$ is not equal to zero, identically, then the invariant distribution $\mathcal{D}$ cannot define a totally geodesic* *foliation on* $ker\pi_*$.

*Proof.* For $V, W \in \mathcal{D}$, and $X \in \mathcal{D}^{\perp}$, using (1.47), (3.5), and (3.189), we get

$$g(\hat{\nabla}_V W, X) = -g(V, JW)g(\text{grad } f, JX). \tag{3.190}$$

Thus, the assertion can be seen from (3.190) and the fact that grad $f \in J\mathcal{D}^{\perp}$.  □

**Theorem 76.** *[261] Let $\pi$ be a Clairaut proper semi-invariant submersion with $r = e^f$ from a Kähler manifold $(M, g, J)$ onto a Riemannian manifold. Then, the fibers of $\pi$ are totally geodesic or the anti-invariant distribution $\mathcal{D}^{\perp}$ is one-dimensional.*

*Proof.* If the fibers of $\pi$ are totally geodesic, it is obvious. Let us assume the opposite. Since $\pi$ is a Clairaut proper semi-invariant submersion, then either $Dim(\mathcal{D}^{\perp}) = 1$ or $Dim(\mathcal{D}^{\perp}) > 1$. If $Dim(\mathcal{D}^{\perp}) > 1$, then we can choose $X, Y \in \mathcal{D}^{\perp}$ such that $\{X, Y\}$ is orthonormal. By using (1.45), (1.46), (3.34), (3.35), and (3.5), we get

$$\mathcal{T}_X JY + \mathcal{H}\nabla_X JY = \nabla_X JY = J\nabla_X Y = \phi\hat{\nabla}_X Y + \omega\hat{\nabla}_X Y + \mathcal{B}\mathcal{T}_X Y + C\mathcal{T}_X Y.$$

So that $g(\mathcal{T}_X JY, X) = g(\phi\hat{\nabla}_X Y, X) + g(\mathcal{B}\mathcal{T}_X Y, X)$. Hence, using (3.5), we obtain

$$g(\mathcal{T}_X JY, X) = -g(\mathcal{T}_X Y, JX). \tag{3.191}$$

Thus, using (3.189) and (3.191), we have

$$g(\text{grad } f, JY) = -g(\mathcal{T}_X X, JY) = g(\mathcal{T}_X JY, X) = -g(\mathcal{T}_X Y, JX).$$

Again, using (3.189), we obtain

$$g(\text{grad } f, JY) = -g(\mathcal{T}_X Y, JX) = g(X, Y)g(\text{grad } f, JX) = 0,$$

since $g(X, Y) = 0$. So, we deduce that

$$\text{grad } f \perp J\mathcal{D}^{\perp}. \tag{3.192}$$

This result contrasts with Lemma 15. Therefore, the dimension of $\mathcal{D}^{\perp}$ must be one.
  □

**Lemma 42.** *[261] Let $\pi$ be a Clairaut proper semi-invariant submersion with $r = e^f$ from a Kähler manifold $(M, g, J)$ onto a Riemannian manifold $(N, g_N)$. If the fibers of $\pi$ are not totally geodesic, then we have*

$$\mathcal{A}_{JX}(ker\pi_*) = 0 \tag{3.193}$$

*and*

$$\mathcal{A}_{\xi}(ker\pi_*) \subset \mu, \tag{3.194}$$

*where $X \in \mathcal{D}^\perp$, $\xi \in \mu$ such that $JX$ and $\xi$ are basic.*

*Proof.* Let $X \in \mathcal{D}^\perp$, $\xi \in \mu$ such that $JX$ and $\xi$ are basic. Then, using (1.47), (3.189) and the fact that $\mathcal{H}\nabla_X JX = \mathcal{A}_{JX}X$, we get

$$g(\mathcal{A}_{JX}X, \xi) = \|X\|^2 g(\text{grad } f, J\xi) = 0, \qquad (3.195)$$

since $\text{grad } f \in J\mathcal{D}^\perp$ and $\mu$ is invariant under the linear isomorphism $J$. Hence, we deduce that

$$\mathcal{A}_{JX}X \in J\mathcal{D}^\perp.$$

Since $J\mathcal{D}^\perp$ is also one-dimensional, using (1.46), we obtain

$$g(\mathcal{A}_{JX}X, JX) = -g(\mathcal{A}_{JX}JX, X) = 0.$$

It means that

$$\mathcal{A}_{JX}X = 0. \qquad (3.196)$$

In a similar way, we find that

$$\mathcal{A}_{JX}V = 0, \qquad (3.197)$$

for any $V \in \mathcal{D}$. Thus, the first assertion in (3.193) follows from (3.196) and (3.197). On the other hand, from (3.195) and the fact that $\mathcal{H}\nabla_X \xi = \mathcal{A}_\xi X$, we get

$$g(\mathcal{A}_\xi X, JX) = -\|X\|^2 g(\text{grad } f, J\xi) = 0.$$

Hence, we deduce that

$$\mathcal{A}_\xi \mathcal{D}^\perp \subset \mu. \qquad (3.198)$$

By a similar method, we get

$$\mathcal{A}_\xi \mathcal{D} \subset \mu. \qquad (3.199)$$

Then, (3.198) and (3.199) give us (3.194). □

**Theorem 77.** *[261] Let $\pi$ be a Clairaut proper anti-holomorphic semi-invariant submersion with $r = e^f$ from a Kähler manifold $(M, g, J)$ onto a Riemannian manifold $(N, g_N)$. If the fibers of $\pi$ are not totally geodesic, then the horizontal distribution $(\ker\pi_*)^\perp$ is always integrable and totally geodesic. In other words, $\mathcal{A}_\xi = 0$ for $\xi \in \Gamma((\ker\pi_*)^\perp)$.*

*Proof.* Let $\pi$ be a Clairaut proper anti-holomorphic semi-invariant submersion with $r = e^f$ from a Kähler manifold $(M, g, J)$ onto a Riemannian manifold $(N, g_N)$. In this case, we have $(\ker\pi_*)^\perp = J\mathcal{D}^\perp$. Since $\mathcal{A}_\xi$ reverses the vertical distribution to the horizontal distribution, by (3.193), we have

$$\mathcal{A}_{\xi}(ker\pi_*)^{\perp} = 0, \tag{3.200}$$

for $\xi \in \Gamma((ker\pi_*)^{\perp})$. Equations (3.193) and (3.200) give us $\mathcal{A}_{\xi} = 0$. □

We now shall study the Clairaut conditions for pointwise slant submersions. We begin to study pointwise slant submersions giving a result that characterizes the geodesics on the total space of such submersions.

**Lemma 43.** *[261] Let $\pi$ be a pointwise slant submersion from a Kähler manifold $(M, g, J)$ onto a Riemannian manifold $(N, g_N)$. If $\alpha : I \to M$ is a regular curve and $V(t)$ and $X(t)$ denote the vertical and horizontal components of its tangent vector field $\dot{\alpha}(t) = E$, then $\alpha$ is a geodesic on $M$ if and only if along $\alpha$*

$$-sin2\theta E[\theta]V + cos^2\theta(\hat{\nabla}_V V + \mathcal{V}\nabla_X V) - (\mathcal{T}_V + \mathcal{A}_X)\omega\phi V \tag{3.201}$$
$$-\mathcal{B}\mathcal{H}((\nabla_V + \nabla_X)\omega V) - \phi((\mathcal{T}_V + \mathcal{A}_X)\omega V) + \mathcal{T}_V X = 0$$

*and*

$$cos^2\theta(\mathcal{T}_V + \mathcal{A}_X)V - \omega((\mathcal{T}_V + \mathcal{A}_X)\omega V) - \mathcal{C}\mathcal{H}((\nabla_V + \nabla_X)\omega V) \tag{3.202}$$
$$+\mathcal{H}\{(\nabla_V + \nabla_X)(\omega\phi V + X)\} = 0$$

*hold.*

*Proof.* Let $\alpha : I \to M$ be a regular curve and $V(t)$ and $X(t)$ denote the vertical and horizontal components of its tangent vector field. Since $M$ is a Kähler manifold, we have

$$\nabla_{\dot{\alpha}(t)}\dot{\alpha}(t) = -J\nabla_{\dot{\alpha}(t)}J\dot{\alpha}(t) = -J\nabla_{\dot{\alpha}(t)}JV(t) - J\nabla_{\dot{\alpha}(t)}JX(t).$$

Using (3.131) and (3.5), we obtain

$$\nabla_{\dot{\alpha}(t)}\dot{\alpha}(t) = -J\nabla_{\dot{\alpha}(t)}(\phi V + \omega V)(t) + \nabla_{\dot{\alpha}(t)}X(t).$$

Again, using (3.131) and (3.5), we get

$$\nabla_{\dot{\alpha}(t)}\dot{\alpha}(t) = -\nabla_{\dot{\alpha}(t)}\phi^2 V - \nabla_{\dot{\alpha}(t)}\omega\phi V - J\nabla_{\dot{\alpha}(t)}\omega V + \nabla_{\dot{\alpha}(t)}X.$$

By Theorem 65, we arrive at

$$\nabla_{\dot{\alpha}(t)}\dot{\alpha}(t) = \nabla_{\dot{\alpha}(t)} cos^2\theta V - \nabla_{\dot{\alpha}(t)}\omega\phi V - J\nabla_{\dot{\alpha}(t)}\omega V + \nabla_{\dot{\alpha}(t)}X.$$

After some calculation, we find that

$$\nabla_{\dot{\alpha}(t)}\dot{\alpha}(t) = \quad -sin2\theta E[\theta(t)]V + cos^2\theta\{\nabla_V V + \nabla_X V\} - \nabla_V\omega\phi V - \nabla_X\omega\phi V$$
$$-J\nabla_V\omega V - J\nabla_X\omega V + \nabla_V X + \nabla_X X.$$

Next, using (1.46)–(1.50) and (3.131), we get along $\alpha$

$$
\begin{aligned}
\nabla_{\dot\alpha}\dot\alpha \;=\; & -\sin2\theta E[\theta]V + \cos^2\theta\{\hat\nabla_V V + \mathcal{T}_V V + \mathcal{V}\nabla_X V + \mathcal{A}_X V\} \\
& -\mathcal{H}\{\nabla_V\omega\phi V + \nabla_X\omega\phi V + \nabla_V X - \nabla_X X\} - (\mathcal{T}_V + \mathcal{A}_X)\omega\phi V \\
& -\mathcal{BH}\{(\nabla_V + \nabla_X)\omega V\} - \phi((\mathcal{T}_V + \mathcal{A}_X)\omega V) - \omega((\mathcal{T}_V + \mathcal{A}_X)\omega V) \\
& +\mathcal{T}_V X - \mathcal{CH}((\nabla_V + \nabla_X)\omega V).
\end{aligned}
$$

Taking the vertical and horizontal parts in the last equation, we find

$$
\begin{aligned}
\mathcal{V}\nabla_{\dot\alpha}\dot\alpha \;=\; & -\sin2\theta E[\theta]V + \cos^2\theta(\hat\nabla_V V + \mathcal{V}\nabla_X V) - (\mathcal{T}_V + \mathcal{A}_X)\omega\phi V \\
& -\mathcal{BH}((\nabla_V + \nabla_X)\omega V) - \phi((\mathcal{T}_V + \mathcal{A}_X)\omega V) + \mathcal{T}_V X
\end{aligned}
$$

and

$$
\begin{aligned}
\mathcal{H}\nabla_{\dot\alpha}\dot\alpha \;=\; & \cos^2\theta(\mathcal{T}_V + \mathcal{A}_X)V - \omega((\mathcal{T}_V + \mathcal{A}_X)\omega V) - \mathcal{CH}((\nabla_V + \nabla_X)\omega V) \\
& +\mathcal{H}\{(\nabla_V + \nabla_X)(\omega\phi V + X)\}.
\end{aligned}
$$

Since $\alpha$ is a geodesic if and only if $\nabla_{\dot\alpha}\dot\alpha = 0$, (3.201) and (3.202) come from the last two equations, respectively. □

**Theorem 78.** *[261] Let $\pi$ be a pointwise slant submersion with connected fibers from a Kähler manifold $(M, g, J)$ onto a Riemannian manifold $(N, g_N)$. Then $\pi$ is a Clairaut submersion with $r = e^f$ if and only if at the point $\alpha(t)$*

$$
\begin{aligned}
[2\tan\theta.d\theta(E) + df(E)]V \;=\; & \sec^2\theta\{(\mathcal{T}_V + \mathcal{A}_X)\omega\phi V + \mathcal{BH}((\nabla_V + \nabla_X)\omega V) \\
& +\phi((\mathcal{T}_V + \mathcal{A}_X)\omega V) - \mathcal{T}_V X\}V
\end{aligned}
$$

*holds, where $\theta$ is the slant function of the submersion $\pi$ and $V$ and $X$ are vertical and horizontal parts of the tangent vector field $\dot\alpha(t) = E$ of the geodesic $\alpha$ on $M$.*

*Proof.* Let $\alpha(t)$ be a geodesic on $M$, then we have

$$
c = \|\dot\alpha(t)\|^2,
$$

where $c$ is a constant. From this equality, we deduce that

$$
g(V(t), V(t)) = c\sin^2\beta(t) \quad and \quad g(X(t), X(t)) = c\cos^2\beta(t),
$$

where $\beta(t)$ is the angle between $\dot\alpha(t)$ and the horizontal space at $\alpha(t)$. Differentiating the first expression in the last equation, we obtain

$$
\frac{d}{dt}g(V(t), V(t)) = 2g(\nabla_{\dot\alpha(t)}V(t), V(t)) = 2c\sin\beta(t)\cos\beta(t)\frac{d\beta(t)}{dt}.
$$

Hence, it follows that along $\alpha$

$$g(\nabla_{\dot{\alpha}(t)} V, V) = c \sin\beta \cos\beta \frac{d\beta}{dt}. \tag{3.203}$$

Using (1.47) and (1.49), we obtain

$$g(\hat{\nabla}_V V + \mathcal{V}\nabla_X V, V) = c \sin\beta \cos\beta \frac{d\beta}{dt} \tag{3.204}$$

from (3.204). By (3.201), we get

$$g(sin2\theta E[\theta]V + (\mathcal{T}_V + \mathcal{A}_X)\omega\phi V + \mathcal{B}\mathcal{H}((\nabla_V + \nabla_X)\omega V) + \phi((\mathcal{T}_V + \mathcal{A}_X)\omega V) - \mathcal{T}_V X, V)$$
$$= \cos^2\theta c \sin\beta \cos\beta \frac{d\beta}{dt}. \tag{3.205}$$

On the other hand, $\pi$ is a Clairaut submersion with $r = e^f$ if and only if the equation

$$\frac{d}{dt}(e^f \sin\beta) = 0$$

is satisfied. Hence, it follows that $\pi$ is a Clairaut submersion with $r = e^f$ if and only if the equation

$$\frac{d(f \circ \alpha)}{dt} \sin\beta + \cos\beta \frac{d\beta}{dt} = 0 \tag{3.206}$$

is satisfied. Multiplying (3.206) by the nonzero factor $c \sin\beta$, we find

$$\frac{d(f \circ \alpha)}{dt} c \sin^2\beta = -c \sin\beta \cos\beta \frac{d\beta}{dt}. \tag{3.207}$$

Thus, using (3.206) and the fact that $g(V, V) = c \sin^2\beta$, we get

$$sin2\theta E[\theta]g(V, V) + \cos^2\theta \frac{d(f \circ \alpha)}{dt} g(V, V)$$
$$= g((\mathcal{T}_V + \mathcal{A}_X)\omega\phi V + \mathcal{B}\mathcal{H}((\nabla_V + \nabla_X)\omega V) + \phi((\mathcal{T}_V + \mathcal{A}_X)\omega V) - \mathcal{T}_V X, V).$$

After some calculation, we arrive at

$$\{2\tan\theta . E[\theta] + g(\text{grad } f, E)\}g(V, V)$$
$$= \sec^2\theta g((\mathcal{T}_V + \mathcal{A}_X)\omega\phi V + \mathcal{B}\mathcal{H}((\nabla_V + \nabla_X)\omega V) + \phi((\mathcal{T}_V + \mathcal{A}_X)\omega V) - \mathcal{T}_V X, V). \tag{3.208}$$

Thus, our assertion follows easily from (3.208). $\qquad \square$

Now, let $\pi$ be a pointwise slant submersion from a Kähler manifold $(M, g, J)$ onto a Riemannian manifold $(N, g_N)$ with slant function $\theta$. If the function $\theta$ is constant, i.e., it is independent of the choice of the point $p \in M$, then $\pi$ is a slant submersion. It is not difficult to prove the following results.

**Lemma 44.** *[261] Let $\pi$ be a slant submersion from a Kähler manifold $(M, g, J)$ onto a Riemannian manifold $(N, g_N)$. If $\alpha : I \to M$ is a regular curve and $V(t)$ and $X(t)$ denote the vertical and horizontal components of its tangent vector field $\dot{\alpha}(t) = E$, then $\alpha$ is a geodesic on $M$ if and only if along $\alpha$*

$$\cos^2\theta(\hat{\nabla}_V V + \mathcal{V}\nabla_X V) - (\mathcal{T}_V + \mathcal{A}_X)\omega\phi V \tag{3.209}$$
$$-\mathcal{B}\mathcal{H}((\nabla_V + \nabla_X)\omega V) - \phi((\mathcal{T}_V + \mathcal{A}_X)\omega V) + \mathcal{T}_V X = 0$$

*and*

$$\cos^2\theta(\mathcal{T}_V + \mathcal{A}_X)V - \omega((\mathcal{T}_V + \mathcal{A}_X)\omega V) - \mathcal{C}\mathcal{H}((\nabla_V + \nabla_X)\omega V) \tag{3.210}$$
$$+\mathcal{H}\{(\nabla_V + \nabla_X)(\omega\phi V + X)\} = 0$$

*hold, where $\theta$ is the slant angle of the submersion $\pi$.*

**Theorem 79.** *[261] Let $\pi$ be a slant submersion with connected fibers from a Kähler manifold $(M, g, J)$ onto a Riemannian manifold $(N, g_N)$. Then $\pi$ is a Clairaut submersion with $r = e^f$ if and only if at the point $\alpha(t)$*

$$df(E)V = \sec^2\theta\{(\mathcal{T}_V + \mathcal{A}_X)\omega\phi V + \mathcal{B}\mathcal{H}((\nabla_V + \nabla_X)\omega V)$$
$$+\phi((\mathcal{T}_V + \mathcal{A}_X)\omega V) - \mathcal{T}_V X\}V$$

*holds, where $\theta$ is the slant angle of the submersion $\pi$ and $V$ and $X$ are vertical and horizontal parts of the tangent vector field $\dot{\alpha}(t) = E$ of the geodesic $\alpha$ on $M$.*

**Remark 10.** Anti-invariant Riemannian submersions have been also studied in the Lie-theoretical approach in [124] as follows. Let $H$ be a closed and connected subgroup of an even-dimensional Lie group $G$. Let $\mathfrak{h}$ and $\mathfrak{g}$ be the associated Lie algebras, respectively. Let $\langle \cdot, \cdot \rangle = \langle \cdot, \cdot \rangle_\mathfrak{g}$ be a non-degenerate symmetric bilinear form on $\mathfrak{g}$ which is invariant under the adjoint action of $H$ and whose restriction to $\mathfrak{h}$ is non-degenerate as well. We use $\langle \cdot, \cdot \rangle_\mathfrak{g}$ to decompose $\mathfrak{g} = \mathfrak{h} \oplus \mathfrak{h}^\perp$ as an orthogonal direct sum. The inner product $\langle \cdot, \cdot \rangle_\mathfrak{g}$ defines a left-invariant pseudo-Riemannian metric on $G$ and, since the inner product is invariant under the adjoint action of $H$, the restriction of $\langle \cdot, \cdot \rangle_\mathfrak{g}$ to $\mathfrak{h}^\perp$ defines a $G$-invariant pseudo-Riemannian metric on the coset manifold $G/H$ so that the natural projection $\pi : G \to G/H$ is a pseudo-Riemannian submersion. Assume $\langle \cdot, \cdot \rangle$ is positive definite; in this case see the discussion at the end of section 1.1 of Chapter 2. Let $J$ be a Hermitian complex structure on $\mathfrak{g}$; $J$ induces a left-invariant Hermitian almost complex structure on $G$. Assume that $J\{\mathfrak{h}\} \subset \mathfrak{h}^\perp$. Then $\pi : G \to G/H$ is an anti-invariant Riemannian submersion; $\pi$ is Lagrangian if and only if $2\dim\{\mathfrak{h}\} = \dim\{\mathfrak{g}\}$. In [124], the authors have produced many new examples of anti-invariant Riemannian submersions. We quote the following example. Let $G = \mathbb{R}^m$ and let $H = \mathbb{R}^n \oplus 0 \subset G$ for $n < m$. We identify $G/H$ with $0 \oplus \mathbb{R}^{m-n}$ and $\pi$ with a projection on the last $m - n$-

coordinates. Take the standard Euclidean inner product on $G$ to obtain a bi-invariant Riemannian metric so that $\pi$ is a Riemannian submersion. In particular, suppose that $m = 2\ell$ and $n = \ell$. Identify $G = \mathbb{C}^\ell$ so that $H$ corresponds to the purely real vectors in $\mathbb{C}^\ell$. We identify $\mathfrak{g}$ with $G$ and $\mathfrak{h}$ with $H$. Then $\sqrt{-1}\mathfrak{h} \perp \mathfrak{h}$ and we obtain a Lagrangian Riemannian submersion; the almost complex structure corresponds to scalar multiplication by $\sqrt{-1}$ and is integrable. For details and more examples, see [124].

**Remark 11.** We conclude this chapter by adding that the concept of anti-invariant Riemannian submersions and their extensions have been also considered from different total bases by various authors.

  (i) Anti-invariant Riemannian submersions from Nearly Kähler manifolds (see Fatima and Ali, [6]).

 (ii) Anti-invariant Riemannian submersions from Kenmotsu manifolds (see Beri, Erken and Murathan [30])

(iii) Slant Riemannian submersions for cosymplectic manifolds (see Erken and Murathan, [105]).

 (iv) Anti-Invariant Riemannian submersions from Cosymplectic manifolds (see Erken and Murathan [106]).

  (v) Anti-invariant semi-Riemannian submersions from almost para-Hermitian manifolds (see Gündüzalp [131]).

 (vi) Anti-invariant Riemannian submersions from almost product manifolds (see Gündüzalp [132]).

(vii) Slant submersions from almost product Riemannian manifolds (see Gündüzalp [133]).

(viii) Slant submersions from Lorentzian almost paracontact manifolds (see Gündüzalp [134]).

 (ix) Slant submersions from almost paracontact Riemannian manifolds (see Gündüzalp [135]).

  (x) Semi-slant submersions from almost product Riemannian manifolds (see Gündüzalp [136]).

 (xi) Anti-invariant $\xi^\perp$-Riemannian submersions from almost contact manifolds (see Lee [171]).

(xii) Slant submersions from Sasakian manifolds (see Erken and Murathan [107]).

(xiii) From quaternion Kähler manifolds (see Park [216], [215], [219]).

(xiv) Pointwise slant submersions from cosymplectic manifolds (see Sepet-Ergüt [265]).

 (xv) Anti-holomorphic semi-invariant submersions (see Taştan [283]).

(xvi) Lagrangian submersions from normal almost contact manifolds (see Taştan [285]).

(xvii) Hemi-slant submersions from almost product Riemannian manifolds (see Akyol-Gündüzalp [3]).

# Riemannian Maps

## Contents

## Abstract

In this chapter, we study Riemannian maps between Riemannian manifolds. In section 1, we define Riemannian maps and give the main properties of such maps. In section 2, we obtain Gauss-Weingarten-like formulas and then we obtain Gauss, Codazzi, and Ricci equations along Riemannian maps. In section 3, we find necessary and sufficient conditions for Riemannian maps to be totally geodesic by using the Bochner identity and generalized divergence theorem. In section 4, we introduce umbilical Riemannian maps and pseudo-umbilical Riemannian maps, and obtain characterizations of such Riemannian maps. In section 5, we discuss the harmonicity and biharmonicity of Riemannian maps. In section 6, we define Clairaut Riemannian maps, give an example, and obtain a characterization. In section 6, we extend the result of Nomizu-Yano to the Riemannian maps. In section 7, we obtain Chen's inequality for Riemannian maps. In the last section, we investigate necessary and sufficient conditions for Riemannian maps to have the Einstein property.

**Keywords:** Riemannian map, umbilical map, umbilical Riemannian map, pseudo-umbilical Riemannian map, Clairaut Riemannian map, circle along Riemannian map, the Bochner identity, generalized divergence theorem, Chen's inequality, Einstein manifold

*Sometimes we stare so long at a door that is closing that we see too late the one that is open.*

*Alexander Graham Bell*

Reimannian Submersions, Reimannian Maps in Hermitian Geometry, and their Applications
http://dx.doi.org/10.1016/B978-0-12-804391-2.50004-X

## 1. Riemannian maps

In 1992, Fischer introduced Riemannian maps between Riemannian manifolds in [117] as a generalization of the notions of isometric immersions and Riemannian submersions. Let $F : (M, g_M) \longrightarrow (N, g_N)$ be a smooth map between Riemannian manifolds such that $0 < rankF < min\{m, n\}$, where $dimM = m$ and $dimN = n$. Then we denote the kernel space of $F_*$ by $\mathcal{V}_p = kerF_{*p}$ at $p \in M$ and consider the orthogonal complementary space $\mathcal{H}_p = (kerF_{*p})^\perp$ to $kerF_{*p}$. Then $T_pM$ of $M$ at $p$ has the following decomposition:

$$T_pM = kerF_{*p} \oplus (kerF_{*p})^\perp = \mathcal{V}_p \oplus \mathcal{H}_p.$$

We denote the range of $F_*$ by $rangeF_{*p}$ at $p \in M$ and consider the orthogonal complementary space $(rangeF_{*p})^\perp$ to $rangeF_{*p}$ in the tangent space $T_{F(p)}N$ of $N$ at $p \in M$. Since $rankF \leq min\{m, n\}$, we have $(rangeF_*)^\perp \neq \{0\}$. Thus the tangent space $T_{F(p)}N$ of $N$ at $F(p) \in N$ has the following decomposition:

$$T_{F(p)}N = (rangeF_{*p}) \oplus (rangeF_{*p})^\perp.$$

Now, a smooth map $F : (M^m, g_M) \longrightarrow (N^n, g_N)$ is called a Riemannian map at $p_1 \in M$ if the horizontal restriction

$$F^h_{*p_1} : (kerF_{*p_1})^\perp \longrightarrow (rangeF_{*p_1})$$

is a linear isometry between the inner product spaces $((kerF_{*p_1})^\perp, g_M(p_1) |_{(kerF_{*p_1})^\perp})$ and $(rangeF_{*p_1}, g_N(p_2) |_{(rangeF_{*p_1})})$, $p_2 = F(p_1)$. Therefore, Fischer stated in [117] that a Riemannian map is a map which is as isometric as it can be. In another words, $F_*$ satisfies the equation

$$g_N(F_*X, F_*Y) = g_M(X, Y) \tag{4.1}$$

for $X, Y$ vector fields tangent to $\mathcal{H}$. It follows that isometric immersions and Riemannian submersions are particular Riemannian maps with $ker F_* = \{0\}$ and $(rangeF_*)^\perp = \{0\}$. It is known that a Riemannian map is a subimmersion, which implies that the rank of the linear map $F_{*p} : T_pM \longrightarrow T_{F(p)}N$ is constant for $p$ in each connected component of $M$ [1] and [117]. A remarkable property of Riemannian maps is that a Riemannian map satisfies the generalized eikonal equation $\| F_* \|^2 = rankF$, which is a bridge between geometric optics and physical optics. Since the left-hand side of this equation is continuous on the Riemannian manifold $M$, and since $rankF$ is an integer valued function, this equality implies that $rankF$ is locally constant and globally constant on connected components. Thus if $M$ is connected, the energy density $e(F) = \frac{1}{2} \| F_* \|^2$ is quantized to integer and half-integer values. The eikonal equation of geometrical optics is solved by using Cauchy's method of characteristics, whereby, for real valued functions $F$, solutions to the partial differential equation $\| dF \|^2 = 1$ are obtained by solving the system of ordinary differential equations $x' = grad f(x)$. Since harmonic maps generalize geodesics, harmonic maps could be used to solve the generalized eikonal equation [117].

In this section, we give formal definition of Riemannian maps, present examples and obtain a characterization of such maps. Since every Riemannian map is a subimmersion, we show the relations among Riemannian maps, isometric immersions and Riemannian submersions, and we also show that Riemannian maps satisfy the eikonal equation.

**Definition 47.** [117] Let $(M^m, g_M)$ and $(N^n, g_N)$ be Riemannian manifolds and

$$F : (M^m, g_M) \longrightarrow (N^n, g_N)$$

a smooth map between them. Then we say that $F$ is a *Riemannian map* at $p_1 \in M$ if $0 \leq rank F_{*p_1} \leq \min\{m, n\}$ and $F_{*p_1}$ maps the horizontal space $\mathcal{H}(p_1) = (ker(F_{*p_1}))^{\perp}$ isometrically onto $range(F_{*p_1})$, i.e.,

$$g_N(F_{*p_1}X, F_{*p_1}Y) = g_M(X, Y) \tag{4.2}$$

for $X, Y \in \mathcal{H}(p_1)$. Also $F$ is called Riemannian if $F$ is Riemannian at each $p_1 \in M$; see Figure 4.1.

We give some examples of Riemannian maps.

**Example 25.** Let $I : (M^m, g_M) \longrightarrow (N^n, g_N)$ be an isometric immersion between Riemannian manifolds. Then $I$ is a Riemannian map with $ker F_* = \{0\}$.

**Example 26.** Let $F : (M^m, g_M) \longrightarrow (N^n, g_N)$ be a Riemannian submersion between Riemannian manifolds. Then $F$ is a Riemannian map with $(range F_*)^{\perp} = \{0\}$.

**Example 27.** Consider the following map defined by

$$F : \qquad \mathbb{R}^5 \qquad \longrightarrow \qquad \mathbb{R}^4$$
$$(x_1, x_2, x_3, x_4, x_5) \qquad (\tfrac{x_1+x_2}{\sqrt{2}}, \tfrac{x_3+x_4}{\sqrt{2}}, x_5, 0).$$

Then we have

$$ker F_* = span\{Z_1 = \frac{\partial}{\partial x_1} - \frac{\partial}{\partial x_2}, Z_2 = \frac{\partial}{\partial x_3} - \frac{\partial}{\partial x_4}\}$$

and

$$(ker F_*)^{\perp} = span\{Z_3 = \frac{\partial}{\partial x_1} + \frac{\partial}{\partial x_2}, Z_4 = \frac{\partial}{\partial x_3} + \frac{\partial}{\partial x_4}, Z_5 = \frac{\partial}{\partial x_5}\}.$$

Hence it is easy to see that

$$g_{\mathbb{R}^4}(F_*(Z_i), F_*(Z_i)) = g_{\mathbb{R}^5}(Z_i, Z_i) = 2, g_{\mathbb{R}^4}(F_*(Z_5), F_*(Z_5)) = g_{\mathbb{R}^5}(Z_5, Z_5) = 1$$

and

$$g_{\mathbb{R}^4}(F_*(Z_i), F_*(Z_j)) = g_{\mathbb{R}^5}(Z_i, Z_j) = 0,$$

$i \neq j$, for $i, j = 3, 4, 5$. Thus $F$ is a Riemannian map.

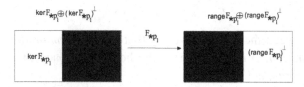

Figure 4.1 The blackscale regions are mapped isometrically to each other by $F$.

If $F$ is Riemannian at $p$ , then there exists orthonormal bases $\{v_1, ..., v_m\}$, $v_i \in T_pM$, $1 \leq i \leq m = dimM$, of $T_pM$, and $\{w_1, ..., w_n\}$, $w_i \in T_{F(p)}N$, $1 \leq i \leq n = dimN$, of $T_{F(p)}N$ such that

$$F_{*p}(v_1, ..., v_r, \underbrace{v_{r+1}, ..., v_m}_{m-r}) = (w_1, ..., w_r, \underbrace{0, ..., 0}_{m-r}),$$

where $\{v_1, ..., v_r\}$ spans $\mathcal{H}_p$, $\{v_{r+1}, ..., v_m\}$ spans $\mathcal{V}_p$, $\{w_1, ..., w_r\}$ spans $rangeF_{*p}$, $\{w_{r+1}, ..., w_n\}$ spans $(rangeF_{*p})^{\perp}$, and where $r$ satisfies $0 \leq r \leq \min\{m, n\}$.

Let $F : (M^m, g_M) \longrightarrow (N^n, g_N)$ be a smooth map between Riemannian manifolds. Define linear transformations

$$\mathcal{P}_{p_1} : T_{p_1}M \longrightarrow T_{p_1}M \quad , \quad \mathcal{P}_{p_1} =^* F_{*p_1} \circ F_{*p_1}$$
$$Q_{p_1} : T_{p_2}N \longrightarrow T_{p_2}N \quad , \quad Q_{p_1} = F_{*p_1} \circ^* F_{*p_1}$$

where $p_2 = F(p_1)$. Using these linear transformations, we obtain the following characterizations of Riemannian maps.

**Theorem 80.** *[117] Let $F : (M, g_M) \longrightarrow (N, g_N)$ be a map between Riemannian manifolds $(M, g_M)$ and $(N, g_N)$. Then the following assertions are equivalent:*
**(1)** *$F$ is Riemannian at $p_1 \in M$.*

**(2)** $\mathcal{P}_{p_1}$ *is a projection, i.e.,* $\mathcal{P}_{p_1} \circ \mathcal{P}_{p_1} = \mathcal{P}_{p_1}$.

**(3)** $Q_{p_1}$ *is a projection, i.e.,* $Q_{p_1} \circ Q_{p_1} = Q_{p_1}$.

*Proof.* $F$ is Riemannian if and only if

$$g_{Np_2}(F_{*p_1}X_1, F_{*p_1}Y_1) = g_{Mp_1}(X_1, Y_1)$$

for $X_1, Y_1 \in T_{p_1}M_1$. Since $g_{Np_2}(F_{*p_1}X_1, F_{*p_1}Y_1) = g_{Mp_1}(X_1, {}^*F_{*p_1} \circ F_{*p_1}Y_1)$, the above statement is equivalent to

$$g_{Np_2}(F_{*p_1}X_1, F_{*p_1} \circ^* F_{*p_1} \circ F_{*p_1}Y_1) = g_{Mp_1}(X_1, {}^*F_{*p_1} \circ F_{*p_1}Y_1)$$

which is equivalent to

$$g_{Mp_1}(X_1, {}^*F_{*p_1} \circ F_{*p_1} \circ^* F_{*p_1} \circ F_{*p_1}Y_1) = g_{Mp_1}(X_1, {}^*F_{*p_1} \circ F_{*p_1}Y_1).$$

Then, since ${}^*F_{*p_1} \circ F_{*p_1} = \mathcal{P}_{p_1}$, we conclude that $F$ is a Riemannian map if and only if

$$g_{Mp_1}(X_1, (\mathcal{P}_{p_1} \circ \mathcal{P}_{p_1}Y_1) = g_{Mp_1}(X_1, \mathcal{P}_{p_1}Y_1).$$

Then the Riemannian metric $g_M$ gives (1) $\Leftrightarrow$ (2). Again, $F$ is Riemannian if and only if

$$g_{Mp_1}(X_1, Y_1) = g_{Np_2}(F_{*p_1}X_1, F_{*p_1}Y_1).$$

Since $range^*F_{*p_1} = \mathcal{H}(p_1)$, this holds if and only if, for $X_2, Y_2 \in T_{p_2}M_2$,

$$g_{Mp_1}({}^*F_{*p_1}X_2, {}^*F_{*p_1}Y_2) = g_{Np_2}(F_{*p_1} \circ^* F_{*p_1}X_2, F_{*p_1} \circ^* F_{*p_1}Y_2),$$

which is equivalent to

$$g_{Np_2}(X_2, F_{*p_1} \circ^* F_{*p_1}Y_2) = g_{Mp_1}({}^*F_{*p_1}X_2, {}^*F_{*p_1} \circ F_{*p_1} \circ^* F_{*p_1}Y_2),$$

that is,

$$g_{Np_2}(X_2, F_{*p_1} \circ^* F_{*p_1}Y_2) = g_{Np_2}(X_2, F_{*p_1} \circ^* F_{*p_1} \circ F_{*p_1} \circ^* F_{*p_1}Y_2).$$

Since $Q_{p_1} = F_{*p_1} \circ^* F_{*p_1}$, we get

$$g_{Np_2}(X_2, Q_{p_1}Y_2) = g_{Np_2}(X_2, (Q_{p_1} \circ Q_{p_1}Y_2).$$

Then, since the above equation holds for all $X_2, Y_2 \in T_{p_2}M_2$, the Riemannian metric $g_N$ implies that $Q_{p_1} \circ Q_{p_1} = Q_{p_1}$. $\qquad\square$

We now recall that a map $F : (M^m, g_M) \longrightarrow N^n$ is called a *subimmersion* at $p_1 \in M$ if there is a neighborhood $U$ of $p_1$, a manifold $P$, a submersion $S : U \longrightarrow P$, and an immersion $I : P \longrightarrow N$ such that $F|_U = F_U = I \circ S$. A map $F : (M^m, g_M) \longrightarrow N^n$ is called a subimmersion if it is a subimmersion at each $p_1 \in M$. It is well known that

$F : (M^m, g_M) \longrightarrow N^n$ is a subimmersion if and only if the rank of the linear map $F_{*p_1} :$ $T_{p_1} M \longrightarrow T_{F(p_1)} N$ is constant for $p_1$ in each connected component of $M$ [1], where $M$ and $N$ are finite dimensional manifolds. Thus, by this definition, a Riemannian map is a subimmersion. Moreover we have the following:

**Theorem 81.** *[117], [122] Let $F : (M^m, g_M) \longrightarrow (N^n, g_N)$ be a Riemannian map between Riemannian manifolds $(M^m, g_M)$ and $(N^n, g_N)$. Let $U$, $P$, $S$ and $I$ be as in the definition of the subimmersion so that $F \mid_U = F_U = I \circ S$. Let $g_{M_U}$ denote the restriction of $g_M$ to $U$ and let $g_P = I^* g_N$. Then $(U, g_{M_U})$ and $(P, g_P)$ are Riemannian manifolds, the submersion $S : (U, g_{M_U}) \longrightarrow (P, g_P)$ is a Riemannian submersion, and the immersion $I : (P, g_P) \longrightarrow (N^n, g_N)$ is an isometric immersion.*

*Proof.* Since $U$ is open in $M$, $(U, g_{M_U})$ is a Riemannian manifold. The Riemannian metric $g_N$ makes $g_P$ also a Riemannian metric. It follows that $(P, g_P)$ is a Riemannian manifold and $I : (P, g_P) \longrightarrow (N^n, g_N)$ is an isometric immersion. Thus, it is enough to show that $S$ is a Riemannian submersion. For each $p_1 \in U$, we define

$$S^h_{*p_1} : \mathcal{H}(p_1) \longrightarrow T_{S(p_1)} P \quad \text{by} \quad S^h_{*p_1} X = S_{*p_1} X$$

and define

$$I^h_{*p_1} : T_{S(p_1)} P \longrightarrow range F_{*p_1} \quad \text{by} \quad I^h_{*p_1} Z = I_{*p_1} Z.$$

Thus we have

$$F^h_{*p_1} = (I_{*S(p_1)} \circ S_{*p_1})^h = I^h_{*S(p_1)} \circ S^h_{*p_1} : \mathcal{H}(p_1) \longrightarrow range F_{*p_1}.$$

Then, for $X, Y \in \mathcal{H}(p_1)$, we have

$$g_N(F^h_{*p_1} X, F^h_{*p_1} Y) = g_N(I^h_{*S(p_1)} \circ S^h_{*p_1} X, I^h_{*S(p_1)} \circ S^h_{*p_1} Y).$$

Since $F : (M^m, g_M) \longrightarrow (N^n, g_N)$ is a proper Riemannian map and $I : (P, g_P) \longrightarrow (M^n, g_M)$ is an isometric immersion, we get

$$g_{M_U}(X, Y) = g_P(S^h_{*p_1} X, S^h_{*p_1} Y),$$

which shows that $S : (U, g_{M_U}) \longrightarrow (P, g_P)$ is a Riemannian submersion.    □

From Theorem 81, we have the following result.

**Proposition 43.** *[117] Let $F_1 : (M_1, g_1) \longrightarrow (M_2, g_2)$ be a Riemannian map between Riemannian manifolds. Then, locally, $F$ is the composition of a surjective Riemannian submersion followed by an injective isometric immersion.*

The next result shows that Riemannian maps satisfy a generalized eikonal equation which is a bridge between physical optics and geometric optics.

**Theorem 82.** *[117], [122] Let $F : (M^m, g_M) \longrightarrow (N^n, g_N)$ be a Riemannian map between Riemannian manifolds $(M^m, g_M)$ and $(N^n, g_N)$. Then,*

$$\| F_* \|^2 = rankF. \tag{4.3}$$

*Proof.* Let $^*F_{*p_1}$ be the adjoint of $F_{*p_1}$ and define a linear transformation

$$G : (\mathcal{H}(p_1), g_{Mp_1 \mathcal{H}(p_1)}) \longrightarrow (\mathcal{H}(p_1), g_{Mp_1 \mathcal{H}(p_1)})$$

by $G_{p_1} =^* F_{*p_1} \circ F_{*p_1}$. Then for $X_1, Y_1 \in \mathcal{H}(p_1)$, we get

$$
\begin{aligned}
g_{Mp_1}(G_{p_1} X_1, Y_1) &= g_{Mp_1}(^*F_{*p_1} \circ F_{*p_1} X_1, Y_1) \\
&= g_{NF(p_1)}(F_{*p_1} X_1, F_{*p_1} Y_1).
\end{aligned}
$$

Since $F$ is a proper Riemannian map, we have

$$g_{Mp_1}(G_{p_1} X_1, Y_1) = g_{Mp_1}(X_1, Y_1).$$

Hence, it follows that $G_{p_1} X_1 = X_1$. Thus, we obtain

$$\| F_* \|^2 (p_1) = \sum_{i=1}^{n} g_{NF(p_1)}(F_{*p_1} e_i, F_{*p_1} e_i) = \sum_{i=1}^{n_1} g_{Mp_1}(e_i, e_i),$$

where $\{e_i\}, i \in \{1, ..., n_1 = dim(\mathcal{H})\}$ is an orthonormal basis of $(ker F_*)^\perp$. Then we have

$$\| F_* \|^2 (p_1) = dim(\mathcal{H}) = rankF.$$

Thus, the proof is complete. $\qquad\qquad\qquad\qquad\qquad\qquad\qquad\qquad$ □

## 2. Geometric structures along Riemannian maps

In differential geometry of submanifolds, there is a set of equations that describe relationships between invariant quantities on the submanifold and ambient manifold when the Riemannian connection is used. These relationships are expressed by the Gauss formula, Weingarten formula, and the equations of Gauss, Codazzi, and Ricci. These equations have such an important role for the Riemannian connection that they are called the fundamental equations for submanifolds. In this section, we extend these equations to Riemannian maps between Riemannian manifolds. To do this, we show that the second fundamental form of a Riemannian map has a special form. Then we define the Gauss formula for Riemannian maps by using the second fundamental form of a Riemannian map. We also obtain Weingarten formula for Riemannian maps by using the linear connection $\nabla^{F\perp}$ on $(F_*(TM))^\perp$ defined in [201]. From the

Gauss-Weingarten formulas, we obtain the Gauss, Ricci and Codazzi equations for Riemannian maps.

## 2.1. Gauss-Weingarten formulas for Riemannian maps

Let $(M, g_M)$ and $(N, g_N)$ be Riemannian manifolds and suppose that $F : M \longrightarrow N$ is a smooth map between them. In this section we construct Gauss-Weingarten-like formulas for Riemannian maps. The following lemma is very useful for Riemannian maps.

**Lemma 45.** *[241], [243]Let $F$ be a Riemannian map from a Riemannian manifold $(M, g_M)$ to a Riemannian manifold $(N, g_N)$. Then, for any $\forall X, Y, Z \in \Gamma((ker F_*)^\perp)$, we have*

$$g_N((\nabla F_*)(X, Y), F_*(Z)) = 0, \tag{4.4}$$

*Proof.* Since $F$ is a Riemannian map, from (1.71) we have

$$g_N((\nabla F_*)(X, Y), F_*(Z)) = g_N(\nabla_X^F F_* Y, F_* Z) - g_M(\nabla_X^M Y, Z). \tag{4.5}$$

On the other hand, since $\nabla^M$ of $M$ is a Levi-Civita connection, from Koszul identity, we have

$$
\begin{aligned}
g_M(\nabla_X^M Y, Z) &= \frac{1}{2}\{X g_M(Y, Z) + Y g_M(X, Z) - Z g_M(X, Y) \\
&+ g_M([X, Y], Z) + g_M([Z, X], Y) - g_M([Y, Z], X)\}.
\end{aligned}
$$

Since $F_*([X, Y]) = [F_* X, F_* Y]$, using $g_M(X, Y) = g_N(F_* X, F_* Y)$, we obtain

$$
\begin{aligned}
g_M(\nabla_X^M Y, Z) &= \frac{1}{2}\{X g_N(F_* Y, F_* Z) + Y g_N(F_* X, F_* Z) - Z g_N(F_* X, F_* Y) \\
&+ g_N([F_* X, F_* Y], F_* Z) + g_N([F_* Z, F_* X], F_* Y) \\
&- g_N([F_* Y, F_* Z], F_* X)\}.
\end{aligned}
$$

Since $\nabla^N$ is also a Levi-Civita connection on $N$, we have

$$g_M(\nabla_X^M Y, Z) = g_N(\nabla_X^F F_* Y, F_* Z). \tag{4.6}$$

Then the proof follows from (4.5) and (4.6). □

As a result of Lemma 45, we have

$$(\nabla F_*)(X, Y) \in \Gamma((range F_*)^\perp), \forall X, Y \in \Gamma((ker F_*)^\perp). \tag{4.7}$$

Thus at $p \in M$, we write

$$\overset{N}{\nabla}{}^F_X F_*(Y)(p) = F_*(\nabla_X^M Y)(p) + (\nabla F_*)(X, Y)(p) \tag{4.8}$$

for $X, Y \in \Gamma((ker F_*)^{\perp})$, where $\overset{N}{\nabla}{}^F_X F_*(Y) \in T_F(p)N$, $F_*(\overset{M}{\nabla}{}^X_X Y)(p) \in F_{*p}(T_pM)$ and $(\nabla F_*)(X, Y)(p) \in (F_{*p}(T_pM))^{\perp}$.

**Remark 12.** For a map $F$ between Riemannian manifolds $(M, g)$ and $(N, g')$, respectively, in [201], the author defined two tensor fields by

$$\tau : TM \times TM \longrightarrow F_*(TM)$$

and

$$\nu : TM \times TM \longrightarrow (F_*(TM))^{\perp}$$

by putting $\tau(X, Y) = $ orthogonal projection of $(\nabla F_*)$ on $F_*(TM)$, $\nu(X, Y) = $ orthogonal projection $(\nabla F_*)$ on $(F_*(TM))^{\perp}$. If $\tau = 0$, $F$ is called a relatively affine map. This notion was introduced in [307]. Thus Lemma 45 implies that a Riemannian map is a relatively affine map. On the other hand, in [119], the author denoted $\overset{2}{\nabla}{}^F_X F_* Y$ by $\overline{\nabla}_X Y$ and $F_*(\overset{M}{\nabla}_X Y)$ by $\mathcal{H}\nabla_X Y$, then as in the isometric immersion theory, the author introduced the following equation:

$$\overline{\nabla}_X Y = \mathcal{H}\nabla_X Y + B(X, Y),$$

where $B(X, Y) \in T^{\perp}M$. In fact, $T^{\perp}M$ was used instead of $(F_*(TM))^{\perp}$ as in the isometric immersion theory. So Lemma 45 shows that $B(X, Y)$ in [119] is actually $(\nabla F_*)(X, Y)$ for $X, Y \in \Gamma((ker F_*)^{\perp})$.

Let $F$ be a Riemannian map from a Riemannian manifold $(M, g_M)$ to a Riemannian manifold $(N, g_N)$. Then we define $\mathcal{T}$ and $\mathcal{A}$ as

$$\mathcal{A}_E F = \mathcal{H}\overset{M}{\nabla}_{\mathcal{H}E} \mathcal{V}F + \mathcal{V}\overset{M}{\nabla}_{\mathcal{H}E} \mathcal{H}F \tag{4.9}$$

$$\mathcal{T}_E F = \mathcal{H}\overset{M}{\nabla}_{\mathcal{V}E} \mathcal{V}F + \mathcal{V}\overset{M}{\nabla}_{\mathcal{V}E} \mathcal{H}F, \tag{4.10}$$

for vector fields $E, F$ on $M$, where $\nabla^M$ is the Levi-Civita connection of $g_M$. In fact, we can see that these tensor fields are O'Neill's tensor fields which were defined for Riemannian submersions. For any $E \in \Gamma(TM)$, $\mathcal{T}_E$ and $\mathcal{A}_E$ are skew-symmetric operators on $(\Gamma(TM), g)$ reversing the horizontal and the vertical distributions. It is also easy to see that $\mathcal{T}$ is vertical, $\mathcal{T}_E = \mathcal{T}_{\mathcal{V}E}$, and $\mathcal{A}$ is horizontal, $\mathcal{A}_E = \mathcal{A}_{\mathcal{H}E}$. We note that the tensor field $\mathcal{T}$ satisfies

$$\mathcal{T}_U W = \mathcal{T}_W U, \forall U, W \in \Gamma(ker F_*). \tag{4.11}$$

On the other hand, from (4.9) and (4.10) we have

$$\nabla_V^M W \;=\; \mathcal{T}_V W + \hat{\nabla}_V W \tag{4.12}$$

$$\nabla_V^M X \;=\; \mathcal{H}\nabla_V^M X + \mathcal{T}_V X \tag{4.13}$$

$$\nabla_X^M V \;=\; \mathcal{A}_X V + \mathcal{V}\nabla_X^M V \tag{4.14}$$

$$\nabla_X^M Y \;=\; \mathcal{H}\nabla_X^M Y + \mathcal{A}_X Y \tag{4.15}$$

for $X, Y \in \Gamma((\ker F_*)^\perp)$ and $V, W \in \Gamma(\ker F_*)$, where $\hat{\nabla}_V W = \mathcal{V}\nabla_V^M W$.

From now on, for simplicity, we denote by $\nabla^N$ both the Levi-Civita connection of $(N, g_N)$ and its pullback along $F$. We denote by $(range F_*)^\perp$ the subbundle of $F^{-1}(TN)$ with fiber $(F_*(T_{p_1}M))^\perp$-orthogonal complement of $F_*(T_{p_1}M)$ for $g_N$ over $p_1$. For any vector field $X$ on $M$ and any section $V$ of $(range F_*)^\perp$, we define $\nabla_X^{F\perp} V$, which is the orthogonal projection of $\nabla_X^N V$ on $(range F_*)^\perp$. Then we have the following result.

**Lemma 46.** *[201] $\nabla^{F\perp}$ is a linear connection on $(range F_*)^\perp$ such that $\nabla^{F\perp} g_N = 0$.*

The proof of the above lemma is similar to the corresponding one for isometric immersions; see, for instance, [64]. Therefore we omit it.

We now suppose that $F$ is a Riemannian map and define $S_V$ as

$$\nabla_{F_*X}^N V = -S_V F_* X + \nabla_X^{F\perp} V, \tag{4.16}$$

where $S_V F_* X$ is the tangential component (a vector field along $F$) of $\nabla_{F_*X}^N V$. Observe that $\nabla_{F_*X}^N V$ is obtained from the pullback connection of $\nabla^N$. Thus, at $p_1 \in M$, we have $\nabla_{F_*X}^N V(p_1) \in T_{F(p_1)}N$, $S_V F_* X(p_1) \in F_{*p_1}(T_{p_1}M)$ and $\nabla_X^{F\perp} V(p_1) \in (F_{*p_1}(T_{p_1}M))^\perp$. It is easy to see that $S_V F_* X$ is bilinear in $V$ and $F_* X$ and $S_V F_* X$ at $p_1$ depends only on $V_{p_1}$ and $F_{*p_1} X_{p_1}$. By direct computations, we obtain

$$g_N(S_V F_* X, F_* Y) = g_N(V, (\nabla F_*)(X, Y)), \tag{4.17}$$

for $X, Y \in \Gamma((\ker F_*)^\perp)$ and $V \in \Gamma((range F_*)^\perp)$. Since $(\nabla F_*)$ is symmetric, it follows that $S_V$ is a symmetric linear transformation of $range F_*$.

As in the case of submanifolds, we call (4.8) and (4.12)-(4.15) the *Gauss formulae* and (4.16) the *Weingarten formula* for the Riemannian map $F$.

## 2.2. The Equations of Ricci, Gauss and Codazzi

In this subsection, we are going to obtain the equations of Gauss, Codazzi, and Ricci for Riemannian maps. This section is taken from [254] and [260]. We first note that if $F$ is a Riemannian map from a Riemannian manifold $(M, g_M)$ to a Riemannian manifold $(N, g_N)$, then considering $F_*^h$ at each $p_1 \in M$ as a linear transformation

$$F_{*p_1}^h : ((ker\, F_*)^\perp(p_1), g_{M\,p_1((ker F_*)^\perp(p_1))}) \longrightarrow (range F_*(p_2), g_{N\,p_2(range F_*)(p_2))}),$$

we will denote the adjoint of $F_*^h$ by $^*F^h{}_{*p_1}$. Let $^*F_{*p_1}$ be the adjoint of

$$F_{*p_1} : (T_{p_1}M, g_{M p_1}) \longrightarrow (T_{p_2}N, g_{N p_2}).$$

Then the linear transformation

$$({}^*F_{*p_1})^h : range F_*(p_2) \longrightarrow (ker\, F_*)^\perp(p_1),$$

defined by $({}^*F_{*p_1})^h y = {}^*F_{*p_1}y$, where $y \in \Gamma(range F_{*p_1})$, $p_2 = F(p_1)$, is an isomorphism and $(F_{*p_1}^h)^{-1} = ({}^*F_{*p_1})^h = {}^*(F_{*p_1}^h)$.

By using (4.8) and (4.16) we have

$$\begin{aligned}
R^N(F_*X, F_*Y)F_*Z &= -\mathcal{S}_{(\nabla F_*)(Y,Z)}F_*X + \mathcal{S}_{(\nabla F_*)(X,Z)}F_*Y \\
&+ F_*(R^M(X,Y)Z) + (\nabla_X(\nabla F_*))(Y,Z) \\
&- (\nabla_Y(\nabla F_*))(X,Z) \qquad (4.18)
\end{aligned}$$

for $X, Y, Z \in \Gamma((ker\, F_*)^\perp)$, where $R^M$ and $R^N$ denote the curvature tensors of $\nabla^M$ and $\nabla^N$ which are metric connections on $M$ and $N$, respectively. Moreover, $(\nabla_X(\nabla F_*))(Y,Z)$ is defined by

$$(\nabla_X(\nabla F_*))(Y,Z) = \nabla_X^{F^\perp}(\nabla F_*)(Y,Z) - (\nabla F_*)(\nabla_X^M Y, Z) - (\nabla F_*)(Y, \nabla_X^M Z).$$

From (4.18), for any vector field $T \in \Gamma((ker\, F_*)^\perp)$, we have

$$\begin{aligned}
g_N(R^N(F_*X, F_*Y)F_*Z, F_*T) &= g_M(R^M(X,Y)Z, T) \\
&+ g_N((\nabla F_*)(X,Z), (\nabla F_*)(Y,T)) \\
&- g_N((\nabla F_*)(Y,Z), (\nabla F_*)(X,T)). \qquad (4.19)
\end{aligned}$$

Taking the $\Gamma((range F_*)^\perp)-$ part of (4.18), we have

$$(R^N(F_*X, F_*Y)F_*Z)^\perp = (\nabla_X(\nabla F_*))(Y,Z) - (\nabla_Y(\nabla F_*))(X,Z). \qquad (4.20)$$

We call (4.19) and (4.20) the *Gauss equation* and the *Codazzi equation*, respectively, for the Riemannian map $F$.

For any vector fields $X, Y$ tangent to $M$ and any vector field $V$ perpendicular to $\Gamma(range F_*)$, we define the curvature tensor field $R^{F\perp}$ of the subbundle $(range F_*)^\perp$ by

$$R^{F\perp}(F_*(X), F_*(Y))V = \nabla_X^{F^\perp}\nabla_Y^{F^\perp}V - \nabla_Y^{F^\perp}\nabla_X^{F^\perp}V - \nabla_{[X,Y]}^{F^\perp}V. \qquad (4.21)$$

Then using (4.8), (4.16), and (4.17), we obtain

$$
\begin{aligned}
R^{N}(F_*(X), F_*(Y))V &= R^{F\perp}(F_*(X), F_*(Y))V \\
&- F_*(\nabla_X^{M*}F_*(\mathcal{S}_V F_*(Y))) + \mathcal{S}_{\nabla_X^{F\perp}V}F_*(Y) \\
&+ F_*(\nabla_Y^{M*}F_*(\mathcal{S}_V F_*(X))) - \mathcal{S}_{\nabla_Y^{F\perp}V}F_*(X) \\
&- (\nabla F_*)(X, {}^*F_*(\mathcal{S}_V F_*(Y))) + (\nabla F_*)(Y, {}^*F_*(\mathcal{S}_V F_*(X))) \\
&- \mathcal{S}_V F_*([X, Y]),
\end{aligned} \tag{4.22}
$$

where

$$
F_*([X, Y]) = \overset{N}{\nabla}{}^F_X F_*(Y) - \overset{N}{\nabla}{}^F_Y F_*(X)
$$

and ${}^*F_*$ is the adjoint map of $F_*$. Then for $F_*(Z) \in \Gamma(range F_*)$, we have

$$
\begin{aligned}
g_N(R^{N}(F_*(X), F_*(Y))V, F_*(Z)) &= g_N((\tilde{\nabla}_Y \mathcal{S})_V F_*(X), F_*(Z)) \\
&- g_N((\tilde{\nabla}_X \mathcal{S})_V F_*(Y)), F_*(Z)),
\end{aligned} \tag{4.23}
$$

where $(\tilde{\nabla}_X \mathcal{S})_V F_*(Y)$ is defined by

$$
(\tilde{\nabla}_X \mathcal{S})_V F_*(Y) = F_*(\nabla_X^{M*}F_*(\mathcal{S}_V F_*(Y))) - \mathcal{S}_{\nabla_X^{F\perp}V}F_*(Y) - \mathcal{S}_V P \overset{N}{\nabla}{}^F_X F_*(Y),
$$

where $P$ denotes the projection morphism on $range F_*$. On the other hand, for $W \in \Gamma((range F_*)^{\perp})$, we get

$$
\begin{aligned}
g_N(R^{N}(F_*(X), F_*(Y))V, W) &= g_N(R^{F\perp}(F_*(X), F_*(Y))V, W) \\
&- g_N((\nabla F_*)(X, {}^*F_*(\mathcal{S}_V F_*(Y))), W) \\
&+ g_N((\nabla F_*)(Y, {}^*F_*(\mathcal{S}_V F_*(X))), W).
\end{aligned} \tag{4.24}
$$

Using (4.17), we obtain

$$
g_N((\nabla F_*)(X, {}^*F_*(\mathcal{S}_V F_*(Y))), W) = g_N(\mathcal{S}_W F_*(X), \mathcal{S}_V F_*(Y))
$$

Since $\mathcal{S}_V$ is self-adjoint, we get

$$
g_N((\nabla F_*)(X, {}^*F_*(\mathcal{S}_V F_*(Y))), W) = g_N(\mathcal{S}_V \mathcal{S}_W F_*(X), F_*(Y)). \tag{4.25}
$$

Then using (4.25) in (4.24), we arrive at

$$
\begin{aligned}
g_N(R^{N}(F_*(X), F_*(Y))V, W) &= g_N(R^{F\perp}(F_*(X), F_*(Y))V, W) \\
&+ g_N([\mathcal{S}_W, \mathcal{S}_V]F_*(X), F_*(Y)),
\end{aligned} \tag{4.26}
$$

where

$$
[\mathcal{S}_W, \mathcal{S}_V] = \mathcal{S}_W \mathcal{S}_V - \mathcal{S}_V \mathcal{S}_W.
$$

We call (4.26) the *Ricci equation* for the Riemannian map $F$.

## 3. Totally geodesic Riemannian maps

In this section, we first give the necessary and sufficient conditions for a Riemannian map to be totally geodesic in terms of (4.9) and (4.16), and then we obtain many results on totally geodesic Riemannian maps by putting appropriate conditions for curvatures of the base manifold and the total manifold, $rankF$, and the tension field. The materials presented in this section, except for Theorem 83, are taken from [122] and [123]. We recall that a differentiable map $F$ between Riemannian manifolds $(M, g_M)$ and $(N, g_N)$ is called a totally geodesic map if $(\nabla F_*)(X, Y) = 0$ for all $X, Y \in \Gamma(TM)$.

**Theorem 83.** *[247] Let $F$ be a Riemannian map from a Riemannian manifold $(M, g_M)$ to a Riemannian manifold $(N, g_N)$. Then $F$ is totally geodesic if and only if*

**(a)** $\mathcal{A}_X Y = 0$, *and*

**(b)** $\mathcal{S}_V F_*(X) = 0$,

**(c)** *the fibers are totally geodesic,*

*for $X, Y \in \Gamma((ker F_*)^\perp)$ and $V \in \Gamma((rangeF_*)^\perp)$.*

*Proof.* First note that a map from a Riemannian manifold $(M, g_M)$ to a Riemannian manifold $(N, g_N)$ is totally geodesic if and only if $(\nabla F_*)(X, Y) = 0$, $(\nabla F_*)(X, U) = 0$ and $(\nabla F_*)(U_1, U_2) = 0$ for $U, U_1, U_2 \in \Gamma(ker F_*)$ and $X, Y \in \Gamma((ker F_*)^\perp)$.
Since $(\nabla F_*)(X, U) \in \Gamma(rangeF_*)$, $(\nabla F_*)(X, U) = 0$ if and only if $g_N((\nabla F_*)(X, U), F_*(Y)) = 0$ for $Y \in \Gamma((kerF_*)^\perp)$. By using (1.71) and (4.9), we have

$$g_M(\mathcal{A}_X U, Y) = -g_N((\nabla F_*)(X, U), F_*(Y)). \qquad (4.27)$$

In a similar way, since $(\nabla F_*)(U, V) \in \Gamma(rangeF_*)$, it follows that $(\nabla F_*)(U, V) = 0$ if and only if $g_N((\nabla F_*)(U, V), F_*(X)) = 0$ for $X \in \Gamma((ker F_*)^\perp)$. Then, from (1.71) and (4.10), we get

$$g_M(\mathcal{T}_U V, X) = -g_N((\nabla F_*)(U, V), F_*(X)). \qquad (4.28)$$

On the other hand, for $X, Y \in \Gamma((ker F_*)^\perp)$, since $(\nabla F_*)(X, Y) \in \Gamma((rangeF_*)^\perp)$, it follows that $(\nabla F_*)(X, Y) = 0$ if and only if $g_N(\nabla F_*)(X, Y), V) = 0$ for $V \in \Gamma((rangeF_*)^\perp)$. Then using (4.17), we obtain

$$g_N((\nabla F_*)(X, Y), V) = g_N(F_*(Y), \mathcal{S}_V F_*(X)). \qquad (4.29)$$

Thus, the assertion comes from (4.13), (4.27) and (4.29). $\qquad\qquad \square$

We now give several results on totally geodesic Riemannian maps obtained by Garcia-Rio and Kupeli. However, we first need the generalized divergence theorem and the Bochner identity.

**Definition 48.** [123] Let $(M, g_M)$ and $(N, g_N)$ be Riemannian manifolds and $F : M \longrightarrow N$ a map between them. Let also $Z$ be a vector field along $F$. Then the divergence of $Z$ is defined by

$$div Z = trace^* F_*(\overset{N}{\nabla})Z, \qquad (4.30)$$

where $^*F_*(\overset{N}{\nabla})Z$ is defined by

$$(^*F_*(\overset{N}{\nabla})Z)(X) =^* F_{*P_1(X)}\overset{N}{\nabla}_X Z,$$

with canonical projection $P_1 = TM \longrightarrow M$ and the pullback connection $\overset{N}{\nabla}$ along $F$. We also note that $^*F_{*P_1(X)}$ is the adjoint of

$$F_{*P_1(X)} : (T_{P_1(X)}M, g_{MP_1(X)}) \longrightarrow T_{F(P_1(X))}M, g_{MF(P_1(X))}).$$

If $\{X_1, ..., X_n\}$ is an orthonormal local frame for $TM$, then we can write

$$div Z = \sum_{i=1}^{n} g_M((^*F_*(\overset{N}{\nabla})Z)(X_i), X_i).$$

Thus we get

$$div Z = \sum_{i=1}^{n} g_M(^*F_{*P_1(X)}\overset{N}{\nabla}_{X_i} Z, X_i)) = \sum_{i=1}^{n} g_M(\overset{N}{\nabla}_{X_i} Z, F_* X_i)).$$

**Lemma 47.** [123] Let $(M^m, g_M)$ and $(N, g_N)$ be Riemannian manifolds and $(F : (M^m, g_M) \to (N^n, g_N)$ a map. If $\{e_1, ..., e_m\}$ is an orthonormal basis for $(T_pM, g_M)$ at $p \in M$ and $\overset{M}{\nabla}\|F_*\|^2$ is the gradient of $\|F_*\|^2$ on $M$, then

$$(\overset{M}{\nabla}\|F_*\|^2)(p_1) = 2 \sum_{i=1}^{m} (^*(\nabla_{e_i} F_*) \circ F_{*p_1})e_i,$$

where $^*(\nabla_{e_i} F_*)$ is the adjoint map of $\nabla_{e_i} F_*$.

*Proof.* Let $\{X_1, ..., X_m\}$ be the extension of $\{e_1, ..., e_m\}$, to an adapted moving frame near $p \in M$. Then, for $X \in \Gamma(TM)$

$$g_M((\overset{M}{\nabla}\|F_*\|^2)(p), X(p)) = X(p)\|F_*\|^2.$$

Thus, Theorem 8 implies that

$$g_M((\overset{M}{\nabla}\|F_*\|^2)(p), X(p)) = 2[\sum_{i=1}^{m} g_2(\overset{N}{\nabla}_X F_* X_i, F_* X_i)](p).$$

Then, from (1.71), and taking into account that the second fundamental form is symmetric, we derive

$$g_M((\overset{M}{\nabla}\|F_*\|^2)(p), X(p)) = 2[\sum_{i=1}^{m} g_2((\nabla F_*)(X_i, X), F_*X_i)](p).$$

Hence

$$g_M((\overset{M}{\nabla}\|F_*\|^2)(p), X(p)) = 2\sum_{i=1}^{m} g_2((\nabla_{e_i} F_*)X(p), F_{*p}e_i).$$

Considering the adjoint map of $(\nabla_{e_i} F_*)$, we find

$$g_M((\overset{M}{\nabla}\|F_*\|^2)(p), X(p)) = 2\sum_{i=1}^{m} g_M(X(p), (^*(\nabla_{e_i} F_*) \circ F_{*p})e_i),$$

which completes the proof.    □

We now give the divergence theorem for a vector field along a map.

**Theorem 84.** *[123] (Generalized divergence theorem) Let $(M, g_M)$ be an oriented Riemannian manifold with boundary $\partial M$ (possibly $\partial M = 0$) and Riemannian volume form $\mathcal{V}_{g_M}$, and let $(N, g_N)$ be a Riemannian manifold. Let $F : ((M, g_M) \longrightarrow (N, g_N)$ be a map and $Z$ be a vector field along $F$ with compact support. Then*

$$\int_M divZ\mathcal{V}_{g_M} + \int_M g_N(Z, \tau(F))\mathcal{V}_{g_M} = \int_{\partial M} g_N(Z, F_*(N))\mathcal{V}_{g_M \partial \tilde{M}} \tag{4.31}$$

*where $N$ is the unit outward normal vector field to $\partial M$ defined on $\partial \tilde{M}$.*

*Proof.* We define a vector field $^*F_*(Z)$ on $M$ by $^*F_*(Z)(p) =^* F_{*p}(Z_p)$ for $p \in M$. If $\{X_1, ..., X_m\}$ is an adapted moving frame near $p$, we have

$$div^*F_*(Z))(p) = \sum_{i=1}^{m} g_M(\overset{M}{\nabla}_{X_i} {}^*F_*(Z), X_i)(p).$$

Since $\overset{M}{\nabla}$ is a metric connection, we have $div^*F_*(Z)(p) = \sum_{i=1}^{m} X_i g_N(Z, F_*(X_i))(p)$. Then Riemannian manifold $(N, g_N)$ and (1.71) imply that

$$div^*F_*(Z)(p) = (\sum_{i=1}^{m} g_N(\overset{N}{\nabla}_{X_i} Z, F_*(X_i)) + g_N(Z, (\nabla F_*)(X_i, X_i))(p).$$

Then from (1.73) we have

$$div^*F_*(Z) = divZ + g_N(Z, \tau(F)). \tag{4.32}$$

Applying Theorem 2 to $^*F_*(Z)$, we have

$$\int_M div^* F_*(Z) \mathcal{V}_g = \int_{\partial M} g_M(N, ^* F_*(Z)) \mathcal{V}_{\tilde{g}}$$

$$= \int_{\partial M} g_M(F_*(N), Z) \mathcal{V}_{\tilde{g}}.$$

Using (4.32) here, we get

$$\int_M div Z \mathcal{V}_{g_M} + \int_M g_N(Z, \tau(F)) \mathcal{V}_{g_M} = \int_{\partial M} g_M(F_*(N), Z) \mathcal{V}_{\tilde{g}}.$$

$\square$

We now give the Bochner identity for a map $F$.

**Lemma 48. (The Bochner Identity)** *[122]) Let $F : (M, g_M) \longrightarrow (N, g_N)$ be a map between Riemannian manifolds $(M, g_M)$ and $(N, g_N)$. Then*

$$-\frac{1}{2}(\overset{M}{\triangle} \| F_* \|^2)(p) = \sum_{i,j=1}^{m} g_N((\nabla F_*)(x_i, x_j), (\nabla F_*)(x_i, x_j)) + div(\tau(F))(p)$$

$$- \sum_{i,j=1}^{m} \{g_N(\overset{N}{R}(F_*(x_i), F_*(x_j))F_*(x_i), F_*(x_j))$$

$$+ g_M((^*F_{*p} \circ F_{*p})x_i, \overset{M}{R}(x_i, x_j)x_j)\} \qquad (4.33)$$

*where $(\overset{M}{\triangle} \| F_* \|^2)$ is the Laplacian of $\| F_* \|^2$ on $M$, $\{x_1, ..., x_m\}$ is an orthonormal basis of $(T_p M, g_{Mp})$, and $\overset{M}{R}$ and $\overset{N}{R}$ are the curvature tensors of $(M, g_M)$ and $(N, g_N)$, respectively.*

*Proof.* Let $\{X_1, ..., X_m\}$ be an adapted moving frame near $p \in M$, i.e., $\{X_1, ..., X_m\}$ is an orthonormal basis frame with $(\overset{M}{\nabla} X_i)_p = 0$ for $i = 1...m$. Then at $p \in M$, using Lemma 47, we have

$$-\frac{1}{2}(\overset{M}{\triangle} \| F_* \|^2)(p) = \{div \sum_{i=1}^{m} (^*(\nabla_{X_i} F_*) \circ F_*)X_i\}(p)\}.$$

Thus metric connection $\overset{M}{\nabla}$ and (1.71) imply

$$-\frac{1}{2}(\overset{M}{\triangle}\parallel F_*\parallel^2)(p) \;=\; \sum_{i,j=1}^{m}\{g_N(\overset{N}{\nabla^F}_{X_j}F_*(X_i),(\nabla_{X_i}F_*)X_j)$$

$$+g_N(F_*(X_i),\overset{N}{\nabla^F}_{X_i}(\nabla_{X_i}F_*)X_j)\}(p).$$

Then from Theorem 11, (1.71), and Theorem 8, we have

$$-\frac{1}{2}(\overset{M}{\triangle}\parallel F_*\parallel^2)(p) = \sum_{i,j=1}^{m}\{g_N((\nabla F_*)(X_j,X_i),(\nabla F_*)(X_j,X_i)$$

$$+g_N(F_*(X_i),\overset{N}{R}(F_*(X_j),F_*(X_i))F_*(X_j)) + X_i g_N(F_*(X_i),\overset{N}{\nabla^F}_{X_j}F_*(X_j))$$

$$-g_N(\overset{N}{\nabla^F}_{X_i}F_*(X_i),\overset{N}{\nabla^F}_{X_j}F_*(X_j) - g_N(F_*(X_i),F_*(\nabla_{X_j}\nabla_{X_i}X_j))\}(p).$$

Using again (1.71) and (1.39), we get

$$-\frac{1}{2}(\overset{M}{\triangle}\parallel F_*\parallel^2)(p) = \sum_{i,j=1}^{m}\{g_N((\nabla F_*)(X_j,X_i),(\nabla F_*)(X_j,X_i))$$

$$-g_N(\overset{N}{R}(F_*(X_i),F_*(X_j))F_*(X_j)F_*(X_i)) + g_N((\nabla_{X_i}F_*)X_i,(\nabla_{X_j}F_*)X_j)$$

$$+g_N(F_*(X_i),F_*(\overset{M}{R}(X_i,X_j)X_i))\}(p).$$

Thus considering the adjoint map of $F_*$ at $p \in M$, we obtain the formula. $\square$

Let $(M,g_M)$ and $(N,g_N)$ be Riemannian manifolds of dimensions of $m$ and $n$, respectively. Let $Ric_M$ and $S_M$ denote the Ricci and scalar curvatures of $M$, respectively. Let also $K_N$ and $S_N$ denote the sectional and scalar curvatures of $N$, respectively. By $S_M \geq A$, we mean $S_M(V) \geq A$ for every $0 \neq V \in \Gamma(TM)$. Also by $K_N \leq B$, we mean $K_N(P) \leq B$ for every plane in $TN$.

We now give several results for totally geodesic Riemannian maps by using the Bochner formula and generalized divergence theorem.

**Theorem 85.** *[123] Let $(M,g_M)$ be a connected Riemannian manifold with Ricci curvature $Ric_M \geq A$ and let $(N,g_N)$ be a Riemannian manifold with sectional curvature $K_N \leq B$. If $F : M \longrightarrow N$ is a Riemannian map with $rankF \geq 2$, $A \geq (rankf - 1)B$ and $div\tau(F) \geq 0$, then $F$ is a totally geodesic map.*

*Proof.* We choose the orthonormal basis $\{x_1,...,x_m\}$ for $T_pM$ such that $\{x_1,...,x_r\}$ is an orthonormal basis for $kerF_{*p}$, where $dim(kerF_{*p}) = r$. Since $^*F_{*p} \circ F_{*p}$ is the identity map and $\| F_* \|^2 = rankF$, from Lemma 48 and (1.18) we have

$$0 = \sum_{i,j=1}^{m} g_N((\nabla F_*)(x_i, x_j), (\nabla F_*)(x_i, x_j))(p) + div(\tau(F))$$

$$- \sum_{i,j=r+1}^{m} \{g_N(\overset{N}{R}(F_*(x_i), F_*(x_j))F_*(x_i), F_*(x_j))\}$$

$$+ \sum_{i=r+1}^{m} Ric_M(x_i, \overset{M}{R}(x_i, x_j)x_j). \tag{4.34}$$

at $p \in M$. By the curvature assumptions in the statement, at each $p \in M$, we get

$$- \sum_{i,j=1}^{m} \{g_N(\overset{N}{R}(F_*(x_i), F_*(x_j))F_*(x_i), F_*(x_j)) \geq -(rankF)(rankF - 1)B$$

and

$$\sum_{i=r+1}^{m} Ric_M(x_i, \overset{M}{R}(x_i, x_j)x_j) \geq rankFA.$$

Thus from (4.34) we find

$$0 \geq \sum_{i,j=1}^{m} g_N((\nabla F_*)(x_i, x_j), (\nabla F_*)(x_i, x_j)) - (rankF)(rankF - 1) + rankFA$$

$$+ div(\tau(F))$$

$$\geq \sum_{i,j=1}^{m} g_N((\nabla F_*)(x_i, x_j), (\nabla F_*)(x_i, x_j)) + div(\tau(F)).$$

Since $div(\tau(F)) \geq$, it follows $\nabla F_* = 0$. □

In particular, if $rankF = 1$, then we have the following result.

**Theorem 86.** *[122] Let $(M, g_M)$ be a connected Riemannian manifold with Ricci curvature $Ric_M \geq 0$ and let $(N, g_N)$ be a Riemannian manifold. If $F : M \longrightarrow N$ is a Riemannian map with $rankF = 1$ and $div\tau(F) \geq 0$, then $F$ is a totally geodesic map, and hence $F$ maps $M$ into a geodesic of $N$.*

*Proof. RankF $= 1$ implies that $\overset{N}{R}(F_*(x_i), F_*(x_j))F_*(x_i), F_*(x_j)) = 0$.* Since $Ric_M \geq 0$ and $div\tau(F) \geq 0$, from (4.34) we find $\nabla F_* = 0$. □

Since isometric immersions and Riemannian submersions are also Riemannian maps, we have the following special cases.

**Corollary 30.** *[122] Let $(M, g_M)$ be a connected Riemannian manifold with $\dim(M) = m \geq 2$ and scalar curvature $S_M \geq A$ and let $(N, g_N)$ be a Riemannian manifold with sectional curvature $K_N \leq B$ such that $A \geq (m-1)B$. If $F : M \longrightarrow N$ is an isometric immersion with $\operatorname{div}\tau(F) \geq 0$, then $F$ is a totally geodesic map.*

*Proof.* Since $F$ is an isometric immersion, at $p \in M$ we have

$$\sum_{i=1}^{m} Ric_M(x_i, x_i) = S_M(p).$$

Hence the proof can be given as in the proof of Theorem 85 by observing the fact that $rankF = m$.  □

For Riemannian submersions, the above theorem gives the following result.

**Corollary 31.** *[122] Let $(M, g_M)$ be a connected Riemannian manifold with Ricci curvature $Ric_M \geq A$ and and let $(N, g_N)$ be a Riemannian with $\dim N = n$ and scalar curvature $r_N \leq B$ such that $nA \geq B$. If $F : M \longrightarrow N$ is a Riemannian submersion with $\operatorname{div}\tau(F) \geq 0$, then $F$ is a totally geodesic map.*

*Proof.* Since $F$ is a submersion, at $p \in M$ we get

$$\sum_{i,j=r+1}^{n} g_N(R(F_*(x_i), F_*(x_j))F_*(x_i), F_*(x_j)) = S_N(F(p)).$$

Hence the proof can be given as in the proof of Theorem 85 by observing the fact that $rankF = n$.  □

## 4. Umbilical Riemannian maps

In this section we study umbilical Riemannian maps. Umbilical maps have been defined in [201] and [273]. In fact, in [201] the author gave many definitions of umbilical maps with respect to the Riemannian metrics of the source manifolds and target manifolds. But we note that the definition given in [273] is the same as the definition of the $g$−umbilicity map given in [201]. We first recall some notions defined on distributions from [20]. For this aim, let $(M, g_M)$ be a Riemannian manifold and $\mathcal{V}$ be a $q$−dimensional distribution on $M$. Denote its orthogonal distribution $\mathcal{V}^\perp$ by $\mathcal{H}$. Then, we

have

$$TM = \mathcal{V} \oplus \mathcal{H}. \tag{4.35}$$

$\mathcal{V}$ is called the vertical distribution and $\mathcal{H}$ is called the horizontal distribution. We use the same letters to denote the orthogonal projections onto these distributions.

By the unsymmetrized second fundamental form of $\mathcal{V}$, we mean the tensor field $A^{\mathcal{V}}$ defined by

$$A_E^{\mathcal{V}} F = \mathcal{H}(\nabla_{\mathcal{V}E} \mathcal{V}F), \quad E, F \in \Gamma(TM), \tag{4.36}$$

where $\nabla$ is the Levi-Civita connection on $M$. The symmetrized second fundamental form $B^{\mathcal{V}}$ of $\mathcal{V}$ is given by

$$B^{\mathcal{V}}(E, F) = \frac{1}{2}\{A_E^{\mathcal{V}} F + A_F^{\mathcal{V}} E\} = \frac{1}{2}\{\mathcal{H}(\nabla_{\mathcal{V}E} \mathcal{V}F) + \mathcal{H}(\nabla_{\mathcal{V}F} \mathcal{V}E)\} \tag{4.37}$$

for any $E, F \in \Gamma(TM)$. The integrability tensor of $\mathcal{V}$ is the tensor field $I^{\mathcal{V}}$ given by

$$I^{\mathcal{V}}(E, F) = A_E^{\mathcal{V}} F - A_F^{\mathcal{V}} E - \mathcal{H}([\mathcal{V}E, \mathcal{V}F]). \tag{4.38}$$

Moreover, the mean curvature of $\mathcal{V}$ is defined by

$$\mu^{\mathcal{V}} = \frac{1}{q} Trace B^{\mathcal{V}} = \frac{1}{q} \sum_{i=1}^{q} \mathcal{H}(\nabla_{e_r} e_r), \tag{4.39}$$

where $\{e_1, ..., e_q\}$ is a local frame of $\mathcal{V}$. In a similar way, the notions of $B^{\mathcal{H}}$, $A^{\mathcal{H}}$ and $I^{\mathcal{H}}$ can be defined similarly. For instance, $B^{\mathcal{H}}$ is defined by

$$B^{\mathcal{H}}(E, F) = \frac{1}{2}\{\mathcal{V}(\nabla_{\mathcal{H}E} \mathcal{H}F) + \mathcal{V}(\nabla_{\mathcal{H}F} \mathcal{H}E)\} \tag{4.40}$$

and, hence we have

$$\mu^{\mathcal{H}} = \frac{1}{m-q} Trace B^{\mathcal{H}} = \frac{1}{m-q} \sum_{s=1}^{m-q} \mathcal{V}(\nabla_{E_s} E_s), \tag{4.41}$$

where $E_1, ..., E_{m-q}$ is a local frame of $\mathcal{H}$. A distribution $\mathcal{D}$ on $M$ is said to be minimal if, for each $x \in M$, the mean curvature vanishes.

We now calculate the tension field of a Riemannian map between Riemannian manifolds.

**Lemma 49.** *[240] Let $F : (M, g_M) \longrightarrow (N, g_N)$ be a Riemannian map between Riemannian manifolds. Then the tension field $\tau$ of $F$ is*

$$\tau = -m_1 F_*(\mu^{ker F_*}) + m_2 H_2, \tag{4.42}$$

*where $m_1 = dim((ker\,F_*))$, and $m_2 = rankF$, $\mu^{ker\,F_*}$ and $H_2$ are the mean curvature vector fields of the distribution $kerF_*$ and $rangeF_*$, respectively.*

*Proof.* Let $\{e_1, ..., e_{m_1}, e_{m_1+1}, ..., e_{m_2}\}$ be an orthonormal basis of $\Gamma(TM)$ such that $\{e_1, ..., e_{m_1}\}$ is an orthonormal basis of $\{ker\,F_*\}$, and $\{e_{m_1+1}, ..., e_{m_2}\}$ is an orthonormal basis of $(ker\,F_*)^{\perp}$. Then the trace of the second fundamental form (restriction to $\{kerF_*\}^{\perp} \times \{ker\,F_*\}^{\perp}$) is given by

$$trace^{\{ker\,F_*\}^{\perp}} \nabla F_* = \sum_{a=m_1+1}^{m_2} (\nabla F_*)(e_a, e_a) = \sum_{a=m_1+1}^{m_2} \sum_{j=1}^{n} g_2((\nabla F_*)(e_a, e_a), \check{Z}_j)\check{Z}_j$$

where $\{\check{Z}_j\}$ is an orthonormal basis of $\Gamma(TN)$. Considering the decomposition of $TN$, we have

$$trace^{\{ker\,F_*\}^{\perp}} \nabla F_* = \sum_{a=m_1+1}^{m_2} \{ \sum_{i=m_1+1}^{m_2} g_2((\nabla F_*)(e_a, e_a), \tilde{e}_i)\tilde{e}_i$$
$$+ \sum_{s=1}^{n_2} g_2((\nabla F_*)(e_a, e_a), \bar{e}_s)\bar{e}_s\},$$

where $\{\bar{e}_s\}$ is an orthonormal basis of $\Gamma(\{rangeF_*\}^{\perp})$. Then from Lemma 45, we get

$$trace^{\{ker\,F_*\}^{\perp}} \nabla F_* = \sum_{a=m_1+1}^{m_2} \sum_{s=1}^{n_2} g_2(\nabla_{e_a}^{F} F_*(e_a), \bar{e}_s)\bar{e}_s.$$

Thus from (4.39) and (4.41), we obtain

$$trace^{\{ker\,F_*\}^{\perp}} \nabla F_* = m_2 \sum_{s=1}^{n_2} g_2(H_2, \bar{e}_s)\bar{e}_s.$$

Hence, we derive

$$trace^{\{ker\,F_*\}^{\perp}} \nabla F_* = m_2 H_2. \tag{4.43}$$

In a similar way, we have

$$trace^{ker\,F_*} \nabla F_* = \sum_{r=n_1+1}^{m} (\nabla F_*)(e_r, e_r) = - \sum_{r=1}^{m_1} F_*(\nabla_{e_r} e_r).$$

Hence, we obtain

$$trace^{ker\,F_*} \nabla F_* = -m_1 F_*(\mu^{ker\,F_*}). \tag{4.44}$$

Then the proof follows from (4.43) and (4.44).    □

We now recall that a map $F : (M^m, g_M) \longrightarrow (N^n, g_N)$ between Riemannian manifolds is called umbilical ([273]) if

$$\nabla F_* = \frac{1}{m} g_M \otimes \tau(F). \tag{4.45}$$

We first show that this definition of umbilical maps has some restrictions for Riemannian maps.

**Theorem 87.** *[244] Every umbilical (in the sense of [273]) Riemannian map between Riemannian manifolds is harmonic. As a result of this, it is totally geodesic.*

*Proof.* Let $F : (M^m, g_M) \longrightarrow (N^n, g_N)$ be an umbilical Riemannian map between Riemannian manifolds $(M^m, g_M)$ and $(N^n, g_N)$. Then from Lemma 45 and (4.45) we have

$$(\nabla F_*)(X, Y) = \frac{1}{m} g_1(X, Y)(-m_1 F_*(\mu^{ker F_*}) + m_2 H_2) \tag{4.46}$$

for $X, Y \in \Gamma((ker F_*)^\perp)$, where $m_1 = dim((ker F_*))$, and $m_2 = rankF$, $\mu^{ker F_*}$ and $H_2$ are the mean curvature vector fields of the distribution $kerF_*$ and $rangeF_*$, respectively. But we know that $(\nabla F_*)(X, Y)$ has no components in $(rangeF_*)$, thus (4.46) implies that $F_*(\mu^{ker F_*}) = 0$, i.e., $\mu^{ker F_*} = 0$. On the other hand, we know that $(\nabla F_*)(U, V)$ has no components in $(rangeF_*)^\perp$ for $U, V \in \Gamma(ker F_*)$. Then

$$(\nabla F_*)(U, V) = \frac{1}{m} g_1(U, V)(m_2 H_2), \tag{4.47}$$

which shows that $H_2 = 0$. Hence it follows that $\tau(F) = 0$, which implies that $(\nabla F_*) = 0$. Thus the proof is complete. □

In fact, the above theorem tells us that the definition of umbilical map does not work for Riemannian maps. Therefore we present the following definition.

**Definition 49.** [244] Let $F$ be a Riemannian map between Riemannian manifolds $(M, g_M)$ and $(N, g_N)$. Then we say that $F$ is an *umbilical Riemannian map* at $p_1 \in M$, if

$$S_V F*p_1(X_{p_1}) = \lambda F_{*p_1}(X_{p_1}) \tag{4.48}$$

for $X \in \Gamma(rangeF_*)$ and $V \in \Gamma((rangeF_*)^\perp)$, where $\lambda$ is a differentiable function on $M$. If $F$ is umbilical for every $p_1 \in M$ then we say that $F$ is an umbilical Riemannian map.

The above definition is the same as that given for isometric immersions. The following lemma gives an useful formula for umbilical Riemannian maps.

**Lemma 50.** *[244] Let F be a Riemannian map between Riemannian manifolds* $(M, g)$ *and* $(N, g_N)$. *Then F is an umbilical Riemannian map if and only if*

$$(\nabla F_*)(X, Y) = g_M(X, Y)H_2 \qquad (4.49)$$

*for* $X, Y \in \Gamma((\ker F_*)^\perp)$ *and* $H_2$ *is, nowhere zero, vector field on* $(\text{range} F_*)^\perp$.

*Proof.* Let $\{e_1, ..., e_{m_1}, e_{m_1+1}, ..., e_{m_2}\}$ be an orthonormal basis of $\Gamma(TM)$ such that $\{e_1, ..., e_{m_1}\}$ is an orthonormal basis of $\{\ker F_*\}$ and $\{e_{m_1+1}, ..., e_{m_2}\}$ is an orthonormal basis of $(\ker F_*)^\perp$. Then Riemannian map $F$ implies that $\{F_*(e_{m_1+1}), ..., F_*(e_{m_2})\}$ is an orthonormal basis of $\text{range} F_*$. Using (4.48), we have

$$\sum_{i=m_1+1}^{m_2} g_N(S_V F_*(e_i), F_*(e_i)) = \lambda m_2.$$

Thus (4.17) and (4.41) imply that

$$\lambda = g_N(H_2, V). \qquad (4.50)$$

On the other hand, from (4.17) we get

$$g_N((\nabla F_*)(X, Y), V) = \lambda g_M(X, Y)$$

for $X, Y \in \Gamma((\ker F_*)^\perp)$. Then the proof comes from (4.50).     □

**Remark 13.** Lemma 50 shows that the notion of umbilicity given in Definition 49 coincides with the notion of strongly $g$–umbilicity given in [201] for an arbitrary map $F$.

The following proposition implies that it is easy to find examples of umbilical Riemannian maps if one has examples of totally umbilical isometric immersions and Riemannian submersions.

**Proposition 44.** *[244] Let* $F_1$ *be a Riemannian submersion from a Riemannian manifold* $(M, g_M)$ *onto a Riemannian manifold* $(N, g_N)$ *and* $F_2$ *a totally umbilical isometric immersion from the Riemannian manifold* $(N, g_N)$ *to a Riemannian manifold* $(\bar{N}, g_{\bar{N}})$. *Then the Riemannian map* $F_2 \circ F_1$ *is an umbilical Riemannian map from* $(M, g_M)$ *to* $(\bar{N}, g_{\bar{N}})$.

*Proof.* Let $F_1$ be a Riemannian submersion from a Riemannian manifold $(M, g_M)$ onto a Riemannian manifold $(N, g_N)$ and $F_2$ a totally umbilical isometric immersion from the Riemannian manifold $(N, g_N)$ to a Riemannian manifold $(\bar{N}, g_{\bar{N}})$. Then it is easy to see that $F_2 \circ F_1$ is a Riemannian map. On the other hand, from (1.72) we have

$$\nabla(F_2 \circ F_1)_*(X, Y) = F_{2*}(\nabla F_1 *(X, Y)) + (\nabla F_{2*})(F_{1*}X, F_{1*}Y)$$

for $X, Y \in \Gamma((ker F_{1*})^\perp)$. Then (3.26) implies that

$$\nabla(F_2 \circ F_1)_*(X, Y) = (\nabla F_{2*})(F_{1*}X, F_{1*}Y).$$

Since $(\nabla F_{2*})$ is the second fundamental form and since $F_2$ is a totally umbilical isometric immersion, we have

$$\nabla(F_2 \circ F_1)_*(X, Y) = g_N(F_{1*}X, F_{1*}Y)H,$$

where $H$ is the mean curvature vector field of $F_2$ which is also the tension field of $F_2$.. Then Riemannian submersion $F_1$ implies that

$$\nabla(F_2 \circ F_1)_*(X, Y) = g_M(X, Y)H,$$

which completes the proof. $\qquad\qquad\qquad\qquad\qquad\qquad\qquad\qquad\qquad$ □

We also introduce a new notion, which generalizes the Riemannian submersions with totally umbilical fibers. Let $F$ be a Riemannian map from a Riemannian manifold to a Riemannian manifold $(M_2, g_2)$. We say that a Riemannian map is a *Riemannian map with totally umbilical fibers* if

$$h_2(X, Y) = g_1(X, Y)H \qquad\qquad\qquad (4.51)$$

for $X, Y \in \Gamma(ker F_*)$, where $h_2$ and $H$ are the second fundamental form and the mean curvature vector field of the distribution $ker F_*$, respectively.

We now define pseudo-umbilical Riemannian maps as a generalization of pseudo-umbilical isometric immersions. Pseudo-umbilical Riemannian maps will be useful when we deal with the biharmonicity of Riemannian maps.

**Definition 50.** [243] Let $F : (M, g_M) \longrightarrow (N, g_N)$ be a Riemannian map between Riemannian manifolds $M$ and $N$. Then we say that $F$ is a *pseudo-umbilical Riemannian map* if

$$S_{H_2}F_*(X) = \lambda F_*(X) \qquad\qquad\qquad (4.52)$$

for $\lambda \in C^\infty(M)$ and $X \in \Gamma((ker F_*)^\perp)$.

Here we present an useful formula for pseudo-umbilical Riemannian maps by using (4.17) and (4.52).

**Proposition 45.** *[243] Let $F : (M, g_M) \longrightarrow (N, g_N)$ be a Riemannian map between Rie-*

*mannian manifolds M and N. Then F is pseudo-umbilical if and only if*

$$g_N((\nabla F_*)(X, Y), H_2) = g_M(X, Y)g_N(H_2, H_2) \tag{4.53}$$

*for* $X, Y \in \Gamma((\ker F_*)^\perp)$.

The following theorem gives a method to find examples of pseudo-umbilical Riemannian maps. It also tells us that if one has an example of pseudo-umbilical submanifolds, it is possible to find an example of pseudo-umbilical Riemannian maps. For examples of pseudo umbilical submanifolds, see [77].

**Theorem 88.** *[243] Let* $F_1 : (M, g_M) \longrightarrow (N, g_N)$ *be a Riemannian submersion and* $F_2 : (N, g_N) \longrightarrow (\bar{N}, g_{\bar{N}})$ *a pseudo-umbilical isometric immersion. Then the map* $F_2 \circ F_1$ *is a pseudo-umbilical Riemannian map.*

## 5. Harmonicity of Riemannian maps

In this section, we give characterizations for Riemannian maps to be harmonic and biharmonic. The notion of biharmonic map was suggested by Eells and Sampson [103]; see also [20]. The first variation formula and, thus, the Euler-Lagrange equation associated to the bienergy was obtained by Jiang in [150] and [151], for translation by H. Urawaka, see [152]. However, biharmonic maps have been extensively studied in the last decade and there are two main research directions. In differential geometry, many authors have obtained classification results and constructed many examples. Biharmonicity of immersions was obtained in [49], [82], and [207] and biharmonic Riemannian submersions were studied in [207]; for a survey on biharmonic maps, see [189] and [208]. From an analytic point of view, biharmonic maps are solutions of fourth-order strongly elliptic semilinear partial differential equations. First from Lemma 49, we have the following.

**Theorem 89.** *[260] Suppose* $F : (M^m, g_M) \longrightarrow (N^n, g_N)$ *is a non-constant Riemannian map between Riemannian manifolds. Then any two conditions below imply the third:*
  **(i)** *F is harmonic,*
 **(ii)** *the distribution* $\ker F_*$ *is minimal, and*
**(iii)** *the distribution* $\mathrm{range} F_*$ *is minimal.*

In fact, by using (1.72), we have the following precisely geometric result.

**Theorem 90.** *[119] A Riemannian map* $F : M \to N$ *is harmonic if and only if* $F = i \circ \pi$ *locally, where i is an injective minimal immersion and* $\pi$ *is a harmonic submersion.*

For the biharmonicity of Riemannian maps, we have the following result.

**Theorem 91.** *[243] Let F be a Riemannian map from a Riemannian manifold $(M, g_M)$ to a space form $(N(c), g_N)$. Then F is biharmonic if and only if*

$$m_1 traceS_{(\nabla F_*)(., \mu^{ker F_*})}F_*(.) - m_1 traceF_*(\nabla_{(.)}\nabla_{(.)}\mu^{ker F_*})$$
$$-m_2 traceF_*(\nabla_{(.)}{}^*F_*(S_{H_2}F_*(.))) - m_2 traceS_{\nabla^{F\perp}_{(.)}H_2}F_*(.)$$
$$-m_1 c(m_2 - 1)F_*(\mu^{ker F_*}) = 0 \tag{4.54}$$

*and*

$$m_1 trace\nabla^{F\perp}_{(.)}(\nabla F_*)(., \mu^{kerF_*}) + m_1 trace(\nabla F_*)(., \nabla_{(.)}\mu^{ker F_*})$$
$$+m_2 trace(\nabla F_*)(., {}^*F_*(S_{H_2}F_*(.))) - m_2\Delta^{R^\perp}H_2$$
$$-m_2^2 cH_2 = 0, \tag{4.55}$$

*where $dim(ker F_*) = m_1$ and $dim((ker F_*)^\perp) = m_2$.*

*Proof.* First of all, from (1.14) and (4.16) we have

$$traceR^2(F_*(.), \tau(F))F_*(.) = m_1 c(m_2 - 1)F_*(\mu^{kerF_*}) - m_2^2 cH_2, \tag{4.56}$$

where $R^2$ is the curvature tensor field of $M_2$. Let $\{\tilde{e}_1, ..., \tilde{e}_{m_1}, e_1, ..., e_{m_2}\}$ be a local orthonormal frame on $M_1$, geodesic at $p \in M_1$ such that $\{\tilde{e}_1, ..., \tilde{e}_{m_1}\}$ is an orthonormal basis of $ker F_*$ and $\{e_1, ..., e_{m_2}\}$ is an orthonormal basis of $(ker F_*)^\perp$. At $p$ we have

$$\Delta\tau(F) = -\sum_{i=1}^{m_2} \nabla^F_{e_i}\nabla^F_{e_i}\tau(F)$$

$$= -\sum_{i=1}^{m_2} \nabla^F_{e_i}\{\nabla^F_{e_i}(-m_1 F_*(\mu^{kerF_*}) + m_2 H_2)\}.$$

Then, using (1.71), (4.4), and (4.16), we get

$$\Delta\tau(F) = -\sum_{i=1}^{m_2} \nabla^F_{e_i}\{-m_1(\nabla F_*)(e_i, \mu^{ker F_*}) - m_1 F_*(\nabla_{e_i}\mu^{ker F_*})$$
$$+ m_2(-S_{H_2}F_*(e_i) + \nabla^{F\perp}_{F_*(e_i)}H_2)\}.$$

Using again (1.71), (4.4), and (4.16), we obtain

$$
\begin{aligned}
\Delta\tau(F) &= m_1 \sum_{i=1}^{m_2} -\mathcal{S}_{(\nabla F_*)(e_i,\mu^{ker F_*})}F_*(e_i) + \nabla^{F\perp}_{F_*(e_i)}(\nabla F_*)(e_i,\mu^{ker F_*}) \\
&\quad + m_1 \sum_{i=1}^{m_2}(\nabla F_*)(e_i,\nabla_{e_i}\mu^{kerF_*}) + F_*(\nabla_{e_i}\nabla_{e_i}\mu^{ker F_*}) \\
&\quad + m_2 \sum_{i=1}^{m_2}\nabla^F_{e_i}\mathcal{S}_{H_2}F_*(e_i) - m_2 \sum_{i=1}^{m_2}-\mathcal{S}_{\nabla^{F\perp}_{F_*(e_i)}H_2}F_*(e_i) \\
&\quad + \nabla^{F\perp}_{F_*(e_i)}\nabla^{F\perp}_{F_*(e_i)}H_2.
\end{aligned}
$$

On the other hand, since $\mathcal{S}_{H_2}F_*(e_i) \in \Gamma(F_*((kerF_*)^\perp))$, we can write

$$
F_*(X) = \mathcal{S}_{H_2}F_*(e_i)
$$

for $X \in \Gamma((ker F_*)^\perp)$, where

$$
X = (F_*)^{-1}(\mathcal{S}_{H_2}F_*(e_i)) = {}^*F_*(\mathcal{S}_{H_2}F_*(e_i)).
$$

Then, using (1.71), we have

$$
\nabla^F_{e_i}\mathcal{S}_{H_2}F_*(e_i) = (\nabla F_*)(e_i,{}^*F_*(\mathcal{S}_{H_2}F_*(e_i))) + F_*(\nabla_{e_i}{}^*F_*(\mathcal{S}_{H_2}F_*(e_i))).
$$

Thus we obtain

$$
\begin{aligned}
\Delta\tau(F) &= m_1 \sum_{i=1}^{m_2} -\mathcal{S}_{(\nabla F_*)(e_i,\mu^{ker F_*})}F_*(e_i) + \nabla^{F\perp}_{F_*(e_i)}(\nabla F_*)(e_i,\mu^{ker F_*}) \\
&\quad + m_1 \sum_{i=1}^{m_2}(\nabla F_*)(e_i,\nabla_{e_i}\mu^{kerF_*}) + F_*(\nabla_{e_i}\nabla_{e_i}\mu^{ker F_*}) \\
&\quad + m_2 \sum_{i=1}^{m_2}(\nabla F_*)(e_i,{}^*F_*(\mathcal{S}_{H_2}F_*(e_i))) + F_*(\nabla_{e_i}{}^*F_*(\mathcal{S}_{H_2}F_*(e_i))) \\
&\quad - m_2 \sum_{i=1}^{m_2}-\mathcal{S}_{\nabla^{F\perp}_{F_*(e_i)}H_2}F_*(e_i) + \nabla^{F\perp}_{F_*(e_i)}\nabla^{F\perp}_{F_*(e_i)}H_2. \qquad (4.57)
\end{aligned}
$$

Thus putting (4.56) and (4.57) in (1.78) and then taking the $F_*((ker F_*)^\perp) = range F_*$ and $(range F_*)^\perp$ parts, we have (4.54) and (4.55). $\qquad\square$

In particular, we have the following.

**Corollary 32.** [243] Let F be a Riemannian map from a Riemannian manifold $(M_1, g_1)$ to a space form $(M_2(c), g_2)$. If the mean curvature vector fields of $range F_*$ and $ker F_*$

*are parallel, then F is biharmonic if and only if*

$$m_1 trace S_{(\nabla F_*)(.,\mu^{ker F_*})} F_*(.) - m_2 trace F_*(\nabla_{(.)}{}^* F_*(S_{H_2} F_*(.)))$$
$$-m_1 c(m_2 - 1) F_*(\mu^{ker F_*}) = 0$$

*and*

$$m_1 trace \nabla^{F\perp}_{F_*(.)}(\nabla F_*)(., \mu^{ker F_*}) + m_2 trace(\nabla F_*)(., {}^* F_*(S_{H_2} F_*(.)))$$
$$-m_2^2 c H_2 = 0.$$

We also have the following result for pseudo-umbilical Riemannian maps.

**Theorem 92.** *[243] Let F be a pseudo-umbilical biharmonic Riemannian map from a Riemannian manifold $(M_1, g_1)$ to a space form $(M_2(c), g_2)$ such that the distribution ker $F_*$ is minimal and the mean curvature vector field $H_2$ is parallel. Then either F is harmonic or $c = \| H_2 \|^2$.*

*Proof.* First note that it is easy to see that $\| H_2 \|^2$ is constant. If $F$ is a biharmonic Riemannian map such that $\mu^{ker F_*} = 0$ and $H_2$ is parallel, then from (4.55) we have

$$m_2 \sum_{i=1}^{m_2} (\nabla F_*)(e_i, {}^* F_*(S_{H_2} F_*(e_i))) - m_2^2 c H_2 = 0.$$

Since $F$ is pseudo-umbilical, we get

$$m_2 \sum_{i=1}^{m_2} (\nabla F_*)(e_i, {}^* F_*(\| H_2 \|^2 F_*(e_i))) - m_2^2 c H_2 = 0.$$

On the other hand, from the linear map ${}^* F_*$ and ${}^* F_* \circ F_* = I$ (identity map), we obtain

$$m_2 \sum_{i=1}^{m_2} (\nabla F_*)(e_i, \| H_2 \|^2 e_i)) - m_2^2 c H_2 = 0.$$

Since the second fundamental form is also linear in its arguments, it follows that

$$m_2 \| H_2 \|^2 \sum_{i=1}^{m_2} (\nabla F_*)(e_i, e_i)) - m_2^2 c H_2 = 0.$$

Hence, we have

$$m_2^2 \| H_2 \|^2 H_2 - m_2^2 c H_2 = 0$$

which implies that

$$(\| H_2 \|^2 - c) H_2 = 0. \tag{4.58}$$

Thus either $H_2 = 0$ or $(\| H_2 \|^2 - c) = 0$. If $H_2 = 0$, then $F$ is harmonic, thus the proof is complete.    $\square$

From (4.58), we have the following result which puts some restrictions on $M_2(c)$.

**Corollary 33.** *[243] There exist no proper biharmonic pseudo-umbilical Riemannian maps from a Riemannian manifold to space forms ($M_2(c)$ with $c \leq 0$ such that the distribution ker $F_*$ is minimal and the mean curvature vector field $H_2$ is parallel.*

We note that harmonicity and biharmonicity of Riemannian maps in certain manifolds equipped with various differential structures may have interesting properties; see [148], [213] and [237]. Therefore it seems that it would be a new research area to investigate the necessary and sufficient conditions for Riemannian maps defined on a manifold equipped with certain differentiable structures.

## 6. Clairaut Riemannian maps

In this section, we introduce Clairaut Riemannian maps as a generalization of Clairaut Riemannian submersions and Clairaut surfaces. For this aim, we first find necessary and sufficient conditions for a curve to be geodesic, then obtain a characterization. We also provide an example of such Riemannian maps.

**Definition 51.** [253], [258] A Riemannian map $F : M \to N$ between Riemannian manifolds $(M, g_M)$ and $(N, g_N)$ is called a *Clairaut Riemannian map* if there is a function $r : M \to \mathbb{R}^+$ such that for every geodesic, making angles $\theta$ with the horizontal subspaces, $r \sin \theta$ is constant.

As we have seen above, the definition involves the notion of geodesic. Therefore we are going to find necessary and sufficient conditions for a curve on $M$ to be geodesic. First suppose that $p \in M$ and $\alpha(s), \mid s \mid < \varepsilon$, is a horizontal curve on $M$. Then $F \circ \alpha : I \longrightarrow N$ is also a curve and for each given vector field $X_s$ along $\alpha$, we can define a vector field $F_*X$ along $F \circ \alpha$ by

$$(F_*X)(s) = F_{*\alpha(s)}X(s). \tag{4.59}$$

Here $s$ is the arc length parameter and the vector field $X_s$ is always the unit tangent vector field along $\alpha$. From (4.59) and (1.47)-(1.50), we obtain the following conditions.

**Corollary 34.** *[253], [258] Let $F : M \to N$ be a Riemannian map. If $c : I \to M$ is regular curve and $U$ and $X$ denote the vertical and the horizontal components of its* *tangent vector field, then c is a geodesic on M if and only if*

$$\hat{\nabla}_U U + \mathcal{T}_U X + \mathcal{V}\nabla_X U = 0$$

*and*

$$\nabla_X^F F_*(X) = -F_*(\mathcal{T}_U U + 2\mathcal{A}_X U) + (\nabla F_*)(X, X) = 0$$

From Corollary 34, we have the following result.

**Corollary 35.** *[253], [258] Let F : M → N be a Riemannian map and c : I → M a geodesic with U(t) = $\mathcal{V}\dot{c}(t)$ and X(t) = $\mathcal{H}\dot{c}(t)$. Then the curve β = F ∘ c is a geodesic on N if and only if*

$$\mathcal{T}_U X + 2\mathcal{A}_X U = 0, (\nabla F_*)(X, X) = 0.$$

*Proof.* Since $(\nabla F_*)(X, X) \in \Gamma((rangeF_{*p})^\perp)$ and $F_*(\mathcal{T}_U U + 2\mathcal{A}_X U) \in \Gamma(rangeF_{*p})$, the assertion follows from Corollary 34.     □

We also have the following result.

**Corollary 36.** *[253], [258] The projection on N of a horizontal geodesic on M is a geodesic if and only if*

$$(\nabla F_*)(X, X) = 0.$$

We note that the assertion of Corollary 36 is valid for a Riemannian submersion without any condition.

Moreover we have the following result.

**Theorem 93.** *[253], [258] Let F : (M, g_M) → (N, g_N) be a Riemannian map with connected fibers. Then F is a Clairaut Riemannian map with r = $e^f$ if and only if each fiber is totally umbilical and has mean curvature vector field H = −grad f.*

*Proof.* Let c : I → M be a geodesic on M with U(t) = $\mathcal{V}\dot{c}(t)$ and X(t) = $\mathcal{H}\dot{c}(t)$ and let ω(t) denote the angle in [0, π] between $\dot{c}(t)$ and X(t). Assuming a = $\| \dot{c}(t) \|^2$, we can obtain

$$g_{c(t)}(X(t), X(t)) = a \cos^2 \omega(t), \ g_{c(t)}(U(t), U(t)) = a \sin^2 \omega(t) \tag{4.60}$$

Thus, by considering the first relation of (4.60) and taking the derivative of it with respect to $t$, we get

$$\frac{d}{dt}g_{c(t)}(X(t), X(t)) = -2a \cos \omega(t) \sin \omega(t)\frac{d\omega(t)}{dt}. \qquad (4.61)$$

On the other hand, since $F$ is a Riemannian map, using (1.71) we have, along $c(t)$,

$$\frac{d}{dt}g_{c(t)}(X, X) = 2g_N(-(\nabla F_*)(\dot{c}, X) + \nabla_{\dot{c}}^F F_*(X), F_*(X)).$$

Since the second fundamental form of $F$ is linear, from (4.4) we derive

$$\frac{d}{dt}g_{c(t)}(X, X) = 2g_N(-(\nabla F_*)(U, X) + \nabla_{\dot{c}}^F F_*(X), F_*(X)).$$

Then, (1.71) and Riemannian map $F$ imply

$$\frac{d}{dt}g_{c(t)}(X, X) = 2g_M(\nabla_X U, X) + 2g_N(\nabla_{\dot{c}}^F F_*(X), F_*(X)).$$

Using (1.49), we obtain

$$\frac{d}{dt}g_{c(t)}(X, X) = 2g_M(\mathcal{A}_X U, X) + 2g_N(\nabla_{\dot{c}}^F F_*(X), F_*(X)).$$

Thus skew-symmetric $\mathcal{A}$ implies that

$$\frac{d}{dt}g_{c(t)}(X, X) = -2g_M(U, \mathcal{A}_X X) + 2g_N(\nabla_{\dot{c}}^F F_*(X), F_*(X)).$$

Hence we obtain

$$\frac{d}{dt}g_{c(t)}(X, X) = 2g_N(\nabla_{\dot{c}}^F F_*(X), F_*(X)). \qquad (4.62)$$

Then, from (4.61) and (4.62), we have

$$g_N(\nabla_{\dot{c}}^F F_*(X), F_*(X)) = -a \cos \omega(t) \sin \omega(t)\frac{d\omega(t)}{dt}. \qquad (4.63)$$

By direct computations, $F$ is a Clairaut Riemannian map with $r = e^f$ if and only if $\frac{d}{dt}(e^{f\circ c} \sin \omega) = 0$. Multiplying this with the nonzero factor $a \sin \omega(t)$, we get

$$-a \sin \omega \cos \omega\frac{d\omega}{dt} = \frac{df}{dt}a \sin^2 \omega. \qquad (4.64)$$

Thus from (4.60), (4.63) and (4.64), we find

$$g_N(\nabla_{\dot{c}}^F F_*(X), F_*(X)) = \frac{df}{dt}g_M(U, U). \qquad (4.65)$$

Since $c(t)$ is geodesic on $M$, from the second equation of Corollary 34 we have

$$g_N(-F_*(\mathcal{T}_U U + 2\mathcal{A}_X U) + (\nabla F_*)(X, X), F_*(X)) = \frac{df}{dt}g_M(U, U).$$

Then Riemannian map $F$ and (4.4) imply

$$-g_M(\mathcal{T}_U X + 2\mathcal{A}_X U, X) = \frac{df}{dt} g_M(U, U).$$

Hence we obtain

$$g_M(\mathcal{T}_U U, X) = \frac{df}{dt} g_M(U, U).$$

The rest of this proof is the same as the calculations given in [111], page 30.    □

We note that the above condition does not imply that the Riemannian map itself is totally umbilical contrary to the Riemannian submersions. We now give an example of Clairaut Riemannian maps.

**Example 28.** [253], [258] Let $(B, g_B)$ and $(F, g_F)$ be two Riemannian manifolds, $f : B \to (0, \infty)$ and $\pi_1 : B \times F \to B$, $\pi_2 : B \times F \to F$ the projection maps given by $\pi_1(p_1, p_2) = p_1$ and $\pi_2(p_1, p_2) = p_2$ for every $(p_1, p_2) \in B \times F$. The warped product (see Definition 10) $M = B \times_f F$ is the manifold $B \times F$ equipped with the Riemannian structure such that

$$g(X, Y) = g_B(\pi_{1*}X, \pi_{1*}Y) + (f \circ \pi_1)^2 g_F(\pi_{2*}X, \pi_{2*}Y)$$

for every $X$ and $Y$ of $M$, where $*$ denotes the tangent map. The function $f$ is called the warping function of the warped product manifold. In particular, if the warping function is constant, then the warped product manifold $M$ is said to be trivial. It is known that the first projection $\pi_1 : B \times F \to B$ is a Riemannian submersion whose vertical and horizontal spaces at any point $p = (p_1, p_2)$ are respectively identified with $T_{p_2}F$, $T_{p_1}B$. Moreover, the fibers of $\pi_1$ are totally umbilical with mean curvature vector field $H = -\frac{1}{2f} grad f$. We now consider the isometric immersion $\pi : B \longrightarrow B \times_f F$; then the composite map $\pi \circ \pi_1$ is a Riemannian map. Moreover, the projection $\pi_1$ and the map $\pi \circ \pi_1$ have the same vertical distribution. Hence $\pi \circ \pi_1$ is a Clairaut Riemannian map with $r = \sqrt{f}$.

We also have another characterization.

**Corollary 37.** *[253], [258] Let $F : (M, g_M) \to (N, g_N)$ be a Riemannian map with connected fibers. Then $F$ is a Clairaut Riemannian map with $r = e^f$ if and only if*

$$g_N(\nabla^F_{\dot{c}(t)} F_*(X), F_*(X)) = \frac{df}{dt} g_M(U, U).$$

## 7. Circles along Riemannian maps

In this section, we are going to extend Nomizu-Yano's result on circles along an immersion to circles along a Riemannian map. We first recall that a smooth curve $\alpha$ on a Riemannian manifold $M$ parametrized by its arc-length is called a *circle* if it satisfies

$$\nabla_{\dot{\alpha}} \nabla_{\dot{\alpha}} \dot{\alpha} = -\kappa^2 \dot{\alpha} \qquad (4.66)$$

with some nonnegative constant $\kappa$, where $\nabla_{\dot{\alpha}}$ denotes the covariant differentiation along $\alpha$ with respect to the Riemannian connection $\nabla$ on $M$ [200]. This condition is equivalent to the condition that there exist a nonnegative constant $\kappa$ and a field of unit vectors $Y$ along this curve which satisfies the following differential equations:

$$\nabla_{\dot{\alpha}} \dot{\alpha} = \kappa Y \qquad (4.67)$$

$$\nabla_{\dot{\alpha}} Y = -\kappa \dot{\alpha}. \qquad (4.68)$$

Here, $\kappa$ is called curvature of $\alpha$. For a given point $p \in M$, an orthonormal pair of tangent vectors $u, v \in T_p M$ and a positive constant $\kappa$, by the existence and uniqueness theorem on solutions for ordinary differential equations we have locally a unique circle $\alpha = \alpha(s)$ with initial condition that $\alpha(0) = p$; $\dot{\alpha}(0) = u$ and $\nabla_{\dot{\alpha}} \dot{\alpha} = \kappa v$. In [200], Nomizu-Yano showed that $\alpha$ is a circle if and only if the following is satisfied

$$\nabla_{\dot{\alpha}}^2 \dot{\alpha} + g(\nabla_{\dot{\alpha}} \dot{\alpha}, \nabla_{\dot{\alpha}} \dot{\alpha}) \dot{\alpha} = 0, \qquad (4.69)$$

where $g$ is the metric and $\nabla_{\dot{\alpha}}^2 \dot{\alpha}$ is $\nabla_{\dot{\alpha}} \nabla_{\dot{\alpha}} \dot{\alpha}$. We also recall that a submanifold $M^n$ of a Riemannian manifold $\bar{M}^m$ is called an *extrinsic sphere* if it is umbilical and has parallel mean curvature vector. Nomizu-Yano also proved that if every circle of radius $\kappa$ in $M^n$ is a circle in $\bar{M}^m$ for some $\kappa > 0$, then $M^n$ is a sphere. Conversely, if $M$ is a sphere in $\bar{M}$, then every circle in $M$ is also a circle or a geodesic in $\bar{M}$. It is shown that it is possible to obtain certain properties of a submanifold by observing the extrinsic shape of circles on this submanifold, see [180] and [158], and the references in there.

Let $F : (M, g_M) \longrightarrow (N, g_N)$ be a Riemannian map and $\alpha : I \longrightarrow M$ a curve, then we say that $\alpha$ is a *horizontal curve* if $\dot{\alpha}(t) \in (Ker F_{*\alpha(t)})^{\perp}$ for any $t \in I$. In this section, we are going to prove the following theorem.

**Theorem 94.** *[258] Let $F$ be a Riemannian map from a connected Riemannian manifold $(M, g_M)$, $dim M \geq 2$, to a Riemannian manifold $(N, g_N)$. For some $\kappa > 0$, let $\alpha$ be a horizontal circle of radius $\kappa$ on $M$, if $F \circ \alpha$ is a circle on $N$ then $F$ is umbilical and the mean curvature vector field $H_2$ is parallel. Conversely, if $F$ is umbilical and the mean curvature vector field $H_2$ is parallel, then for every horizontal circle $\alpha$ on $M$, $F \circ \alpha$ is a circle on $N$.*

*Proof.* Suppose that $p \in M$ and $\alpha(s), \mid s \mid < \varepsilon$, is a horizontal curve on $M$. Then $F \circ \alpha : I \longrightarrow N$ is also a curve and for each given vector field $X_s$ along $\alpha$, we can define a vector field $F_* X$ along $F \circ \alpha$ by

$$(F_* X)(s) = F_{*\alpha(s)} X(s).$$

Here, $s$ is the arc length parameter and the vector field $X_s$ is always the unit tangent vector field along $\alpha$. Now suppose that $F \circ \alpha$ is a circle on $N$. Then, from (4.69), we have

$$(\nabla_{X_s}^F)^2 F_*(X_s) + g_N(\nabla_{X_s}^F F_*(X_s), \nabla_{X_s}^F F_*(X_s)) F_*(X_s) = 0. \tag{4.70}$$

On the other hand, using (1.71), (4.4), and (4.16), we derive

$$(\nabla_{X_s}^F)^2 F_*(X_s) = -S_{(\nabla F_*)(X_s, X_s)} F_*(X_s) + \nabla_{X_s}^\perp (\nabla F_*)(X_s, X_s)$$
$$+ (\nabla F_*)(X_s, \nabla_{X_s}^1 X_s) + F_*(\nabla^{1^2}_{X_s} X_s). \tag{4.71}$$

Substituting (1.71) and (4.71) in (4.70) and using (4.67) and (4.68), we obtain

$$(\nabla F_*)(X_s, \nabla_{X_s}^1 X_s) \quad + \quad g_N((\nabla F_*)(X_s, X_s), (\nabla F_*)(X_s, X_s)) F_*(X_s)$$
$$- S_{(\nabla F_*)(X_s, X_s)} F_*(X_s) \quad + \quad \nabla_{X_s}^\perp (\nabla F_*)(X_s, X_s) = 0. \tag{4.72}$$

due to $\alpha$ being a circle. By looking at the tangential and normal components, we have

$$(\nabla F_*)(X_s, \nabla_{X_s}^1 X_s) + \nabla_{X_s}^\perp (\nabla F_*)(X_s, X_s) = 0 \tag{4.73}$$

and

$$- S_{(\nabla F_*)(X_s, X_s)} F_*(X_s) + g_N((\nabla F_*)(X_s, X_s), (\nabla F_*)(X_s, X_s)) F_*(X_s) = 0. \tag{4.74}$$

We now define

$$(\tilde{\nabla}_X (\nabla F_*))(Y, Z) = \nabla_X^\perp (\nabla F_*)(Y, Z) - (\nabla F_*)(\nabla_X^1 Y, Z) - (\nabla F_*)(Y, \nabla_X^1 Z)$$

for $X, Y, Z \in \Gamma(TM)$. Then we can write

$$(\tilde{\nabla}_{X_s} (\nabla F_*))(X_s, X_s) = \nabla_{X_s}^\perp (\nabla F_*)(X_s, X_s) - 2(\nabla F_*)(\nabla_{X_s}^1 X_s, X_s). \tag{4.75}$$

Using (4.75) in (4.73), we arrive at

$$3(\nabla F_*)(\nabla_{X_s}^1 X_s, X_s) = -(\tilde{\nabla}_{X_s} (\nabla F_*))(X_s, X_s). \tag{4.76}$$

Thus at $s = 0$, we get

$$(\nabla F_*)(X, Y) = \frac{-1}{3\kappa} (\tilde{\nabla}_{X_s} (\nabla F_*))(X_s, X_s), \forall X, Y \in \Gamma((\ker F_*)^\perp). \tag{4.77}$$

This equation shows that given a unit vector $X \in T_p M$, $(\nabla F_*)(X, Y)$ is independent of a unit vector $Y \in T_p M$, provided that $Y$ is orthogonal to $X$. Changing $Y$ into $-Y$, we

have

$$(\nabla F_*)(X, Y) = 0. \tag{4.78}$$

On the other hand, $\frac{1}{\sqrt{2}}(X + Y)$ and $\frac{1}{\sqrt{2}}(X - Y)$ are orthogonal, hence we derive

$$(\nabla F_*)(\frac{1}{\sqrt{2}}(X + Y), \frac{1}{\sqrt{2}}(X - Y)) = 0. \tag{4.79}$$

Since $\nabla F_*$ is linear, we obtain

$$(\nabla F_*)(X, X) = (\nabla F_*)(Y, Y). \tag{4.80}$$

Now let $\{X_1, ..., X_n\}$ be an orthonormal basis in $\Gamma((ker\, F_*)^\perp)$, then we have

$$(\nabla F_*)(X_1, X_1) = (\nabla F_*)(X_2, X_2) = ...$$

Thus we have

$$H_2 = (\nabla F_*)(X_1, X_1).$$

Moreover, choosing $X = \sum a_i X_i$, $Y = \sum b_j X_j$, we get

$$(\nabla F_*)(X, Y) = \sum_{i,j} a_i b_j (\nabla F_*)(X_i, X_j) = g_M(X, Y) H_2,$$

which shows that $F$ is umbilical. On the other hand, from (4.73) we have

$$\nabla^\perp_{X_s} (\nabla F_*)(X_s, X_s) = 0$$

due to $(\nabla F_*)(X_s, \nabla^1_{X_s} X_s) = 0$. This implies that $H_2$ is parallel. Conversely suppose that $F$ is an umbilical Riemannian map and $H_2$ is parallel. From (1.71), we have

$$g_N(\nabla^F_{X_s} F_*(X_s), \nabla^F_{X_s} F_*(X_s)) = g_N((\nabla F_*)(X_s, X_s), (\nabla F_*)(X_s, X_s))$$
$$+ g_N(F_*(\nabla^1_{X_s} X_s), F_*(\nabla^1_{X_s} X_s)).$$

Then Riemannian map $F$ and (4.49) imply that

$$g_N(\nabla^F_{X_s} F_*(X_s), \nabla^F_{X_s} F_*(X_s)) = \| H_2 \|^2 + g_M(\nabla^1_{X_s} X_s, \nabla^1_{X_s} X_s). \tag{4.81}$$

On the other hand, since $F$ is umbilical Riemannian map and $H_2$ is parallel, we have

$$\nabla^\perp_{X_s}(\nabla F_*)(X_s, X_s) = 0 \quad \text{and} \quad (\nabla F_*)(X_s, \nabla^1_{X_s} X_s) = 0.$$

Then, using (4.71) and (4.49), we get

$$(\nabla^F_{X_s})^2 F_*(X_s) = - \| H_2 \|^2 F_*(X_s) + F_*(\nabla^{1^2}_{X_s} X_s). \tag{4.82}$$

Thus from (4.81) and (4.82) we obtain

$$(\nabla^F_{X_s})^2 F_*(X_s) \quad + \quad g_N(\nabla^F_{X_s} F_*(X_s), \nabla^F_{X_s} F_*(X_s)) F_*(X_s) = F_*(\nabla^1_{X_s} X_s)$$
$$+ g_M(\nabla^1_{X_s} X_s, \nabla^1_{X_s} X_s) F_*(X_s).$$

Since $\alpha$ is a circle on $M$, using (4.66) and (4.67) we arrive at

$$(\nabla^F_{X_s})^2 F_*(X_s) + g_N(\nabla^F_{X_s} F_*(X_s), \nabla^F_{X_s} F_*(X_s)) F_*(X_s) = 0,$$

which shows that $\gamma = F \circ \alpha$ is a circle on $N$.                                  □

## 8. Chen first inequality for Riemannian maps

In [72] and [73], B Y. Chen established the sharp inequality for a submanifold in a real space form involving intrinsic invariants of the submanifolds and squared mean curvature, the main extrinsic invariant. In this section we are going to obtain a Chen inequality for Riemannian maps. We first recall the following lemma, which will be very important for our computations.

**Lemma 51.** *[72] Let $n \geq 2$ and $a_1, a_2, ..., a_n, b$ be real numbers such that*

$$(\sum_{i=1}^n a_i)^2 = (n-1)(\sum_{i=1}^n a_i^2 + b), \tag{4.83}$$

*then $2a_1 a_2 \geq b$, with equality holds if and only if*

$$a_1 + a_2 = a_3 = ... = a_n.$$

Let $F$ be a Riemannian map from a Riemannian manifold $(M, g_M)$ to a Riemannian manifold $(N, g_N)$. Now suppose that $N$ is a space form $N(c)$. Since $F$ is a Riemannian map we have

$$g_M(R^M(X, Y)Z, T) = c(g_M(Y, , Z) g_M(X, T) - g_M(X, Z) g_M(Y, T))$$
$$- g_N((\nabla F_*)(X, Z), (\nabla F_*)(Y, T))$$
$$+ g_N((\nabla F_*)(Y, Z), (\nabla F_*)(X, T)) \tag{4.84}$$

for $X, Y, Z, T \in \Gamma(TM)$.

**Theorem 95.** *[257] Let $F$ be a Riemannian map from a Riemannian manifold $(M, g_M)$ to a space form $(N(c), g_N)$ with $\text{rank} F = r \geq 3$. Then for each point $p \in M$ and each plane section $\pi \subset T_pM$, we have*

$$K(\pi) \geq \rho_{\mathcal{H}} - \frac{r-2}{2}((r+1)c + \frac{1}{r-1} \| \tau^{\mathcal{H}} \|^2), \tag{4.85}$$

where $\rho_{\mathcal{H}}$ is the scalar curvature defined on $\mathcal{H} = (\ker F_*)^{\perp}$ and $\tau^{\mathcal{H}}$ is defined by

$$\tau^{\mathcal{H}} = \sum_{i=1}^{r} g_N((\nabla F_*)(e_i, e_i), (\nabla F_*)(e_i, e_i)).$$

Equality holds if and only if there exists an orthonormal basis $\{e_1, e_2, ..., e_r\}$ of $(\ker F_{*p})^{\perp}$ and orthonormal basis $\{V_{r+1}, V_{r+2}, ..., V_{r+d}\}$ of $(range F_{*p})^{\perp}$ such that the shape operator takes the following forms:

$$S_{r+1} = \begin{pmatrix} a & 0 & 0 & \cdots & 0 \\ 0 & b & 0 & \cdots & 0 \\ 0 & 0 & \mu & \cdots & 0 \\ \vdots & \vdots & \vdots & \ddots & \vdots \\ 0 & 0 & 0 & \cdots & \mu \end{pmatrix}, a + b = \mu$$

and

$$S_\alpha = \begin{pmatrix} B_{11}^\alpha & B_{12}^\alpha & 0 & \cdots & 0 \\ B_{12}^\alpha & -B_{11}^\alpha & 0 & \cdots & 0 \\ 0 & 0 & 0 & \cdots & 0 \\ \vdots & \vdots & \vdots & \ddots & \vdots \\ 0 & 0 & 0 & \cdots & 0 \end{pmatrix}, \alpha = r + 2, ..., r + d.$$

Proof. Taking $X = T = e_i$ and $Y = Z = e_j$ in (4.84), we get

$$r(r - 1)c = 2\rho_{\mathcal{H}} + \sum_{i,j=1}^{r} g_N((\nabla F_*)(e_i, e_j), (\nabla F_*)(e_i, e_j)) - \| \tau^{\mathcal{H}} \| . \qquad (4.86)$$

Set

$$\varepsilon = -r(r-1)c + 2\rho_{\mathcal{H}} - \frac{(r-2)}{(r-1)} \| \tau^{\mathcal{H}} \|^2, \qquad (4.87)$$

then we have

$$\| \tau^{\mathcal{H}} \| = (r-1) \sum_{i,j=1}^{r} (g_N((\nabla F_*)(e_i, e_j), (\nabla F_*)(e_i, e_j)) + \varepsilon). \qquad (4.88)$$

If we use the notion given in (4.105), we get

$$\| \tau^{\mathcal{H}} \| = (r-1)(\sum_{i,j=1}^{r} g_N(B(e_i, e_j), B(e_i, e_j)) + \varepsilon). \qquad (4.89)$$

Now for $p \in M$, consider a plane $\pi \subset T_p M$ spanned by $\{e_1, e_2\}$. Also from Lemma 45 we know that $\tau^{\mathcal{H}}$ belongs to $(range F_*)^{\perp}$. Thus we take an orthonormal frame

$\{V_{r+1}, ..., V_d\}$ of $(rangeF_*)^\perp$ such that $V_{r+1}$ is parallel to $\tau^\mathcal{H}$. Also for convenience, we set $B_{ij}^n = g_N((\nabla F_*)(e_i, e_j), V_n)$. Then we get

$$(\sum_{i=1}^{r} B_{ii}^{r+1})^2 = (r-1) \sum_{i,j=1}^{r} \sum_{\alpha=r+1}^{r+d} (B_{ij}^\alpha)^2 + \varepsilon) \tag{4.90}$$

or

$$(\sum_{i=1}^{r} B_{ii}^{r+1})^2 = (r-1)\{\sum_{i=1}^{r}(B_{ii}^{r+1})^2 + \sum_{i \neq j=1}^{r}(B_{ij}^{r+1})^2$$

$$+ \sum_{\alpha=r+2}^{r+d} \sum_{i,j=1}^{r}(B_{ij}^\alpha)^2 + \varepsilon)\}. \tag{4.91}$$

Applying Lemma 51, we get

$$2B_{11}^{r+1} B_{22}^{r+1} \geq \sum_{i \neq j}^{r}(B_{ij}^{r+1})^2 + \sum_{\alpha=r+2}^{r+d} \sum_{i,j=1}^{r}(B_{ij}^\alpha)^2 + \varepsilon. \tag{4.92}$$

Thus we obtain

$$2B_{11}^{r+1} B_{22}^{r+1} \geq 2(B_{12}^{r+1})^2 + \sum_{i \neq j>2}^{r}(B_{ij}^{r+1})^2 + 2\sum_{j>2}^{r}((B_{1j}^{r+1})^2 + (B_{2j}^{r+1})^2)$$

$$+ 2\sum_{\alpha=r+2}^{r+d}(B_{12}^\alpha)^2 + \sum_{\alpha=r+2}^{r+d} \sum_{i,j>2}^{r}(B_{ij}^\alpha)^2 + 2\sum_{\alpha=r+2}^{r+d} \sum_{j>2}^{r}((B_{1j}^\alpha)^2 + (B_{2j}^\alpha)^2)$$

$$+ \sum_{\alpha=r+2}^{r+d}((B_{11}^\alpha)^2 + (B_{22}^\alpha)^2) + \varepsilon. \tag{4.93}$$

Hence we have

$$2B_{11}^{r+1} B_{22}^{r+1} - 2(B_{12}^{r+1})^2 - 2\sum_{\alpha=r+2}^{r+d}(B_{12}^\alpha)^2 + 2\sum_{\alpha=r+2}^{r+d} B_{11}^\alpha B_{22}^\alpha \geq \sum_{i \neq j>2}^{r}(B_{ij}^{r+1})^2$$

$$+2\sum_{j>2}^{r}((B_{1j}^{r+1})^2 + (B_{2j}^{r+1})^2) + \sum_{\alpha=r+2}^{r+d} \sum_{i,j>2}^{r}(B_{ij}^\alpha)^2$$

$$+2\sum_{\alpha=r+2}^{r+d} \sum_{j>2}^{r}((B_{1j}^\alpha)^2 + (B_{2j}^\alpha)^2) + \sum_{\alpha=r+2}^{r+d}(B_{11}^\alpha + B_{22}^\alpha)^2 + \varepsilon. \tag{4.94}$$

Taking $i = 1$ and $j = 2$ in (4.19) and using them in (4.94), Riemannian map $F$ implies that

$$K(\pi) \geq \sum_{\alpha=r+1}^{r+d} \sum_{j>2}^{r} ((B_{1j}^\alpha)^2 + (B_{2j}^\alpha)^2) + \frac{1}{2} \sum_{i \neq j > 2}^{r} (B_{ij}^{r+1})^2$$

$$\frac{1}{2} \sum_{\alpha=r+2}^{r+d} \sum_{i,j>2} (B_{ij}^\alpha)^2 + \frac{1}{2} \sum_{\alpha=r+2}^{r+d} (B_{11}^\alpha + B_{22}^\alpha)^2 + c + \frac{\varepsilon}{2} \geq c + \frac{\varepsilon}{2}. \qquad (4.95)$$

Thus we arrive at (4.85). If the equality in (4.85) at a point $p$ holds, then the inequality (4.95) becomes equality. In this case, from (4.95), we have

$$\begin{cases} B_{1j}^{r+1} = B_{2j}^{r+1} = B_{ij}^{r+1} = 0, & i \neq j > 2, \\ B_{ij}^\alpha = 0, \forall i \neq j, & i, j = 3, ..., r, \quad \alpha = r+2, ..., r+d, \\ B_{11}^\alpha + B_{22}^\alpha = 0, \forall \alpha = r+2, ..., r+d, \\ B_{11}^{r+2} + B_{22}^{r+2} = ... = B_{11}^{r+d} + B_{22}^{r+d} = 0. \end{cases}$$

Now, if we choose $e_1, e_2$ such that $B_{12}^{r+1} = 0$ and we denote by $a = B_{11}^\alpha, b = B_{22}^\alpha, \mu = B_{33}^{r+1} = ... = B_{33}^\alpha$. Therefore, by choosing the suitable orthonormal basis, the shape operators $S_V$ take the desired forms. $\qquad \square$

From Theorem 95, we have the following corollary.

**Corollary 38.** *[257] Let F be a harmonic Riemannian map from a Riemannian manifold $(M, g_M)$ to an Euclidean space $\mathbb{E}^n$ with $rankF = r \geq 3$. Then for each point $p \in M$ and each plane section $\pi \subset T_pM$, we have*

$$K(\pi) \geq \rho_{\mathcal{H}},$$

*where $\rho_{\mathcal{H}}$ is the scalar curvature defined on $\mathcal{H} = (\ker F_*)^\perp$.*

From Lemma 50, we have the following result.

**Corollary 39.** *[257] An umbilical Riemannian map F from a Riemannian manifold $(M, g_M)$ to a space form $(N(c), g_N)$ with $rankF = r \geq 3$ satisfies the equality in (4.85).*

**Remark 14.** We can see that if $r = dim(M)$, then Riemannian map becomes an isometric immersion and Theorem 95 gives the immersion case Theorem 3.

Let $\mathbb{C}^{m+1}$ denote the complex Euclidean $(m+1)$-space and let

$$S^{2m+1} = \{z = (z_1, ..., z_{m+1}) \in \mathbb{C}^{m+1} \mid <z, z> = 1\}$$

be the unit hypersphere of $\mathbb{C}^{m+1}$. Then consider the Hopf fibration $\pi : S^{2m+1} \longrightarrow \mathbb{CP}^m(4c)$. It is well known that this map is a Riemannian submersion with totally geodesic fibers. Let $N$ be an n-dimensional submanifold of $\mathbb{CP}^m(4c)$. Denote the pre-image $\pi^{-1}(N)$ of $N$ in $M$ by $\tilde{N}$. Then $\bar{\pi} : \tilde{N} \longrightarrow N$ is also a Riemannian submersion with totally geodesic fibers, where $\bar{\pi}$ is the restriction $\pi_N$. For a horizontal 2-plane $P_x \subset T_x\tilde{N}$, we denote $dim(S^{2m+1}) - dim(\mathbb{CP}^m(4c)) + 2-$ subspaces panned by $P_x$ and and the vertical $\mathcal{V}_x$ by $\tilde{P}_x$. In [4], Alegre, Chen, and Munteanu found the following result for the Hopf submersion: let $\pi : S^{2m+1} \longrightarrow \mathbb{CP}^m(4)$ be the Hopf fibration and let $N$ be an n-dimensional submanifold of $\mathbb{CP}^m(4)$. Then we have

$$\rho_{\tilde{N}}(x) - \inf_{\tilde{P}_x} \rho_{\tilde{N}}\tilde{P}_x \le \frac{n^2(n-2)}{2(n-1)}\|H\|^2 + \|\mathcal{P}\|^2 + \frac{1}{2}(n+1)(n-2)c,$$

where $\tilde{P}_x$ runs over $(m+3)-$subspaces associated with all horizontal 2−planes $P_x$ at $x \in \tilde{N}$, $\mathcal{P}$ is the projection from $\mathbb{CP}^m$ to $TN$, and $\|H\|^2$ is the squared mean curvature of $N$ in $\mathbb{CP}^m$. Equality holds if and only if there exists an orthonormal basis $\{e_1, e_2, ..., e_m\}$ such that:

(a)  the shape operator $A$ of $N$ in $\mathbb{CP}^m(4)$ satisfies

$$A_{e_s} = \begin{pmatrix} B_s & 0 \\ 0 & \mu_s I \end{pmatrix}, s = n+1...m$$

where $I$ is an identity $(n-2) \times (n-2)$ matrix and $B_s$ are symmetric $2 \times 2$ submatrices satisfying $\mu_s = traceBs$, $s = n+1, ..., 2m$, and

(b)  $\mathcal{P}e_1 = \mathcal{P}e_2 = 0$.

Since the base space is a complex manifold for the Hopf map, two inequalities seem different due to extra terms, but they they have some relations.

Moreover, in [79] and [78], Chen obtained another inequality for Riemannian submersions and he found interesting result about non-existence of certain immersions defined on the same total space. Let $\pi : M \to B$ be a Riemannian submersion with totally geodesic fibers and $\phi : M \to \bar{M}$ an isometric immersion into a Riemannian manifold $\bar{M}$. Another inequality in terms of O'Neill's tensor fields was obtained in [79] as

$$\check{A}_\pi \le \frac{n^2}{4}\|H\|^2 + b(n-b)max\bar{K},$$

where $\check{A}_\pi$ is the submersion invariant defined by $\check{A}_\pi = \sum_{i=1}^b \sum_{s=b+1}^n \|A_{e_i}e_s\|^2$ and $max\bar{K}(p)$ denotes the maximum value of the sectional curvature function of $\bar{M}^m$ restricted to plane sections in $T_pM$. By using this inequality, he proves that $\pi$ cannot be isometrically immersed into any Riemannian manifold of non-positive sectional curvature as a minimal submanifold.

From Proposition 44, we have the following example.

**Example 29.** [257] We consider the Hopf fibration $\pi : \mathbf{S}^7 \to \mathbf{S}^4$. This map is a Riemannian submersion with totally geodesic fibers and it has fibers $\mathbf{S}^3$. We also consider the isometric immersion $i : \mathbf{S}^4 \to \mathbb{R}^5$ as a hypersurface of $\mathbb{R}^5$. Then $i$ is a totally umbilical isometric immersion. Thus $i \circ \pi$ is a totally umbilical Riemannian map and therefore it satisfies (4.85).

## 9. Einstein metrics on the total space of a Riemannian map

In this section, we are going to find necessary and sufficient conditions for the total space of a Riemannian map to be an Einstein manifold. By using (1.59), (1.64), (4.19), (1.60), and (1.62), we have the following result for the Ricci tensor.

**Lemma 52.** *[256] Let $F : (M, g) \longrightarrow (B, g_B)$ be a Riemannian map. Then, we have*

$$
\begin{aligned}
\rho(W_1, W_2) &= \hat{\rho}(W_1, W_2) - rg(H, T_{W_1} W_2) \\
&+ \sum_{j=r+1}^{m_1} g((\nabla_{e_j} T)_{W_1} W_2, e_j) + g(A_{e_j} W_1, A_{e_j} W_2),
\end{aligned}
\tag{4.96}
$$

*where $\hat{\rho}(W_1, W_2)$ and $H$ are Ricci tensor and the mean curvature vector field of any fiber,*

$$
\begin{aligned}
\rho(X, Y) &= \sum_{i=1}^{r} g((\nabla_X T)_{u_i} u_i, Y) + g((\nabla_{u_i} A)_X Y, u_i) - g(T_{u_i} X, T_{u_i} Y) + g(A_X u_i, A_Y u_i) \\
&+ \rho^{(rangeF_*)}(F_*(X), F_*(Y)) - \sum_{r+1=1}^{m} g_N((\nabla F_*)(e_j, Y), (\nabla F_*)(X, e_j)) \\
&+ g_N((\nabla F_*)(X, Y), \tau^{((ker F_*)^\perp)}),
\end{aligned}
\tag{4.97}
$$

*where $\rho^{(rangeF_*)}(F_*(X), F_*(Y))$ and $\tau^{((ker F_*)^\perp)}$ are the Ricci tensor of $rangeF_*$ and $((ker F_*)^\perp)$– component of the tension field $\tau$,*

$$
\begin{aligned}
\rho(X, U) &= \sum_{i=1}^{r} g((\nabla_U T)_{u_i} u_i, X) - g((\nabla_{u_i} T)_U u_i, X) + \sum_{j=r+1}^{m_1} g((\nabla_{e_j} A)_{e_j} X, U) \\
&+ 2g(T_U e_j, A_{e_j} X)
\end{aligned}
\tag{4.98}
$$

*for $W_1, W_2, U \in \Gamma(ker F_*)$ and $X, Y \in \Gamma((ker F_*)^\perp)$, where $\{u_1, ..., u_r\}$ and $\{e_{r+1}, ..., e_m\}$ are orthonormal frames of $(ker F_*)$ and $(ker F_*)^\perp = \mathcal{H}$.*

**Proposition 46.** *[256] Let $F : (M, g) \longrightarrow (B, g_B)$ be a Riemannian map with totally geodesic fibers. Then, $(M, g_M)$ is Einstein if and only if the following conditions are* *satisfied:*

$$\hat{\rho}(W_1, W_2) + \sum_{j=r+1}^{m_1} g(A_{e_j} W_1, A_{e_j} W_2) = \frac{r}{m} g(U, V), \tag{4.99}$$

$$\sum_{i=1}^{r} g(A_X u_i, A_Y u_i) + \rho^{(rangeF_*)}(F_*(X), F_*(Y)) - \sum_{j=r+1}^{m_1} g_N((\nabla F_*)(e_j, Y), (\nabla F_*)(X, e_j))$$

$$+ g_N((\nabla F_*)(X, Y), \tau^{((ker F_*)^\perp)}) = \frac{r}{m} g(X, Y), \tag{4.100}$$

*and*

$$\sum_{j=r+1}^{m_1} g((\nabla_{e_j} A)_{e_j} X, U) = 0. \tag{4.101}$$

We note that the above conditions (4.99) and (4.101) are the same as the conditions given for Riemannian submersions in [111, Page 144]. The only difference is the condition (4.100). Using (4.96) and (4.97), we have the scalar curvature of the total space as follows.

**Theorem 96.** *[256] Let $F : (M, g) \longrightarrow (B, g_B)$ be a Riemannian map. Then, we have*

$$s = \hat{s} + s^{(rangeF_*)} + \| \tau^{((ker F_*)^\perp)} \|^2 - r^2 \| H \|^2 - \sum_{j,l=r+1}^{m_1} \| \nabla F_*)(e_j, e_l) \|^2$$

$$+ 2 \sum_{j=r+1}^{m_1} \sum_{k=1}^{r} g((\nabla_{e_j} T)_{u_k} u_k, e_j) + \| A_{e_j} u_k \|^2$$

$$+ \sum_{l=r+1}^{m_1} \sum_{i=1}^{r} g((\nabla_{u_i} A)_{e_l} e_l, u_i) - \| T_{u_i} e_l \|^2 \tag{4.102}$$

*where $s$, $\hat{s}$, and $s^{(rangeF_*)}$ denote the scalar curvature of M, the scalar curvature of the fiber and the scalar curvature $(rangeF_*)$, respectively.*

Using (4.34), we have also the following result.

**Corollary 40.** *[256] Let $F : (M^m, g_M) \longrightarrow (B, g_B)$ be a Riemannian map. We then have*

$$s = \hat{s} + s^{\mathcal{H}} + div\tau(F) + \sum_{i,j=1}^{m} g_B((\nabla F_*)(\mu_\alpha, \mu_\beta), (\nabla F_*)(e_\alpha, e_\beta))$$

$$+ \quad \| \tau^{((ker F_*)^\perp)} \|^2 - r^2 \| H \|^2 - \sum_{j,l=r+1}^{m_1} \| \nabla F_*)(e_j, e_l) \|^2$$

$$+ \quad 2 \sum_{j=r+1}^{m_1} \sum_{k=1}^{r} g((\nabla_{e_j} T)_{u_k} u_k, e_j) + \| A_{e_j} u_k \|^2$$

$$+ \quad \sum_{l=r+1}^{m_1} \sum_{i=1}^{r} g((\nabla_{u_i} A)_{e_l} e_l, u_i) - \| T_{u_i} e_l \|^2 \qquad (4.103)$$

*at $p \in M$, where $s^{\mathcal{H}}$ denote the scalar curvature of $\mathcal{H}$ and $\{\mu_1, ..., \mu_m\}$ is an orthonormal frame of $M$.*

From Theorem 96, we have the following results.

**Corollary 41.** *[256] Let $F : (M, g) \longrightarrow (B, g_B)$ be a Riemannian map with totally geodesic fibers. Then*

$$s \leq \hat{s} + s^{(range F_*)} + 2 \sum_{j=r+1}^{m_1} \sum_{k=1}^{r} \| A_{e_j} u_k \|^2 + \| \tau^{((ker F_*)^\perp)} \|^2,$$

*and the equality is satisfied if and only if $F$ is totally geodesic.*

**Corollary 42.** *[256] Let $F : (M, g) \longrightarrow (B, g_B)$ be a Riemannian map with totally geodesic fibers. Then*

$$s \geq \hat{s} + s^{(range F_*)} - \| \nabla F_*)(e_j, e_l) \|^2,$$

*and the equality is satisfied if and only if $F$ is harmonic and the horizontal distribution is integrable.*

From Corollary 40, we have the following result.

**Corollary 43.** *[256] Let $F : (M, g) \longrightarrow (B, g_B)$ be a Riemannian map with totally geodesic* *fibers. Then, we have*

$$s = \hat{s} + s^{\mathcal{H}} + div\tau(F) + \sum_{\alpha,\beta=1}^{m} g_{B}((\nabla F_*)(\mu_\alpha, \mu_\beta), (\nabla F_*)(\mu_\alpha, \mu_\beta))$$

$$+ \| \tau^{((ker F_*)^\perp)} \|^2 - \sum_{j,l=r+1}^{m_1} \| \nabla F_*)(e_j, e_l) \|^2 + \| A_{e_j} u_k \|^2$$

$$+ \sum_{l=r+1}^{m_1} \sum_{i=1}^{r} g((\nabla_{u_i} A)_{e_l} e_l, u_i)$$

*at $p \in M$.*

From Corollary 43, we have the following results.

**Corollary 44.** *[256]Let $F : (M, g) \longrightarrow (B, g_B)$ be a Riemannian map with totally geodesic fibers. We then have*

$$s \geq \hat{s} + s^{\mathcal{H}} + div\tau(F) - \sum_{j,l=r+1}^{m_1} \| \nabla F_*)(e_j, e_l) \|^2 + \| A_{e_j} u_k \|^2$$

$$+ \sum_{l=r+1}^{m_1} \sum_{i=1}^{r} g((\nabla_{u_i} A)_{e_l} e_l, u_i)$$

*at $p \in M$. The equality is satisfied if and only if $F$ is totally geodesic. In the equality case, it takes the following form*

$$s = \hat{s} + s^{\mathcal{H}} + \| A_{e_j} u_k \|^2 + \sum_{l=r+1}^{m_1} \sum_{i=1}^{r} g((\nabla_{u_i} A)_{e_l} e_l, u_i)$$

*at $p \in M$.*

**Corollary 45.** *[256]Let $F : (M, g) \longrightarrow (B, g_B)$ be a Riemannian map with totally geodesic fibers. We then have*

$$s \geq \hat{s} + s^{\mathcal{H}} + div\tau(F) - \sum_{j,l=r+1}^{m_1} \| \nabla F_*)(e_j, e_l) \|^2 + \sum_{l=r+1}^{m_1} \sum_{i=1}^{r} g((\nabla_{u_i} A)_{e_l} e_l, u_i)$$

*at $p \in M$, the equality is satisfied if and only if $F$ is totally geodesic and the horizontal distribution is integrable. In the equality case, it takes the following form:*

$$s = \hat{s} + s^{\mathcal{H}}. \tag{4.104}$$

**Corollary 46.** *[256]Let $F : (M, g) \longrightarrow (B, g_B)$ be a Riemannian map with totally geodesic fibers. We then have*

$$s \geq \hat{s} + s^{\mathcal{H}} + div\tau(F) + \sum_{i,j=1}^{r} g_B((\nabla F_*)(u_i, u_j), (\nabla F_*)(u_i, u_j))$$

$$+ \| A_{e_j} u_k \|^2 + \sum_{l=r+1}^{m_1} \sum_{i=1}^{r} g((\nabla_{u_i} A)_{e_l} e_l, u_i)$$

*at $p \in M$, the equality is satisfied if and only if $F$ is harmonic. In this case, the equality takes the following form:*

$$s = \hat{s} + s^{\mathcal{H}} + \sum_{i,j=1}^{r} g_B((\nabla F_*)(u_i, u_j), (\nabla F_*)(u_i, u_j)) + \| A_{e_j} u_k \|^2$$

$$+ \sum_{l=r+1}^{m_1} \sum_{i=1}^{r} g((\nabla_{u_i} A)_{e_l} e_l, u_i)$$

*at $p \in M$.*

**Remark 15.** We note that Riemannian maps have been also considered in different constructions. For instance, in [271], the author introduces the second fundamental form $B$ for Riemannian map as follows. Let $\varphi : (M, g) \to (\bar{M}, \bar{g})$ be a Riemannian map. The second fundamental form $B$ of $\varphi$ is a bilinear bundle map $B : \oplus^2(ker(\varphi_*)^\perp) \to T\bar{M}$ over $\varphi$ given by

$$B(u, x) = \bar{\nabla}_{\varphi_*(u)} x - \varphi_*(\nabla_u x) \qquad (4.105)$$

for basic vector fields. He shows that $B$ vanishes identically if and only if the image $\varphi(M)$ is a totally geodesic submanifold of $\bar{M}$. He also considers Riemannian maps between Riemannian manifolds and asks the following question: what Riemannian metrics on tangent bundles are preserved under Riemannian maps? In particular, he considers the Sasaki metric on the tangent bundle (See Definition 38). He obtains the following theorem.

**Theorem 97.** *[271] Let $\varphi : (M, g) \to (\bar{M}, \bar{g})$ be a Riemannian map and let their tangent bundles be given their corresponding Sasaki metrics $G(g)$ and $G(\bar{g})$, respectively. Suppose further that the fibers and the image of $\varphi$ are totally geodesic. Let $u \in T_pM$ and let $X, Y (\in ker\varphi_{*p})^\perp$. Then, the pullback metric on the orthogonal complement to*

$ker\varphi_{**p}$ *is given by*

$$(\varphi_*)^*G(\bar{g})(X^V, Y^V) \ = \ G(g)(X^V, Y^V)$$
$$(\varphi_*)^*G(\bar{g})(X^H, Y^V) = 0$$
$$(\varphi_*)^*G(\bar{g})(X^H, Y^H) \ = \ G(g)(X^H, Y^H)$$

*and thus, $\varphi_*$ is also a Riemannian map.*

# CHAPTER 5

# Riemannian Maps From Almost Hermitian Manifolds

Contents

## Abstract

In this chapter, we study Riemannian maps from almost Hermitian manifolds to Riemannian manifolds. In section 1, we study holomorphic Riemannian maps as a generalization of holomorphic submersions and obtain a characterization of such maps. In section 2, we investigate anti-invariant Riemannian maps as a generalization of anti-invariant submersion, investigate the geometry of leaves of distributions defined by such maps, and give necessary and sufficient conditions for anti-invariant Riemannian maps to be totally geodesic. We also find necessary and sufficient conditions for the total manifold of anti-invariant Riemannian maps to be an Einstein manifold. In section 3, as a generalization of the semi-invariant submersion, we introduce semi-invariant Riemannian maps, give examples, and obtain the main properties of such maps. In section 4, we introduce generic Riemannian maps from almost Hermitian manifolds to Riemannian manifolds as a generalization of generic submersions and we give examples. We also find new conditions for Riemannian maps to be totally geodesic and harmonic. In section 5, we study slant Riemannian maps as a generalization of slant submersion, give examples, and obtain the harmonicity of such maps. We also find new necessary and sufficient conditions for such maps to be totally geodesic. Moreover, we obtain a decomposition theorem by slant Riemannian maps. In section 6, we define semi-slant Riemannian maps and give examples. In section 7, we introduce hemi-slant Riemannian maps as a generalization of hemi-slant submersions and slant Riemannian maps, give examples, obtain integrability conditions for distribution, and investigate the geometry of these distributions. We also obtain conditions for such maps to be harmonic and totally geodesic, respectively.

**Keywords:** Kähler manifold, holomorphic Riemannian map, invariant Riemannian map, anti-invariant Riemannian map, semi-invariant Riemannian map, generic Riemannian map, slant Riemannian map, semi-slant Riemannian map, hemi-slant Riemannian map, holomorphic

Reimannian Submersions, Reimannian Maps in Hermitian Geometry, and their Applications
http://dx.doi.org/10.1016/B978-0-12-804391-2.50005-1

submersion, anti-invariant submersion, semi-invariant submersion, slant submersion, Einstein manifold

*Many of life's failures are people who did not realize how close they were to success when they gave up.*

**Thomas A. Edison**

## 1. Holomorphic Riemannian maps from almost Hermitian manifolds

In this section, we define holomorphic Riemannian maps and obtain a geometric characterization of harmonic holomorphic Riemannian maps from a Kähler manifold to an almost Hermitian manifold.

**Definition 52.** [254] Let $F$ be a Riemannian map from an almost Hermitian manifold $(M_1, g_1, J_1)$ to an almost Hermitian manifold $(M_2, g_2, J_2)$. Then we say that $F$ is a holomorphic Riemannian map at $p \in M_1$ if

$$J_2 F_* = F_* J_1. \tag{5.1}$$

If $F$ is a holomorphic Riemannian map at every point $p \in M_1$, then we say that $F$ is a holomorphic Riemannian map between $M_1$ and $M_2$.

It is known that vertical and horizontal distributions of an almost Hermitian submersion are invariant with respect to the complex structure of the total manifold. Next, we show that this is true for a holomorphic Riemannian map.

**Lemma 53.** *Let $F$ be a holomorphic Riemannian map between almost Hermitian manifolds $(M_1, g_1, J_1)$ and $(M_2, g_2, J_2)$. Then the distributions $\ker F_*$ and $(\ker F_*)^\perp$ are invariant with respect to $J_1$.*

*Proof.* For $X \in \Gamma(\ker F_*)$, from (5.1) we have $F_*(J_1 X) = J_2 F_*(X) = 0$, which implies that $J_1 X \in \Gamma(\ker F_*)$. In a similar way, we show that $(\ker F_*)^\perp$ is invariant. □

In a similar way, it is easy to see that $(range F_*)^\perp$ is invariant under the action of $J_2$. Since $F$ is a subimmersion, it follows that the rank of $F$ is constant on $M_1$, then the rank theorem for functions implies that $\ker F_*$ is an integrable subbundle of $TM_1$. Thus, it follows from the above definition that the leaves of the distribution $\ker F_*$ of a holomorphic Riemannian map are holomorphic submanifolds of $M_1$. We now give examples of holomorphic Riemannian maps.

**Example 30.** Every holomorphic submersion between almost Hermitian manifolds is a holomorphic Riemannian map with $(rangeF_*)^\perp = \{0\}$.

**Example 31.** Every Kählerian submanifold of a Kähler manifold is a holomorphic Riemannian map with $kerF_* = \{0\}$.

In the following, $\mathbb{R}^{2m}$ denotes the Euclidean $2m-$ space with the standard metric. An almost complex structure $J$ on $\mathbb{R}^{2m}$ is said to be compatible if $(\mathbb{R}^{2m}, J)$ is complex analytically isometric to the complex number space $C^m$ with the standard flat Kählerian metric. We denote by $J$ the compatible almost complex structure on $\mathbb{R}^{2m}$ defined by

$$J(a^1, ..., a^{2m}) = (-a^2, a^1, ..., -a^{2m}, a^{2m-1}).$$

**Example 32.** [254] Consider the following Riemannian map given by

$$F: \quad \mathbb{R}^4 \quad \longrightarrow \quad \mathbb{R}^4$$
$$(x_1, x_2, x_3, x_4) \quad (\tfrac{x_1+x_3}{\sqrt{2}}, \tfrac{x_2+x_4}{\sqrt{2}}, 0, 0).$$

Then $F$ is a holomorphic Riemannian map.

**Remark 16.** We note that the notion of invariant Riemannian map has been introduced in [241] as a generalization of invariant immersion of almost Hermitian manifolds and holomorphic Riemannian submersions. We can see that every holomorphic Riemannian map is an invariant Riemannian map, but the converse is not true. In other words, an invariant Riemannian map may not be a holomorphic Riemannian map.

We now investigate the harmonicity of holomorphic Riemannian maps. We first note that if $M_1$ and $M_2$ are Kähler manifolds and $F : M_1 \longrightarrow M_2$ is a holomorphic map, then from Theorem 18, $F$ is harmonic. But there is no guarantee when $M_1$ or $M_2$ is an almost Hermitian manifold.

**Theorem 98.** *[254] Let F be a holomorphic Riemannian map from a Kähler manifold $(M_1, g, J_1)$ to almost Hermitian manifold $(M_2, g_2, J_2)$. Then F is harmonic if and only if the distribution $F_*((ker F_*)^\perp)$ is minimal.*

*Proof.* Since $TM_1 = ker F_* \oplus (ker F_*)^\perp$, we can write $\tau = \tau^1 + \tau^2$, where $\tau^1$ and $\tau^2$ are the components of $\tau$ in $ker F_*$ and $(ker F_*)^\perp$, respectively. First we compute $\tau^1 = \sum_{i=1}^{n_1}(\nabla F_*)(e_i, e_i)$, where $\{e_1, ..., e_{n_1}\}$ is an orthonormal basis of $ker F_*$. From (1.71), we

have

$$\tau^1 = -\sum_{i=1}^{n_1} F_*(\nabla^1_{e_i} e_i). \tag{5.2}$$

We note that, since $(ker\, F_*)$ is an invariant space with respect to $J_1$, then $\{J_1 e_i\}_{i=1}^{n_1}$ is also an orthonormal basis of $ker\, F_*$. Thus we can write

$$\tau^1 = \sum_{i=1}^{n_1} (\nabla F_*)(J_1 e_i, J_1 e_i) = -\sum_{i=1}^{n_1} F_*(\nabla^1_{J_1 e_i} J_1 e_i).$$

Since $M_1$ is a Kähler manifold and $ker\, F_*$ is integrable, using (5.1), we obtain

$$\tau^1 = -\sum_{i=1}^{n_1} J_2 F_*(\nabla^1_{e_i} J_1 e_i).$$

Using again (5.1), we derive

$$\tau^1 = \sum_{i=1}^{n_1} F_*(\nabla^1_{e_i} e_i). \tag{5.3}$$

Thus (5.2) and (5.3) imply that $\tau^1 = 0$. On the other hand, using Lemma 45 and (1.71), we obtain

$$\tau^2 = g_2 \Big( \sum_{s=1}^{m_2} \sum_{a=1}^{n_1} (\nabla^F_{e_a} F_*(e_a), \mu_s) \mu_s = H_{(rangeF_*)},$$

where $H_{(rangeF_*)}$ is the mean curvature vector field of $(rangeF_*)$. Then our assertion follows from the above equation and (5.3).    □

Next, by using (3.2) and (3.5), we have the following.

**Lemma 54.** *[254] Let F be a holomorphic Riemannian map from an almost Hermitian manifold $(M_1, g_1, J_1)$ to a Kähler manifold $(M_2, g_2, J_2)$. Then we have*

$$(\nabla F_*)(X, J_1 Y) = (\nabla F_*)(Y, J_1 X) = J_2(\nabla F_*)(X, Y), \tag{5.4}$$

*for $X, Y \in \Gamma((ker\, F_*)^\perp)$.*

**Lemma 55.** *[254] Let F be a holomorphic Riemannian map from an almost Hermitian manifold $(M_1, g_1, J_1)$ to a Kähler manifold $(M_2, g_2, J_2)$. Then we have*

$$\begin{aligned}
g_1(R^1(X, J_1 X)J_1 X, X) &= g_2(R^2(F_*X, J_2 F_*X)J_2 F_*X, F_*X) \\
&\quad - 2\|(\nabla F_*)(X, X)\|^2
\end{aligned} \tag{5.5}$$

*for $X \in \Gamma((ker\, F_*)^\perp)$.*

*Proof.* Putting $Y = J_1X$, $Z = J_1X$ and $T = X$ in (4.19) and by using (5.1) and (5.4), we obtain (5.5). □

As a result of Lemma 55, we have the following result for the leaves of $(\ker F_*)^\perp$.

**Theorem 99.** *[254] Let F be a holomorphic Riemannian map from an almost Hermitian manifold $(M_1, g_1, J_1)$ to a complex space form $(M_2(c), g_2, J_2)$ of constant holomorphic sectional curvature c such that $(\ker F_*)^\perp$ is integrable. Then the integral manifold of $(\ker F_*)^\perp$ is a complex space form $M'(c)$ if and only if $(\nabla F_*)(X, X) = 0$ for $X \in \Gamma((\ker F_*)^\perp)$.*

**Remark 17.** We note that we can introduce the notion of invariant Riemannian maps from almost Hermitian manifolds to Riemannian manifolds. It is easy to see that holomorphic Riemannian maps and invariant Riemannian maps are different, even if both the source and target manifolds are almost Hermitian manifolds. Again, it is clear that every holomorphic Riemannian map is an invariant Riemannian map, but the converse is not true.

## 2. Anti-invariant Riemannian maps from almost Hermitian manifolds

In this section, we define anti-invariant Riemannian maps, provide examples and investigate the geometry of such maps. We first present the following definition.

**Definition 53.** [246] Let F be a Riemannian map from an almost Hermitian manifold $(M_1, g_1, J_1)$ to a Riemannian manifold $(M_2, g_2)$. We say that F is an *anti-invariant Riemannian map* if the following condition is satisfied:

$$J_1(\ker F_*) \subset (\ker F_*)^\perp.$$

We denote the orthogonal complementary subbundle to $J_1(\ker F_*)$ in $(\ker F_*)^\perp$ by $\mu$. Then it is easy to see that $\mu$ is invariant with respect to $J_1$. Since F is a subimmersion, it follows that the rank of F is constant on $M_1$, then the rank theorem for functions implies that $\ker F_*$ is an integrable subbundle of $TM_1$. Thus it follows from the above definition that the leaves of the distribution $\ker F_*$ of an anti-invariant Riemannian map are anti-invariant (totally real) submanifolds of $M_1$.

**Proposition 47.** *[246] Let F be anti-invariant Riemannian map from an almost Hermitian manifold $(M_1^{2m}, g_1, J_1)$ to a Riemannian manifold $(M_2^n, g_2)$ with $\dim(\ker F_*) = r$. Then $\dim(\mu) = 2(m - r)$.*

*Proof.* Since $dim(ker F_*) = r$ and $F$ is an anti-invariant Riemannian map, it follows that $dim(J_1(kerF_*)) = r$. Hence $dim(\mu) = 2(m - r)$.   □

Let $F$ be an anti-invariant Riemannian map from an almost Hermitian manifold $(M_1^{2m}, g_1, J_1)$ to a Riemannian manifold $(M_2^n, g_2)$, then, we say that $F$ is a *Lagrangian Riemannian map* from $M_1$ to $M_2$ if $J_1(ker F_*) = (ker F_*)^\perp$. Hence an anti-invariant Riemannian map $F$ is a Lagrangian Riemannian map if and only if $\frac{dim(M_1)}{2} = dim(ker F_*)$. It is also easy to see that an anti-invariant Riemannian map is Lagrangian if and only if $\mu = \{0\}$. Moreover, we have the following.

**Proposition 48.** *[246] Let F be anti-invariant Riemannian map from an almost Hermitian manifold $(M_1, g_1, J_1)$ to a Riemannian manifold $(M_2, g_2)$ with $dim(ker F_*) = r$. Then F is a Lagrangian map if and only if $\frac{dim(M_1)}{2} = dim(M_2)$.*

We now give some examples for anti-invariant Riemannian maps.

**Proposition 49.** *Every proper Riemannian map F (F is neither an isometric immersion nor a Riemannian submersion) from an almost Hermitian manifold $(M_1^2, J, g_1)$ to a Riemannian manifold $(M_2^n, g_2)$ is an anti-invariant Riemannian map.*

**Example 33.** Every anti-invariant Riemannian submersion from an almost Hermitian manifold to a Riemannian manifold is an anti-invariant Riemannian map with $(rangeF_*)^\perp = \{0\}$.

**Example 34.** Let $F$ be a map defined by

$$F: \quad \mathbb{R}^4 \quad \longrightarrow \quad \mathbb{R}^3$$
$$(x_1, x_2, x_3, x_4) \quad (\tfrac{x_1-x_3}{\sqrt{2}}, 0, \tfrac{x_2+x_4}{\sqrt{2}}).$$

Then it follows that

$$(ker F_*) = Span\{Z_1 = \frac{\partial}{\partial x_1} + \frac{\partial}{\partial x_3}, Z_2 = \frac{\partial}{\partial x_2} - \frac{\partial}{\partial x_4}\}$$

and

$$(ker F_*)^\perp = Span\{Z_3 = \frac{\partial}{\partial x_1} - \frac{\partial}{\partial x_3}, Z_2 = \frac{\partial}{\partial x_2} + \frac{\partial}{\partial x_4}\}.$$

By direct computations, we can see that

$$F_*(Z_3) = \frac{2}{\sqrt{2}} \frac{\partial}{\partial x_1}, \quad F_*(Z_4) = \frac{2}{\sqrt{2}} \frac{\partial}{\partial x_3}.$$

Thus $F$ is a Riemannian map with $(range F_*)^\perp = Span\{\frac{\partial}{\partial x_2}\}$. Moreover it is easy to see that $JZ_1 = Z_4$, $JZ_2 = -Z_3$. As a result, $F$ is an anti-Riemannian map.

Let $F$ be an anti-invariant Riemannian map from a Kähler manifold $(M_1, g_1, J)$ to a Riemannian manifold $(M_2, g_2)$. Then, for $X \in \Gamma((ker F_*)^\perp)$, we have

$$JX = \mathcal{B}X + CX, \tag{5.6}$$

where $\mathcal{B}X \in \Gamma(ker F_*)$ and $CX \in \Gamma(\mu)$.

We now investigate the geometry of leaves of the distributions $(ker F_*)$ and $(ker F_*)^\perp$.

**Proposition 50.** *[246] Let $F$ be an anti-invariant Riemannian map from a Kähler manifold $(M_1, g_1, J_1)$ to a Riemannian manifold $(M_2, g_2)$. Then $(ker F_*)$ defines a totally geodesic foliation on $M_1$ if and only if*

$$g_2((\nabla F_*)(X, \mathcal{B}Z), F_*(J_1 Y)) = g_2((\nabla F_*)(J_1 Y, X), F_*(CZ)) \tag{5.7}$$

*for $X, Y \in \Gamma(ker F_*)$ and $Z \in \Gamma((ker F_*)^\perp)$.*

*Proof.* For $X, Y \in \Gamma(ker F_*)$ and $Z \in \Gamma((ker F_*)^\perp)$, by using (3.2), (3.5) and (5.6), we have

$$g_1(\nabla^1_X Y, Z) = g_1(\nabla^1_X J_1 Y, \mathcal{B}Z) + g_1(\nabla^1_X J_1 Y, CZ),$$

where $\nabla^1$ is the Levi-Civita connection on $M_1$. Since $\mathcal{B}Z \in \Gamma(ker F_*)$, we get

$$g_1(\nabla^1_X Y, Z) = -g_1(J_1 Y, \nabla^1_X \mathcal{B}Z) + g_1(\nabla^1_X J_1 Y, CZ).$$

Since $F$ is a Riemannian map, using (1.71) we obtain

$$g_1(\nabla^1_X Y, Z) = g_2(F_*(J_1 Y), (\nabla F_*)(X, \mathcal{B}Z)) - g_2((\nabla F_*)(X, J_1 Y), F_*(CZ)),$$

which proves the assertion. $\qquad\qquad\square$

For the distribution $(ker F_*)^\perp$, we have the following.

**Proposition 51.** *[246]Let $F$ be an anti-invariant Riemannian map from a Kähler manifold $(M_1, g_1, J_1)$ to a Riemannian manifold $(M_2, g_2)$. Then $(ker F_*)^\perp$ defines a totally geodesic foliation on $M_1$ if and only if*

$$g_2((\nabla F_*)(Z_1, \mathcal{B}Z_2), F_*(J_1 X)) = -g_2(\nabla^F_{Z_1} F_*(J_1 X), F_*(CZ_2)) \tag{5.8}$$

*for $Z_1, Z_2 \in \Gamma((ker F_*)^\perp)$ and $X \in \Gamma(ker F_*)$.*

*Proof.* For $Z_1, Z_2 \in \Gamma((ker F_*)^\perp)$ and $X \in \Gamma(ker F_*)$, we have $g_1(\nabla^1_{Z_1} Z_2, X) = -g_1(Z_2, \nabla^1_{Z_1} X)$. Using (3.2), (3.5), and (5.6), we obtain

$$g_1(\nabla^1_{Z_1} Z_2, X) = -g_1(\nabla^1_{Z_1} J_1 X, \mathcal{B} Z_2) - g_1(\nabla^1_{Z_1} J_1 X, C Z_2).$$

Hence, we get

$$g_1(\nabla^1_{Z_1} Z_2, X) = g_1(J_1 X, \nabla^1_{Z_1} \mathcal{B} Z_2) - g_1(\nabla^1_{Z_1} J_1 X, C Z_2).$$

Then Riemannian map $F$ and (1.71) imply that

$$\begin{aligned}
g_1(\nabla^1_{Z_1} Z_2, X) &= -g_2(F_*(J_1 X), (\nabla F_*)(Z_1, \mathcal{B} Z_2)) + g_2((\nabla F_*)(Z_1, J_1 X), F_*(C Z_2)) \\
&\quad - g_2(\nabla^F_{Z_1} F_*(J_1 X), F_*(CZ)).
\end{aligned}$$

From (4.4), we know that there are no components of $(\nabla F_*)(Z_1, J_1 X)$ in $(range F_*)$, hence we get the result.  □

From Proposition 50 and Proposition 51, we have the following decomposition theorem.

**Theorem 100.** *[246] Let F be an anti-invariant Riemannian map from a Kähler manifold $(M_1, g_1, J_1)$ to a Riemannian manifold $(M_2, g_2)$. Then $M_1$ is a locally product Riemannian manifold in the form $M_{(ker F_*)} \times M_{(ker F_*)^\perp}$ if and only if (5.7) and (5.8) are satisfied.*

We now give necessary and sufficient conditions for an anti-invariant Riemannian map from an almost Hermitian manifold to a Riemannian manifold to be a totally geodesic map.

**Theorem 101.** *[246] Let F be an anti-invariant Riemannian map from an almost Hermitian manifold $(M_1, g_1, J_1)$ to a Riemannian manifold $(M_2, g_2)$. Then F is totally geodesic if and only if*

$$^*F_*(S_v F_*(J_1 X)) \in \Gamma(\mu) \tag{5.9}$$

$$^*F_*(S_v F_*(Z_1)) \in \Gamma(J_1(ker F_*)) \tag{5.10}$$

$$g_2(F_*(J_1 Y), (\nabla F_*)(X, \mathcal{B} Z)) = g_2((\nabla F_*)(X, J_1 Y), F_*(CZ)) \tag{5.11}$$

$$\begin{aligned}
g_1(\nabla_X \mathcal{B} Z, \mathcal{B} Z_2) &= g_2((\nabla F_*)(X, \mathcal{B} Z) + (\nabla F_*)(X, CZ), F_*(C Z_1)) \\
&\quad - g_2(F_*(CZ), (\nabla F_*)(X, \mathcal{B} Z_2)) \tag{5.12}
\end{aligned}$$

*for $X, Y \in \Gamma(ker F_*)$, $Z_1, Z_2 \in \Gamma(\mu)$ and $V \in \Gamma((range F_*)^\perp)$.*

*Proof.* From (1.71), (3.2), and (3.5), we get

$$g_2((\nabla F_*)(X, Z), F_*(\bar{Z})) = -g_1(\nabla_X J_1 Z, J_1 \bar{Z})$$

for $X \in \Gamma(ker\, F_*)$ and $Z, \bar{Z} \in \Gamma((ker\, F_*)^\perp)$. Using (5.6), we have

$$
\begin{aligned}
g_2((\nabla F_*)(X, Z), F_*(\bar{Z})) &= -g_1(\nabla_X \mathcal{B}Z, \mathcal{B}\bar{Z}) - g_1(\nabla_X \mathcal{B}Z, CZ_2) \\
&\quad - g_1(\nabla_X CZ, \mathcal{B}\bar{Z}) - g_1(\nabla_X CZ, C\bar{Z}).
\end{aligned}
$$

Since $F$ is a Riemannian map, we obtain

$$
\begin{aligned}
g_2((\nabla F_*)(X, Z), F_*(\bar{Z})) &= -g_1(\nabla_X \mathcal{B}Z, \mathcal{B}\bar{Z}) - g_2(F_*(\nabla_X \mathcal{B}Z), F_*(CZ_2)) \\
&\quad + g_1(CZ, \nabla_X \mathcal{B}\bar{Z}) - g_2(F_*(\nabla_X CZ), F_*(C\bar{Z})).
\end{aligned}
$$

Then Riemannian map $F$ and (1.71) imply that

$$
\begin{aligned}
g_2((\nabla F_*)(X, Z), F_*(\bar{Z})) &= -g_1(\nabla_X \mathcal{B}Z, \mathcal{B}\bar{Z}) + g_2((\nabla F_*)(X, \mathcal{B}Z), F_*(CZ_2)) \\
&\quad - g_2(F_*(CZ), (\nabla F_*)(X, \mathcal{B}\bar{Z})) \\
&\quad + g_2((\nabla F_*)(X, CZ), F_*(C\bar{Z})). \tag{5.13}
\end{aligned}
$$

In a similar way, we can obtain

$$
\begin{aligned}
g_2((\nabla F_*)(X, Y), F_*(Z)) &= -g_2(F_*(J_1 Y), (\nabla F_*)(X, \mathcal{B}Z)) \\
&\quad + g_2((\nabla F_*)(X, J_1 Y), F_*(CZ)) \tag{5.14}
\end{aligned}
$$

for $X, Y \in \Gamma(ker\, F_*)$ and $Z \in \Gamma((ker\, F_*)^\perp)$. On the other hand, from (1.71) we have

$$g_2((\nabla F_*)(Z_1, Z_2), V) = g_2(\nabla^F_{Z_1} F_*(Z_2), V)$$

for $Z_1, Z_2 \in \Gamma(\mu)$ and $V \in \Gamma((range F_*)^\perp)$. Hence, we get

$$g_2((\nabla F_*)(Z_1, Z_2), V) = -g_2(F_*(Z_2), \nabla^2_{F_*(Z_1)} V).$$

Then, using (4.16) we obtain

$$g_2((\nabla F_*)(Z_1, Z_2), V) = g_2(F_*(Z_2), S_v F_*(Z_1)).$$

Thus, we have

$$g_2((\nabla F_*)(Z_1, Z_2), V) = g_1(Z_2, {}^*F_*(S_v F_*(Z_1))). \tag{5.15}$$

In a similar way, we obtain

$$g_2((\nabla F_*)(J_1 X, J_1 Y), V) = g_1(J_1 Y, {}^*F_*(S_v F_*(J_1 X))) \tag{5.16}$$

for $X, Y \in \Gamma(ker\, F_*)$ and $V \in \Gamma((range F_*)^\perp)$. Thus proof comes from (5.13)-(5.16).
$$\square$$

We also have the following result on pluriharmonic maps.

**Theorem 102.** *[246] Let $(M_1, g_1, J)$ be a connected Kähler manifold and $M_2$ a Riemannian manifold. If a Lagrangian Riemannian map from $M_1$ to $M_2$ is pluriharmonic, then it is totally geodesic.*

*Proof.* Suppose that $F$ is a pluriharmonic Lagrangian Riemannian map from $M_1$ to $M_2$. Then we have

$$(\nabla F_*)(X, Y) + (\nabla F_*)(JX, JY) = 0$$

for $X, Y \in \Gamma(ker\, F_*)$. On the other hand, from Lemma 45, we know that

$$(\nabla F_*)(JX, JY) \in \Gamma((rangeF_*)^\perp), \quad (\nabla F_*)(X, Y) \in \Gamma(rangeF_*).$$

Since $T_{F(p)}M_2 = rangeF_{*p} \oplus (rangeF_{*p})^\perp$, we get

$$(\nabla F_*)(JX, JY) = 0 \tag{5.17}$$

and

$$(\nabla F_*)(X, Y) = 0. \tag{5.18}$$

We also claim that $(\nabla F_*)(X, JY) = 0$ for $X, Y \in \Gamma(ker\, F_*)$. Suppose that $F$ is pluriharmonic and $(\nabla F_*)(X, JY) \neq 0$. But since $F$ is pluriharmonic, we have $(\nabla F_*)(X, JY) - (\nabla F_*)(JX, Y) = 0$. Then, using (1.71), we have

$$-F_*(\nabla^1_X JY) + F_*(\nabla^1_Y JX) = 0.$$

Since $M_1$ is a Kähler manifold, we derive

$$-F_*(J\nabla^1_X Y) + F_*(J\nabla^1_Y X) = 0.$$

Hence we obtain

$$F_*(J[X, Y]) = 0.$$

Hence we conclude that $J[X, Y] \in \Gamma(ker\, F_*)$, which implies that $[X, Y] \in \Gamma((ker\, F_*)^\perp)$. This tells us that $ker\, F_*$ is not integrable which contradicts the rank theorem. Since $F$ is a subimmersion, it follows that the rank of $F$ is constant on $M_1$, then the rank theorem for functions implies that $ker\, F_*$ is an integrable subbundle of $TM_1$. Thus we should have

$$(\nabla F_*)(X, JY) = 0 \tag{5.19}$$

for $X, Y \in \Gamma(ker\, F_*)$. Then the proof comes from (5.17)-(5.19).    $\square$

In the sequel we show that anti Riemannian map with totally umbilical fibers puts some restrictions on the geometry of the distribution $ker\, F_*$.

**Theorem 103.** *[246] Let F be a Lagrangian Riemannian map with totally umbilical fibers from a Kähler manifold $(M_1, g_1, J)$ to a Riemannian manifold $(M_2, g)$. If $dim(kerF_{*p}) > 1$, $p \in M_1$, then the distribution $\ker F_*$ defines a totally geodesic foliation on $M_1$.*

*Proof.* Since $F$ is a Lagrangian map with totally umbilical fibers, for $X, Y \in \Gamma(\ker F_*)$ and $Z \in \Gamma((\ker F_*)^\perp)$, we have $g_1(\nabla_X^1 Y, Z) = g_1(X, Y)g_1(H, Z)$. Using (3.2) we obtain $g_1(\nabla_X^1 JY, JZ) = g_1(X, Y)g_1(H, Z)$. Hence we get

$$-g_1(\nabla_X^1 JZ, JY) = g_1(X, Y)g_1(H, Z).$$

Then (4.51) implies that

$$-g_1(X, JZ)g_1(H, JY) = g_1(X, Y)g_1(H, Z).$$

Thus we have

$$g_1(H, JY)JX = g_1(X, Y)H.$$

Taking inner product both sides with $JX$ and using (3.2), we arrive at

$$g_1(H, JY)g_1(X, X) = g_1(X, Y)g_1(H, JX).$$

Since $dim(\ker F_*) > 1$, we can choose unit vector fields $X$ and $Y$ such that $g_1(X, Y) = 0$, thus we derive

$$g_1(H, JY)g_1(X, X) = 0.$$

Since $F$ is Lagrangian, the above equation implies that $H = 0$, which shows that $\ker F_*$ is totally geodesic.    □

Thus from the remark of section 3 of Chapter 3 and Theorem 103, we obtain the following result.

**Corollary 47.** *[246] Let $M_1$ be a Kähler manifold and $M_2$ a Riemannian manifold. Then there do not exist a Lagrangian Riemannian map F from $M_1$ to $M_2$ such that $M_1$ is a locally twisted product manifold of the form $M_{(\ker F_*)^\perp} \times_f M_{(\ker F_*)}$, where $M_{(\ker F_*)^\perp}$ and $M_{(\ker F_*)}$ are the integral manifolds of $(\ker F_*)^\perp$ and $\ker F_*$, and $f$ is the twisting function.*

In the rest of this section, we are going to check conditions for the total manifold of anti-invariant Riemannian maps to be an Einstein manifold. We first give the following result whose proof is exactly the same as Theorem 25, therefore we omit it.

**Lemma 56.** *[246] Let $F : M \longrightarrow N$ be a Lagrangian Riemannian from a Kähler manifold $M$ to a Riemannian manifold $N$. Then the horizontal distribution $(\ker F_*)^{\perp}$ is integrable and totally geodesic. As a result of this, we have $A_X = 0$ for $X \in \Gamma((\ker F_*)^{\perp})$.*

We also note that one can see that those curvature relations of Riemannian submersions (1.60)-(1.64) are valid for Riemannian maps.

**Lemma 57.** *[246] Let $F$ be an anti-invariant Riemannian map from a Kähler manifold $(M, g_1, J)$ to a Riemannian manifold $(N, g_N)$. Then we have*

$$Ric(U, V) = \sum_{i=1}^{r} -g_1((\nabla_J U\mathcal{T})(e_i, e_i), JV) - g_1((\nabla_{e_i}\mathcal{A})(JU, JV), e_i) \tag{5.20}$$

$$+ g_1(\mathcal{T}_{e_i} JU, \mathcal{T}_{e_i} JV) - g_1(\mathcal{A}_{JU} e_i, \mathcal{A}_{JV} e_i) + Ric^{(rangeF_*)}(F_*(JU), F_*(JV))$$

$$+ g_N((\nabla F_*)(JU, JV), \tau^{(\ker F_*)^{\perp}}) - \sum_{j=r+1}^{m_1} g_N((\nabla F_*)(E_i, JV), (\nabla F_*)(JU, E_i)),$$

$$Ric(U, X) = \sum_{i=1}^{r} \{g_1((\nabla_{\mathcal{B}X}\mathcal{T})(e_i, e_i), JU) - g_1((\nabla_{e_i}\mathcal{T})(\mathcal{B}X, e_i), JU)$$

$$+ g_1((\nabla_{JU}\mathcal{T})(e_i, e_i), CX) + g_1((\nabla_{e_i}\mathcal{A})(JU, CX), e_i)$$

$$- g_1(\mathcal{T}_{e_i} JU, \mathcal{T}_{e_i} CX) + g_1(\mathcal{A}_{JU} e_i, \mathcal{A}_{CX} e_i)\}$$

$$+ Ric^{(rangeF_*)}(F_* JU, F_* CX) + g_N((\nabla F_*)(JU, CX), \tau^{(\ker F_*)^{\perp}})$$

$$+ \sum_{j=r+1}^{m_1} g_1((\nabla_{E_j}\mathcal{A})(E_j, JU), \mathcal{B}X) + 2g_1(\mathcal{A}_{E_j} JU, \mathcal{T}_{\mathcal{B}X} E_j)$$

$$- g_N((\nabla F_*)(E_j, CX), (\nabla F_*)(JU, E_j)), \tag{5.21}$$

$$Ric(X, Y) = \sum_{i=1}^{r} \{g_1((\nabla_Y\mathcal{T})(e_i, e_i), X) + g_1((\nabla_{e_i}\mathcal{A})(Y, X), e_i) - g_1(\mathcal{T}_{e_i} Y, \mathcal{T}_{e_i} X)$$

$$+ g_1(\mathcal{A}_Y e_i, \mathcal{A}_X e_i)\} + Ric^{(rangeF_*)}(F_*(X), F_*(Y)) + g_N((\nabla F_*)(X, Y), \tau^{(\ker F_*)^{\perp}})$$

$$- \sum_{j=r+1}^{m_1} g_N((\nabla F_*)(E_j, Y), (\nabla F_*)(X, E_j)), \tag{5.22}$$

*for $U, V \in \Gamma(\ker F_*)$ and $X, Y \in \Gamma((\ker F_*)^{\perp})$, where $Ric^{(rangeF_*)}(F_*(JU), F_*(JV))$ and $\tau^{(\ker F_*)^{\perp}}$ denote the Ricci tensor of $rangeF_*$ and the $(\ker F_*)^{\perp})-$ part of tension field $\tau(F)$. Moreover, $\{e_1, ..., e_r\}$ and $\{E_1, ..., E_{r+s}\}$ are orthonormal frames of $(\ker F_*)$, $J(\ker F_*) \oplus \mu$, $r = dim(\ker F_*)$, $m_1 = dim(M)$.*

*Proof.* Since $M$ is Kähler, from Theorem 19 we have $Ric(X, Y) = Ric(JX, JY)$. Thus we have

$$
\begin{aligned}
Ric(U, V) \;=\;& Ric(JU, JV) \\
=\;& \sum_{i=1}^{r} g(R(e_i, JU)JV, e_i) + g(R(Je_i, JU)JV, Je_i) + \sum_{j=1}^{s} g(R(\mu_j, JU)JV, \mu_j).
\end{aligned}
$$

Using the Gauss equations (4.19) and (1.64), we obtain (5.20). The other expressions can be obtained in a similar way by using (4.19) and (1.60)-(1.63). $\qquad\square$

For the total manifold $M$ of an anti-invariant Riemannian map, we have the following conditions for $M$ to be an Einstein manifold.

**Proposition 52.** *[246] Let $F$ be an anti-invariant Riemannian map from a Kähler manifold $(M, g_1, J)$ to a Riemannian manifold $(N, g_N)$. Then, $(M, g_1)$ is Einstein if and only if the following relations hold:*

$$
\begin{aligned}
& -\frac{r}{m} g_1(U, V) + \sum_{i=1}^{r} -g_1((\nabla_{JU}\mathcal{T})(e_i, e_i), JV) - g_1((\nabla_{e_i}\mathcal{A})(JU, JV), e_i) \\
& +g_1(\mathcal{T}_{e_i}JU, \mathcal{T}_{e_i}JV) - g_1(\mathcal{A}_{JU}e_i, \mathcal{A}_{JV}e_i) + Ric^{(rangeF_*)}(F_*(JU), F_*(JV)) \quad (5.23) \\
& +g_N((\nabla F_*)(JU, JV), \tau^{(kerF_*)^\perp}) - \sum_{j=r+1}^{m_1} g_N((\nabla F_*)(E_i, JV), (\nabla F_*)(JU, E_i)) = 0,
\end{aligned}
$$

$$
\begin{aligned}
& \sum_{i=1}^{r} \{g_1((\nabla_{\mathcal{B}X}\mathcal{T})(e_i, e_i), JU) - g_1((\nabla_{e_i}\mathcal{T})(\mathcal{B}X, e_i), JU) \\
& +g_1((\nabla_{JU}\mathcal{T})(e_i, e_i), CX) + g_1((\nabla_{e_i}\mathcal{A})(JU, CX), e_i) - g_1(\mathcal{T}_{e_i}JU, \mathcal{T}_{e_i}CX) \\
& +g_1(\mathcal{A}_{JU}e_i, \mathcal{A}_{CX}e_i)\} + Ric^{(rangeF_*)}(F_*JU, F_*CX) + g_N((\nabla F_*)(JU, CX), \tau^{(kerF_*)^\perp}) \\
& + \sum_{j=r+1}^{m_1} g_1((\nabla_{E_j}\mathcal{A})(E_j, JU), \mathcal{B}X) + 2g_1(\mathcal{A}_{E_j}JU, \mathcal{T}_{\mathcal{B}X}E_j) \\
& - g_N((\nabla F_*)(E_j, CX), (\nabla F_*)(JU, E_j)) = 0 \quad\quad (5.24)
\end{aligned}
$$

$$
\begin{aligned}
& -\frac{r}{m} g_1(X, Y) + \sum_{i=1}^{r} \{g_1((\nabla_Y\mathcal{T})(e_i, e_i), X) + g_1((\nabla_{e_i}\mathcal{A})(Y, X), e_i) \\
& -g_1(\mathcal{T}_{e_i}Y, \mathcal{T}_{e_i}X) + g_1(\mathcal{A}_Y e_i, \mathcal{A}_X e_i)\} + Ric^{(rangeF_*)}(F_*(X), F_*(Y)) \\
& +g_N((\nabla F_*)(X, Y), \tau^{(kerF_*)^\perp}) - \sum_{j=r+1}^{m_1} g_N((\nabla F_*)(E_j, Y), (\nabla F_*)(X, E_j)) = 0, (5.25)
\end{aligned}
$$

*where r is the scalar curvature of M.*

Denoting the scalar curvature of ($rangeF_*$) of an anti-invariant Riemannian map from a Kähler manifold to a Riemannian manifold, using (5.23) and (5.25) we have the following theorem.

**Theorem 104.** *[246] Let F be a Lagrangian Riemannian map from a Kähler manifold* $(M, g_1, J)$ *to a Riemannian manifold* $(N, g_N)$. *The following relation is valid:*

$$\frac{r}{2} = r^{(rangeF_*)} + \parallel \tau^{(kerF_*)^\perp} \parallel^2 + \sum_{i,l=1}^{r}\{-g_1((\nabla_{Je_i}\mathcal{T})(e_i, e_i), Je_l)$$

$$+ g_1(\mathcal{T}_{e_i}Je_l, \mathcal{T}_{e_i}Je_l) - g_N((\nabla F_*)(E_i, E_l), (\nabla F_*)(E_l, E_i))\}.$$

From Theorem 104 and Lemma 56, we have the following result.

**Corollary 48.** *[246] Let F be a Lagrangian Riemannian map from a Kähler manifold* $(M, g_1, J)$ *to a Riemannian manifold* $(N, g_N)$ *with totally geodesic fibers. Then*

$$\frac{r}{2} \leq r^{(rangeF_*)} + \parallel \tau^{(kerF_*)^\perp} \parallel^2 .$$

*The equality holds if and only if F is totally geodesic.*

From Theorem 104, we also have the following result.

**Corollary 49.** *[246] Let F be a Lagrangian Riemannian map from a Kähler manifold* $(M, g_1, J)$ *to a Riemannian manifold* $(N, g_N)$ *with totally geodesic fibers. Then*

$$\frac{r}{2} \geq r^{(rangeF_*)} - \sum_{i,l=1}^{r}\{g_N((\nabla F_*)(E_i, E_l), (\nabla F_*)(E_l, E_i))\}.$$

*The equality holds if and only if F is harmonic.*

## 3. Semi-invariant Riemannian maps from almost Hermitian manifolds

In this section, we introduce semi-invariant Riemannian maps from almost Hermitian manifolds, give examples, investigate the geometry of leaves of the distributions defined by semi-invariant Riemannian maps, and obtain decomposition theorems.

**Definition 54.** [247] Let $F$ be a Riemannian map from an almost Hermitian manifold $(M, g_M, J)$ to a Riemannian manifold $(M_2, g_N)$. Then we say that $F$ is a *semi-invariant Riemannian map* if the following conditions are satisfied:

**(A)**  There exists a subbundle of $ker\, F_*$ such that

$$J(D_1) = D_1.$$

**(B)**  There exists a complementary subbundle $D_2$ to $D_1$ in $ker\, F_*$ such that

$$J(D_2) \subsetneq (ker\, F_*)^{\perp}.$$

From the definition, we have

$$ker\, F_* = D_1 \oplus D_2. \tag{5.26}$$

Now, we denote the orthogonal complementary subbundle of $(ker\, F_*)^{\perp}$ to $J(D_2)$ by $\mu$. Then it is easy to see that $\mu$ is invariant. Since $F$ is a subimmersion, it follows that the rank of $F$ is constant on $M_1$, then the rank theorem for functions implies that $ker\, F_*$ is an integrable subbundle of $TM_1$. Thus it follows from the above definition that the leaves of the distribution $ker\, F_*$ of a semi-invariant Riemannian map are CR-submanifolds of $M_1$. We now provide some examples of semi-invariant Riemannian maps.

**Example 35.** Every holomorphic submersion between almost Hermitian manifolds is a semi-invariant Riemannian map with $D_1 = ker\, F_*$ and $(range F_*)^{\perp} = \{0\}$.

**Example 36.** Every anti-invariant Riemannian submersion from an almost Hermitian manifold to a Riemannian manifold is a semi-invariant Riemannian map $D_2 = ker\, F_*$ and $(range F_*)^{\perp} = \{0\}$.

**Example 37.** Every semi-invariant submersion from an almost Hermitian manifold to a Riemannian manifold is a semi-invariant Riemannian map with $(range F_*)^{\perp} = \{0\}$.

In the following $\mathbb{R}^{2m}$ denotes the Euclidean $2m-$ space with the standard metric. An almost complex structure $J$ on $\mathbb{R}^{2m}$ is said to be compatible if $(\mathbb{R}^{2m}, J)$ is complex analytically isometric to the complex number space $C^m$ with the standard flat Kählerian metric. We denote by $J$ the compatible almost complex structure on $\mathbb{R}^{2m}$ defined by

$$J(a^1, ..., a^{2m}) = (-a^2, a^1, ..., -a^{2m}, a^{2m-1}).$$

We say that a semi-invariant Riemannian map is proper if $D_1 \neq \{0\}$, $D_2 \neq \{0\}$ and $\mu \neq \{0\}$. Here is an example of a proper semi-invariant Riemannian map.

**Example 38.** [247] Consider the following map defined by

$$F: \qquad \mathbb{R}^6 \qquad \longrightarrow \qquad \mathbb{R}^4$$
$$(x^1, x^2, x^3, x^4, x^5, x^6) \qquad (\tfrac{x^1-x^3}{\sqrt{2}}, \tfrac{x^2-x^4}{\sqrt{2}}, \tfrac{x^5+x^6}{\sqrt{2}}, 0).$$

Then we have

$$\ker F_* = span\{Z_1 = \frac{\partial}{\partial x^1} + \frac{\partial}{\partial x^3}, Z_2 = \frac{\partial}{\partial x^2} + \frac{\partial}{\partial x^4}, Z_3 = \frac{\partial}{\partial x^5} - \frac{\partial}{\partial x^6}\}$$

and

$$(\ker F_*)^\perp = span\{Z_4 = \frac{\partial}{\partial x^1} - \frac{\partial}{\partial x^3}, Z_5 = \frac{\partial}{\partial x^2} - \frac{\partial}{\partial x^4}, Z_6 = \frac{\partial}{\partial x^5} + \frac{\partial}{\partial x^6}\}.$$

Hence it is easy to see that

$$g_{\mathbb{R}^4}(F_*(Z_i), F_*(Z_i)) = g_{\mathbb{R}^6}(Z_i, Z_i) = 2$$

and

$$g_{\mathbb{R}^4}(F_*(Z_i), F_*(Z_j)) = g_{\mathbb{R}^6}(Z_i, Z_j) = 0,$$

$i \neq j$, for $i, j = 4, 5, 6$. Thus $F$ is a Riemannian map. On the other hand, we have $JZ_1 = Z_2$ and $JZ_3 = Z_6 \in \Gamma((rangeF_*)^\perp)$, where $J$ is the complex structure of $\mathbb{R}^6$. Thus $F$ is a semi-invariant Riemannian map with $D_1 = span\{Z_1, Z_2\}$, $D_2 = span\{Z_3\}$ and $\mu = span\{Z_4, Z_5\}$.

Let $F$ be a semi-invariant Riemannian map from an almost Hermitian manifold $(M, g_M, J)$ to a Riemannian manifold $(N, g_N)$. Then for $U \in \Gamma(\ker F_*)$, we write

$$JU = \phi U + \omega U, \tag{5.27}$$

where $\phi U \in \Gamma(D_1)$ and $\omega U \in \Gamma(JD_2)$. Also for $X \in \Gamma((\ker F_*)^\perp)$, we write

$$JX = \mathcal{B}X + \mathcal{C}X, \tag{5.28}$$

where $\mathcal{B}X \in \Gamma(D_2)$ and $\mathcal{C}X \in \Gamma(\mu)$.

Since $F$ is a subimmersion, it follows that the rank of $F$ is constant on $M_1$, then the rank theorem for functions implies that $\ker F_*$ is an integrable subbundle of $TM_1$. For the integral manifolds of $\ker F_*$, we have the following.

**Theorem 105.** [247] *Let $F$ be a semi-invariant Riemannian map from a Kähler manifold $(M, g_M, J)$ to a Riemannian manifold $(N, g_N)$. Then the distribution $\ker F_*$ defines* *a totally geodesic foliation if and only if*

$$g_N((\nabla F_*)(U, \omega V), F_*(CX)) = g_M(\hat{\nabla}_U \phi V, \mathcal{B}X) + g_M(\mathcal{T}_U \phi V, CX)$$
$$-g_M(\omega V, \mathcal{T}_U \mathcal{B}X)$$

*for $U, V \in \Gamma(ker F_*)$ and $X \in \Gamma((ker F_*)^\perp)$.*

*Proof.* It is clear that *ker $F_*$* defines a totally geodesic foliation if and only if $g_M(\nabla_U V, X) = 0$ for $U, V \in \Gamma(ker F_*)$ and $X \in \Gamma((ker F_*)^\perp)$. Using (5.27), (5.28), and (4.12), we have

$$g_M(\nabla_U V, X) = g_M(\hat{\nabla}_U \phi V, \mathcal{B}X) + g_M(\mathcal{T}_U \phi V, CX))$$
$$-g_M(\omega V, \mathcal{T}_U \mathcal{B}X) + g_M(\nabla_U \omega V, CX)$$

for $U, V \in \Gamma(ker F_*)$ and $X \in \Gamma((ker F_*)^\perp)$. Since $F$ is a Riemannian map, by using (1.71) we get

$$g_M(\nabla_U V, X) = g_M(\hat{\nabla}_U \phi V, \mathcal{B}X) + g_M(\mathcal{T}_U \phi V, CX))$$
$$-g_M(\omega V, \mathcal{T}_U \mathcal{B}X) - g_N((\nabla_U F_*)(U, \omega V), F_*(CX)),$$

which proves the assertion. $\square$

For the integrability of the distribution $(ker F_*)^\perp$, we have the following result.

**Theorem 106.** *[247] Let $F$ be a semi-invariant Riemannian map from a Kähler manifold $(M, g_M, J)$ to a Riemannian manifold $(N, g_N)$. Then the distribution $(ker F_*)^\perp$ is integrable if and only if*

$$g_M(\mathcal{B}Z_2, \mathcal{A}_{Z_1} \omega U) - g_M(\mathcal{B}Z_1, \mathcal{A}_{Z_2} \omega U) = g_M(\nabla_{Z_1} \mathcal{B}Z_2 - \nabla_{Z_2} \mathcal{B}Z_1, \phi U)$$
$$+g_N((\nabla F_*)(Z_1, \phi U) - \nabla^F_{Z_1} F_*(\omega U), F_*(CZ_2))$$
$$-g_N((\nabla F_*)(Z_2, \phi U) - \nabla^F_{Z_2} F_*(\omega U), F_*(CZ_1))$$

*for $Z_1, Z_2 \in \Gamma((ker F_*)^\perp)$ and $U \in \Gamma(ker F_*)$.*

*Proof.* By direct computations, we have

$$g_M([Z_1, Z_2], U) = -g_M(Z_2, \nabla_{Z_1} U) + g_M(Z_1, \nabla_{Z_2} U)$$

for $Z_1, Z_2 \in \Gamma((ker F_*)^\perp)$ and $U \in \Gamma(ker F_*)$. Then, using (3.2), (3.5), (5.27), (5.28), and (4.15) we get

$$
\begin{aligned}
g_M([Z_1, Z_2], U) &= -g_M(\mathcal{B}Z_2, \nabla_{Z_1}\phi U) - g_M(\mathcal{B}Z_2, \mathcal{A}_{Z_1}\omega U) \\
&\quad - g_M(\mathcal{C}Z_2, \nabla_{Z_1}\phi U) - g_M(\mathcal{C}Z_2, \nabla_{Z_1}\omega U) \\
&\quad + g_M(\mathcal{B}Z_1, \nabla_{Z_2}\phi U) + g_M(\mathcal{B}Z_1, \mathcal{A}_{Z_2}\omega U) \\
&\quad + g_M(\mathcal{C}Z_1, \nabla_{Z_2}\phi U) + g_M(\mathcal{C}Z_1, \nabla_{Z_2}\omega U).
\end{aligned}
$$

Since $F$ is a Riemannian map, from (1.71) and (4.7), we obtain

$$
\begin{aligned}
g_M([Z_1, Z_2], U) &= -g_M(\mathcal{B}Z_2, \nabla_{Z_1}\phi U) - g_M(\mathcal{B}Z_2, \mathcal{A}_{Z_1}\omega U) \\
&\quad + g_N(F_*(\mathcal{C}Z_2), (\nabla F_*)(Z_1, \phi U)) - g_N(F_*(\mathcal{C}Z_2), \nabla^F_{Z_1}F_*(\omega U)) \\
&\quad + g_M(\mathcal{B}Z_1, \nabla_{Z_2}\phi U) + g_M(\mathcal{B}Z_1, \mathcal{A}_{Z_2}\omega U) \\
&\quad - g_N(F_*(\mathcal{C}Z_1), (\nabla F_*)(Z_2, \phi U)) + g_N(F_*(\mathcal{C}Z_1), \nabla^F_{Z_2}F_*(\omega U)),
\end{aligned}
$$

which gives us the assertion. □

For the leaves of the distribution $(\ker F_*)^\perp$, we have the following result.

**Theorem 107.** *[247] Let $F$ be a semi-invariant Riemannian map from a Kähler manifold $(M, g_M, J)$ to a Riemannian manifold $(N, g_N)$. Then the distribution $(\ker F_*)^\perp$ defines a totally geodesic foliation if and only if*

$$
g_N(\nabla^F_X F_*(\mathcal{C}Y), F_*(\omega U)) = g_M(\mathcal{A}_X \phi U, \mathcal{C}Y) - g_M(\mathcal{A}_X \mathcal{B}Y, \omega U) - g_M(\nabla_X \mathcal{B}Y, \phi U)
$$

*for $X, Y \in \Gamma((\ker F_*)^\perp)$ and $U \in \Gamma(\ker F_*)$.*

*Proof.* From (3.2), (3.5), (4.14), (5.27), and (5.28), we obtain

$$
\begin{aligned}
g_M(\nabla_X Y, U) &= g_M(\nabla_X \mathcal{B}Y, \phi U) + g_M(\mathcal{A}_X \mathcal{B}Y, \omega U) \\
&\quad - g_M(\mathcal{C}Y, \mathcal{A}_X \phi U) + g_M(\nabla_X \mathcal{C}Y, \omega U)
\end{aligned}
$$

for $X, Y \in \Gamma((\ker F_*)^\perp)$ and $U \in \Gamma(\ker F_*)$. Since $F$ is a semi-invariant Riemannian map, using (1.71) we get the assertion. □

From Theorem 105 and Theorem 107, we have the following decomposition theorem.

**Theorem 108.** *[247] Let $F$ be a semi-invariant Riemannian map from a Kähler manifold $(M, g_M, J)$ to a Riemannian manifold $(N, g_N)$. Then $M$ is a locally product Riemannian manifold if and only if*

$$
g_N(\nabla^F_X F_*(\mathcal{C}Y), F_*(\omega U)) = g_M(\mathcal{A}_X \phi U, \mathcal{C}Y) - g_M(\mathcal{A}_X \mathcal{B}Y, \omega U) - g_M(\nabla_X \mathcal{B}Y, \phi U)
$$

*and*

$$g_N((\nabla F_*)(U, \omega V), F_*(CX)) = g_M(\hat{\nabla}_U \phi V, \mathcal{B} X) + g_M(\mathcal{T}_U \phi V, CX)$$
$$-g_M(\omega V, \mathcal{T}_U \mathcal{B} X)$$

*for $U, V \in \Gamma(ker F_*)$ and $X, Y \in \Gamma((ker F_*)^\perp)$.*

Since $(ker F_*)^\perp = J(D_2) \oplus \mu$ and $F$ is a Riemannian map from an almost Hermitian manifold $(M, g_M, J)$ to a a Riemannian manifold $(N, g_N)$, for $X \in \Gamma(D_2)$ and $Y \in \Gamma(\mu)$, we have

$$g_N(F_*(JX), F_*(Y)) = g_M(JX, Y) = 0.$$

This implies that the distributions $F_*(JD_2)$ and $F_*(\mu)$ are orthogonal. Thus, if we denote $F_*(JD_2)$ and $F_*(\mu)$ by $\bar{D}_2$ and $\bar{\mu}$, we have the following decomposition for $(range F_*)$:

$$range F_* = \bar{D}_2 \oplus \bar{\mu}. \tag{5.29}$$

We now investigate the geometry of the leaves of the distributions $D_1$ and $D_2$.

**Theorem 109.** *[247] Let $F$ be a semi-invariant Riemannian map from a Kähler manifold $(M, J, g_M)$ to a Riemannian manifold $(N, g_N)$. Then $D_1$ defines a totally geodesic foliation on $M$ if and only if*

$$(\nabla F_*)(X_1, JY_1) \in \Gamma(\bar{\mu})$$

*and*

$$g_M(\hat{\nabla}_{X_1} JY_1, \mathcal{B} X) = g_N((\nabla F_*)(X_1, JY_1), F_*(CX))$$

*for $X_1, Y_1 \in \Gamma(D_1)$, $X_2 \in \Gamma(D_2)$ and $X \in \Gamma((ker F_*)^\perp)$.*

*Proof.* From the definition of a semi-invariant Riemannian map, it follows that the distribution $D_1$ defines a totally geodesic foliation on $M$ if and only if $g_M(\nabla_{X_1} Y_1, X_2) = 0$ and $g_M(\nabla_{X_1} Y_1, X) = 0$ for $X_1, Y_1 \in \Gamma(D_1)$, $X_2 \in \Gamma(D_2)$ and $X \in \Gamma((ker F_*)^\perp)$. Since $F$ is a Riemannian map, from (1.71) we have

$$g_M(\nabla_{X_1} Y_1, X_2) = -g_N((\nabla F_*)(X_1, JY_1), F_*(JX_2)). \tag{5.30}$$

On the other hand, by using (5.28), we derive

$$g_M(\nabla_{X_1} Y_1, X) = g_M(\nabla_{X_1} JY_1, \mathcal{B} X) + g_M(\nabla_{X_1} JY_1, CX).$$

Then Riemannian map $F$ and (1.71) and (4.12) imply that

$$g_M(\nabla_{X_1} Y_1, X) = g_M(\hat{\nabla}_{X_1} JY_1, \mathcal{B}X) - g_N((\nabla F_*)(X_1, JY_1), F_*(CX)). \qquad (5.31)$$

Thus proof follows from (5.29), (5.30) and (5.31).    □

For the leaves of $D_2$, we have the following result.

**Theorem 110.** *[247] Let F be a semi-invariant Riemannian map from a Kähler manifold $(M, J, g_M)$ to a Riemannian manifold $(N, g_N)$. Then $D_2$ defines a totally geodesic foliation on M if and only if*

$$(\nabla F_*)(X_2, JY_1) \in \Gamma(\bar{\mu})$$

*and*

$$g_M(\mathcal{T}_{X_2} \mathcal{B}X, JY_2) = -g_N((\nabla F_*)(X_2, JY_2), F_*(CX))$$

*for $X_1 \in \Gamma(D_1)$, $X_2, Y_2 \in \Gamma(D_2)$ and $X \in \Gamma((ker F_*)^\perp)$.*

*Proof.* From the definition of a semi-invariant Riemannian map, it follows that the distribution $D_2$ defines a totally geodesic foliation on $M$ if and only if $g_M(\nabla_{X_2} Y_2, X_1) = 0$ and $g_M(\nabla_{X_2} Y_2, X) = 0$ for $X_1 \in \Gamma(D_1)$, $X_2, Y_2 \in \Gamma(D_2)$ and $X \in \Gamma((ker F_*)^\perp)$. Since we have $g_M(\nabla_{X_2} Y_2, X_1) = -g_M(\nabla_{X_2} X_1, Y_2)$, from (1.71) we have

$$g_M(\nabla_{X_2} Y_2, X_1) = g_N((\nabla F_*)(X_2, JX_1), F_*(JY_2)). \qquad (5.32)$$

In a similar way, by using (5.28), we derive

$$g_M(\nabla_{X_2} Y_2, X) = -g_M(\nabla_{X_2} \mathcal{B}X, JY_2) + g_M(\nabla_{X_2} JY_2, CX).$$

Then Riemannian map $F$ and (1.71) and (4.12) imply that

$$g_M(\nabla_{X_1} Y_1, X) = -g_M(JY_2, \mathcal{T}_{X_2} \mathcal{B}X) - g_N((\nabla F_*)(X_2, JY_2), F_*(CX)). \qquad (5.33)$$

The proof follows from (5.29), (5.32), and (5.33).    □

From (5.30) and (5.32) we have the following result.

**Corollary 50.** *[247] Let F be a semi-invariant Riemannian map from a Kähler manifold $(M, J, g_M)$ to a Riemannian manifold $(N, g_N)$. Then the fibers are locally product Riemannian manifolds if and only if*

$$(\nabla F_*)(U, JY_1) \in \Gamma(\bar{\mu})$$

*for $U \in \Gamma(ker F_*)$ and $Y_1 \in \Gamma(D_1)$.*

By using the adjoint map $^*F_*$ of $F_*$, we now give a characterization for semi-invariant Riemannian maps to be totally geodesic in terms of $\mathcal{A}, \mathcal{T}$ and $\mathcal{S}$.

**Theorem 111.** *[247] Let F be a semi-invariant Riemannian map from a Kähler manifold $(M, J, g_M)$ to a Riemannian manifold $(N, g_N)$. Then F is totally geodesic if and only if*

$$g_M(\hat{\nabla}_X \mathcal{B}Z_1, \mathcal{B}Z_2) = g_M(X, \mathcal{A}_{CZ_1}CZ_2) - g_M(\mathcal{T}_X\mathcal{B}Z_1, CZ_2)$$
$$-g_M(\mathcal{T}_XCZ_1, \mathcal{B}Z_2),$$

$$g_M(\hat{\nabla}_X\mathcal{B}Z, \phi Y) = -\{g_M(\phi Y, \mathcal{T}_XCZ) + g_M(\omega Y, \mathcal{T}_X\mathcal{B}Z)$$
$$+g_M(\mathcal{A}_{\omega Y}CZ, X)\}$$

*and*

$$^*F_*(\mathcal{S}_V F_*(JU)) = 0, \; ^*F_*(\mathcal{S}_V F_*(Z_3)) \in \Gamma(J(D_2))$$

*for $Z_1, Z_2, Z \in \Gamma((ker F_*)^\perp), X, Y \in \Gamma(kerF_*), U \in \Gamma(D_2), Z_3 \in \Gamma(\mu)$ and $V \in \Gamma((rangeF_*)^\perp)$.*

*Proof.* From the definition of a Riemannian map, $F$ is totally geodesic if and only if $(\nabla F_*)(X, Y) = 0, (\nabla F_*)(X, Z) = 0$ and $(\nabla F_*)(Z_1, Z_2) = 0$ for $X, Y \in \Gamma(kerF_*), Z, Z_1, Z_2 \in \Gamma((ker F_*)^\perp)$. From the above information, decomposition of the tangent bundle of $M$, and (4.7), a semi-invariant Riemannian map $F$ is totally geodesic if and only if

$$g_N((\nabla F_*)(X, Z_1), F_*(Z_2)) = 0, \; g_N((\nabla F_*)(X, Y), F_*(Z)) = 0$$

and

$$g_N((\nabla F_*)(JU, Z), V) = 0, \; g_N((\nabla F_*)(Z_3, Z_4), V) = 0$$

for $X, Y \in \Gamma(ker F_*), U \in \Gamma(D_2), Z, Z_1, Z_2 \in \Gamma((ker F_*)^\perp), Z_3, Z_4 \in \Gamma(\mu)$. Since $F$ is a Riemannian map, by using (1.71), (3.2), (3.5), and (5.28) we have

$$g_N((\nabla F_*)(X, Z_1), F_*(Z_2)) = -g_M(\nabla_X\mathcal{B}Z_1, \mathcal{B}Z_2) - g_M(\nabla_XCZ_1, \mathcal{B}Z_2)$$
$$-g_M(\nabla_X\mathcal{B}Z_1, CZ_2) - g_M(\nabla_XCZ_1, CZ_2).$$

Then from (4.12), (4.13), and (1.71) we get

$$g_N((\nabla F_*)(X, Z_1), F_*(Z_2)) = -g_M(\hat{\nabla}_X\mathcal{B}Z_1, \mathcal{B}Z_2) - g_M(\mathcal{T}_XCZ_1, \mathcal{B}Z_2)$$
$$-g_M(\mathcal{T}_X\mathcal{B}Z_1, CZ_2) + g_N((\nabla F_*)(X, CZ_1), F_*(CZ_2)). \tag{5.34}$$

On the other hand, since the second fundamental form is symmetric, we get

$$g_N((\nabla F_*)(X, CZ_1), F_*(CZ_2)) = g_M(X, \nabla_{CZ_1}CZ_2).$$

Thus from (4.15) we obtain

$$g_N((\nabla F_*)(X, CZ_1), F_*(CZ_2)) = g_M(X, \mathcal{A}_{CZ_1} CZ_2). \tag{5.35}$$

Using (5.35) in (5.34) we have

$$\begin{aligned} g_N((\nabla F_*)(X, Z_1), F_*(Z_2)) &= -g_M(\hat{\nabla}_X \mathcal{B} Z_1, \mathcal{B} Z_2) - g_M(\mathcal{T}_X CZ_1, \mathcal{B} Z_2) \\ &\quad - g_M(\mathcal{T}_X \mathcal{B} Z_1, CZ_2) + g_M(X, \mathcal{A}_{CZ_1} CZ_2). \end{aligned} \tag{5.36}$$

In a similar way, we get

$$\begin{aligned} g_N((\nabla F_*)(X, Y), F_*(Z)) &= g_M(\hat{\nabla}_X \mathcal{B} Z, \phi Y) + g_M(\mathcal{T}_X CZ, \phi Y) \\ &\quad + g_M(\mathcal{T}_X \mathcal{B} Z, \omega Y) + g_M(\mathcal{A}_{\omega Y} CZ, X). \end{aligned} \tag{5.37}$$

On the other hand, from (4.17) we have

$$g_N((\nabla F_*)(JU, Z), V) = g_N(F_*(Z), S_V F_*(JU)).$$

Then, using the adjoint map, we obtain

$$g_N((\nabla F_*)(JU, Z), V) = g_M(Z, {}^* F_* S_V F_*(JU)). \tag{5.38}$$

In a similar way, we have

$$g_N((\nabla F_*)(Z_3, Z_4), V) = g_M(Z_4, {}^* F_* S_V F_*(Z_3)). \tag{5.39}$$

Thus, the proof follows from (5.36),(5.37),(5.38) and (5.39).    □

In the rest of this section, we study semi-invariant Riemannian maps with totally umbilical fibers and obtain a characterization for such semi-invariant Riemannian maps. We first note that (4.51) is equivalent to the following equation:

$$\mathcal{T}_U V = g_M(U, V) H \tag{5.40}$$

for $U, V \in \Gamma(ker\, F_*)$, where $H$ is the mean curvature vector field of $ker\, F_*$.

**Lemma 58.** *[247] Let F be a semi-invariant Riemannian map F from a Kähler manifold $(M, J, g_M)$ to a Riemannian manifold $(N, g_N)$ with totally umbilical fibers. Then $H \in \Gamma(JD_2)$.*

*Proof.* Since $M$ is a Kähler manifold, we have $\nabla_X JY = J\nabla_X Y$ for $X, Y \in \Gamma(D_1)$. Then, using (4.12), (5.27), (5.28), and (5.40), for $W \in \Gamma(\mu)$ we get

$$g_M(X, JY)g_M(H, W) = -g_M(X, Y)g_M(H, JW). \tag{5.41}$$

Interchanging the role of $X$ and $Y$, we also have

$$g_M(Y, JX)g_M(H, W) = -g_M(X, Y)g_M(H, JW). \tag{5.42}$$

Thus, from (5.41) and (5.42) we obtain

$$2g_M(JY, X)g_M(H, W) = 0,$$

which shows that $H \in \Gamma(JD_2)$ due to the fact that $g_M$ is a Riemannian metric.    ⊔

**Theorem 112.** *[247] Let F be a semi-invariant Riemannian map F from a Kähler manifold $(M, J, g_M)$ to a Riemannian manifold $(N, g_N)$ with totally umbilical fibers. Then either $D_2$ is one-dimensional or the fibers are totally geodesic.*

*Proof.* Since the fibers are totally umbilical, we have

$$g_M(\mathcal{T}_V U, JV) = g_M(H, JV)g_M(U, V)$$

for $U, V \in \Gamma(D_2)$. Then, from (4.12) we get

$$-g_M(U, \nabla_V JV) = g_M(H, JV)g_M(U, V).$$

Hence, we have

$$g_M(JU, \nabla_V V) = g_M(H, JV)g_M(U, V).$$

Using (4.12) and (5.40), we obtain

$$g_M(V, V)g(H, JU) = g(U, V)g(H, JV). \tag{5.43}$$

Interchanging the role of $U$ and $V$, we have

$$g_M(U, U)g(H, JV) = g(U, V)g(H, JU). \tag{5.44}$$

Then, from (5.43) and (5.44), we arrive at

$$g(H, JV) = \frac{g_M(U, V)^2}{g_M(U, U)g_M(V, V)}g_M(JV, H). \tag{5.45}$$

Equation (5.45) implies that $U$ and $V$ are linearly dependent or $H = 0$ due to Lemma 58.    □

From Theorem 112 and Lemma 49, we have the following result.

**Corollary 51.** *Let F be a semi-invariant Riemannian map F from a Kähler manifold $(M, J, g_M)$ to a Riemannian manifold $(N, g_N)$ with totally umbilical fibers. If $dim(D_2) > 1$, then F is harmonic if and only if the distribution $rangeF_*$ is minimal.*

## 4. Generic Riemannian maps from almost Hermitian manifolds

In this section, we are going to define and study generic Riemannian maps from almost Hermitian manifolds to Riemannian manifolds. Let $F$ be a Riemannian map from an almost Hermitian manifold $(M, g_M, J)$ to a Riemannian manifold $(N, g_N)$. Define

$$\mathcal{D}_p = (ker F_{*p} \cap J(ker F_{*p})), p \in M,$$

the complex subspace of the vertical subspace $\mathcal{V}_p$.

**Definition 55.** [255] Let $F$ be a Riemannian map from an almost Hermitian manifold $(M, g_M, J)$ to a Riemannian manifold $(N, g_N)$. If the dimension $\mathcal{D}_p$ is constant along $M$ and it defines a differentiable distribution on $M$, then we say that $F$ is a *generic Riemannian map*

A generic Riemannian map is purely real (respectively, complex) if $\mathcal{D}_p = \{0\}$ (respectively, $\mathcal{D}_p = ker F_{*p}$). For a generic Riemannian map, the orthogonal complementary distribution $\mathcal{D}^\perp$, called purely real distribution, satisfies

$$ker F_* = \mathcal{D} \oplus \mathcal{D}^\perp \tag{5.46}$$

and

$$\mathcal{D} \cap \mathcal{D}^\perp = \{0\}. \tag{5.47}$$

Let $F$ be a generic Riemannian map from an almost Hermitian manifold $(M, g_M, J)$ to a Riemannian manifold $(N, g_N)$. Then for $U \in \Gamma(ker F_*)$, we write

$$JU = \phi U + \omega U, \tag{5.48}$$

where $\phi U \in \Gamma(ker F_*)$ and $\omega U \in \Gamma((ker F_*)^\perp)$. Now we consider the complementary orthogonal distribution $\mu$ to $\omega \mathcal{D}^\perp$ in $(ker F_*)^\perp$. It is obvious that we have

$$\phi \mathcal{D}^\perp \subseteq \mathcal{D}^\perp, (ker F_*)^\perp = \omega \mathcal{D}^\perp \oplus \mu.$$

Also, for $X \in \Gamma((ker F_*)^\perp)$, we write

$$JX = \mathcal{B}X + CX, \tag{5.49}$$

where $\mathcal{B}X \in \Gamma(\mathcal{D}^\perp)$ and $CX \in \Gamma(\mu)$. Then it is clear that we get

$$\mathcal{B}((ker F_*)^\perp) = \mathcal{D}^\perp. \tag{5.50}$$

Considering (5.46), for $U \in \Gamma(ker F_*)$, we can write

$$JU = P_1 U + P_2 U + \omega U, \tag{5.51}$$

where $P_1$ and $P_2$ are the projections from $ker F_*$ to $\mathcal{D}$ and $\mathcal{D}^\perp$, respectively.

Since $F$ is a subimmersion, it follows that the rank of $F$ is constant on $M_1$, then the rank theorem for functions implies that $ker F_*$ is an integrable subbundle of $TM_1$. Thus it follows from the above definition that the leaves of the distribution $ker F_*$ of a generic Riemannian map are generic submanifolds of $M_1$.

We now give some examples of generic Riemannian maps from almost Hermitian manifolds to Riemannian manifolds.

**Example 39.** Every semi-invariant submersion $F$ is a generic Riemannian map with $(range F_*)^\perp = \{0\}$ and $\mathcal{D}^\perp$ is a totally real distribution.

**Example 40.** Every generic submersion $F$ is a generic Riemannian map with $(range F_*)^\perp = \{0\}$.

**Example 41.** Every semi-invariant Riemannian map $F$ from an almost Hermitian manifold to a Riemannian manifold is a generic Riemannian map such that $\mathcal{D}^\perp$ is a totally real distribution.

Since semi-invariant Riemannian maps include invariant Riemannian maps and anti-invariant Riemannian maps, such Riemannian maps are also examples of generic Riemannian maps. We say that a generic Riemannian map is *proper* if $\mathcal{D}^\perp$ is neither complex nor purely real. We now present an example of proper generic Riemannian maps.

**Example 42.** [255] Consider the following map defined by

$$F: \quad \mathbb{R}^8 \quad \longrightarrow \quad \mathbb{R}^5$$
$$(x^1, x^2, x^3, x^4, x^5, x^6, x^7, x^8) \quad (x^2, x^1, \tfrac{x^5+x^6+x^4}{\sqrt{3}}, 0, \tfrac{x^5-x^6}{\sqrt{2}}).$$

Then we have

$$ker F_* = span\{U_1 = \frac{\partial}{\partial x^8}, U_2 = \frac{\partial}{\partial x^7}, U_3 = \frac{\partial}{\partial x^3}, U_4 = -2\frac{\partial}{\partial x^4} + \frac{\partial}{\partial x^5} + \frac{\partial}{\partial x^6}\}$$

and

$$(ker F_*)^\perp = span\{Z_1 = \frac{\partial}{\partial x^1}, Z_2 = \frac{\partial}{\partial x^2}, Z_3 = \frac{\partial}{\partial x^4} + \frac{\partial}{\partial x^5} + \frac{\partial}{\partial x^6}, Z_4 = \frac{\partial}{\partial x^5} - \frac{\partial}{\partial x^6}\}.$$

Hence it is easy to see that

$$g_{\mathbb{R}^5}(F_*(Z_i), F_*(Z_i)) = g_{\mathbb{R}^8}(Z_i, Z_i), i = 1, 2, 3, 4$$

and

$$g_{\mathbb{R}^5}(F_*(Z_i), F_*(Z_j)) = g_{\mathbb{R}^8}(Z_i, Z_j) = 0,$$

$i \neq j$. Thus $F$ is a Riemannian map. On the other hand, we have $JU_1 = U_2$ and $JU_3 = -\frac{1}{3}U_4 + \frac{1}{3}Z_3$ and $JU_4 = 2U_3 - Z_4$, where $J$ is the complex structure of $\mathbb{R}^6$. Thus $F$ is a generic Riemannian map with $\mathcal{D} = span\{U_1, U_2\}$, $\mathcal{D}^\perp = span\{U_3, U_4\}$

and $\mu = span\{Z_1, Z_2\}$.

We now investigate the effect of a generic Riemannian map on the geometry of the total manifold, the base manifold, and the map itself.

**Lemma 59.** *[255] Let F be a generic Riemannian map from a Kähler manifold $(M, J, g_M)$ to a Riemannian manifold $(N, g_N)$. Then the distribution $\mathcal{D}$ is integrable if and only if the following expression is satisfied:*

$$\mathcal{T}_X JY = \mathcal{T}_Y JX \tag{5.52}$$

*for $X, Y \in \Gamma(\mathcal{D})$.*

*Proof.* From (3.5), (4.12), (5.49), and (5.52), we have

$$\hat{\nabla}_X JY + \mathcal{T}_X JY = P_1 \hat{\nabla}_X Y + P_2 \hat{\nabla}_X Y + \omega \hat{\nabla}_X Y + \mathcal{B}\mathcal{T}_X Y + \mathcal{C}\mathcal{T}_X Y \tag{5.53}$$

for $X, Y \in \Gamma(\mathcal{D})$. Taking the vertical parts and the horizontal parts of (5.53), we get

$$\hat{\nabla}_X JY = P_1 \hat{\nabla}_X Y + P_2 \hat{\nabla}_X Y + \mathcal{B}\mathcal{T}_X Y \tag{5.54}$$

$$\mathcal{T}_X JY = \omega \hat{\nabla}_X Y + \mathcal{C}\mathcal{T}_X Y \tag{5.55}$$

Interchanging the role of $X$ and $Y$ in (5.55), and taking into account that $\mathcal{T}$ is symmetric on the vertical distribution, we derive

$$\mathcal{T}_X JY - \mathcal{T}_Y JX = \omega[X, Y],$$

which gives the proof. $\qquad\square$

In a similar way, we have the following lemma.

**Lemma 60.** *[255]Let F be a generic Riemannian map from a Kähler manifold $(M, J, g_M)$ to a Riemannian manifold $(N, g_N)$. Then the distribution $\mathcal{D}^\perp$ is integrable if and only if the following expression is satisfied:*

$$\hat{\nabla}_{Z_1} P_2 Z_2 - \hat{\nabla}_{Z_2} P_2 Z_1 = P_2[Z_1, Z_2] + \mathcal{T}_{Z_1} \omega Z_2 - \mathcal{T}_{Z_2} \omega Z_1 \tag{5.56}$$

*for $Z_1, Z_2 \in \Gamma(\mathcal{D}^\perp)$.*

We now investigate the geometry of leaves of $\mathcal{D}$ and $\mathcal{D}^\perp$.

**Lemma 61.** *[255] Let F be a generic Riemannian map from a Kähler manifold $(M, J, g_M)$ to a Riemannian manifold $(N, g_N)$. Then the distribution $\mathcal{D}$ defines a totally geodesic* *foliation in M if and only if:*

**(1)** $\hat{\nabla}_X P_2 Z + \mathcal{T}_X \omega Z$ *has no components in* $\mathcal{D}$ *for* $X \in \Gamma(\mathcal{D})$ *and* $Z \in \Gamma(\mathcal{D}^\perp)$,

**(2)** $\hat{\nabla}_X \mathcal{B}W + \mathcal{T}_X CW$ *has no components in* $\mathcal{D}$ *for* $X \in \Gamma(\mathcal{D})$ *and* $W \in \Gamma((\ker F_*)^\perp)$, *and*

*Proof.* From (3.5), (4.12), (5.49), and (5.52), we have

$$g_M(\nabla_X Y, Z) = -g_M(\hat{\nabla}_X P_2 Z + \mathcal{T}_X \omega Z, JY)$$

for $X, Y \in \Gamma(\mathcal{D})$ and $Z \in \Gamma(\mathcal{D}^\perp)$. This gives us (1). Also, from (3.5), (5.49), and (5.52) we get

$$g_M(\nabla_X Y, W) = -g_M(\nabla_X \mathcal{B}W + CW, JY)$$

for $X, Y \in \Gamma(\mathcal{D})$ and $W \in \Gamma((\ker F_*)^\perp)$. Now, using (4.12) and (4.13) we obtain

$$g_M(\nabla_X Y, W) = -g_M(\hat{\nabla}_X \mathcal{B}W + \mathcal{T}_X CW, JY),$$

which gives us (2). □

In a similar way, we have the following result.

**Lemma 62.** *[255] Let F be a generic Riemannian map from a Kähler manifold* $(M, J, g_M)$ *to a Riemannian manifold* $(N, g_N)$. *Then the distribution* $\mathcal{D}^\perp$ *defines a totally geodesic foliation in M if and only if*

**(1)** $\hat{\nabla}_{Z_1} P_2 Z_2 + \mathcal{T}_{Z_1} \omega Z_2 = 0$ *for* $Z_1, Z_2 \in \Gamma(\mathcal{D}^\perp)$.

**(2)** $C\mathcal{H}\nabla_{Z_1} \omega Z_2 + C\mathcal{T}_{Z_1} P_2 Z_2$ *has no components in* $\mu$.

From Lemma 61 and Lemma 62; we obtain the following decomposition theorem.

**Theorem 113.** *[255]Let F be a generic Riemannian map from a Kähler manifold* $(M, J, g_M)$ *to a Riemannian manifold* $(N, g_N)$. *Then the fibers are locally product Riemannian manifold of the form* $M_{\mathcal{D}} \times M_{\mathcal{D}^\perp}$ *if:*

**1.** $\hat{\nabla}_Y P_2 Z_2 + \mathcal{T}_Y \omega Z_2 = 0$ *for* $Y \in \Gamma(\ker F_*)$ *and* $Z_2 \in \Gamma(\mathcal{D}^\perp)$,

**2.** $C\mathcal{H}\nabla_{Z_1} \omega Z_2 + C\mathcal{T}_{Z_1} P_2 Z_2$ *has no components in* $\mu$ $Z_1 \in \Gamma(\mathcal{D}^\perp)$, *and*

**3.** $\hat{\nabla}_X \mathcal{B}W + \mathcal{T}_X CW$ *has no components in* $\mathcal{D}$ *for* $X \in \Gamma(\mathcal{D})$ *and* $W \in \Gamma((\ker F_*)^\perp)$.

**Lemma 63.** *[255] Let F be a generic Riemannian map from a Kähler manifold* $(M, J, g_M)$ *to a Riemannian manifold* $(N, g_N)$. *Then the distribution* $\ker F_*$ *defines a totally geodesic foliation in M if and only if:*

**(1)** $\hat{\nabla}_X P_2 Z + \mathcal{T}_X \omega Z$ *has no components in* $\mathcal{D}$ *for* $X \in \Gamma(\mathcal{D})$ *and* $Z \in \Gamma(\mathcal{D}^\perp)$, *and*

**(2)** $\hat{\nabla}_X \mathcal{B}W + \mathcal{T}_X CW$ *has no components in* $\mathcal{D}$ *for* $X \in \Gamma(\mathcal{D})$ *and* $W \in \Gamma((\ker F_*)^\perp)$.

*Proof.* From (3.5), (5.49), and (5.52), we have

$$
\begin{aligned}
g_M(\nabla_U V, \xi) &= g_M(\hat{\nabla}_U \phi V, \mathcal{B}\xi) + g_M(\mathcal{T}_U \omega V, \mathcal{B}\xi) \\
&\quad + g_M(\mathcal{T}_U \omega V, C\xi) + g_M(\langle \nabla_U \phi V, C\xi)
\end{aligned}
$$

for $U, V \in \Gamma(\ker F_*)$ and $\xi \in \Gamma((\ker F_*)^\perp)$. Also, from (3.5), (5.49), and (5.52) we get

$$
g_M(\nabla_X Y, W) = -g_M(\nabla_X \mathcal{B}W + CW, JY)
$$

for $X, Y \in \Gamma(\mathcal{D})$ and $W \in \Gamma((\ker F_*)^\perp)$. Now, using (4.12) and (4.13), we obtain

$$
g_M(\nabla_X Y, W) = -g_M(\hat{\nabla}_X \mathcal{B}W + \mathcal{T}_X CW, JY)
$$

which gives us (2).  □

In a similar way, we have the following result for distribution $(\ker F_*)^\perp$.

**Lemma 64.** *[255] Let F be a generic Riemannian map from a Kähler manifold $(M, J, g_M)$ to a Riemannian manifold $(N, g_N)$. Then the distribution $(\ker F_*)^\perp$ defines a totally geodesic foliation in M if and only if:*

$$
\mathcal{B}\mathcal{A}_X \mathcal{B}Y + \mathcal{B}H\nabla_X CY = -\phi V \nabla_X \mathcal{B}Y - \phi \mathcal{A}_X CY,
$$

*for $X, Y \in \Gamma((\ker F_*)^\perp)$.*

From Lemma 63 and Lemma 64, we have the following corollary.

**Corollary 52.** *[255] Let F be a generic Riemannian map from a Kähler manifold $(M, J, g_M)$ to a Riemannian manifold $(N, g_N)$. Then M is a locally product Riemannian manifold if and only if:*
**(1)** $\hat{V}_U P_2 Z + \mathcal{T}_U \omega Z$ *has no components in $\mathcal{D}$ for $U \in \Gamma(\mathcal{D})$ and $Z \in \Gamma(\mathcal{D}^\perp)$,*
**(2)** $\hat{V}_U \mathcal{B}W + \mathcal{T}_U CW$ *has no components in $\mathcal{D}$ for $U \in \Gamma(\mathcal{D})$ and $W \in \Gamma((\ker F_*)^\perp)$, and*
**(3)** $\mathcal{B}\mathcal{A}_X \mathcal{B}Y + \mathcal{B}H\nabla_X CY = -\phi V \nabla_X \mathcal{B}Y - \phi \mathcal{A}_X CY$
*for $X, Y \in \Gamma((\ker F_*)^\perp)$.*

In the rest of this section, we are going to find necessary and sufficient conditions for generic Riemannian maps from Kähler manifolds to Riemannian manifolds to be totally geodesic and harmonic, respectively. We recall again that a map between Riemannian manifolds is totally geodesic if $(\nabla F_*) = 0$ on the total manifold. A geometric interpretation of a totally geodesic map is that it maps every geodesic in the total manifold into a geodesic in the base manifold in proportion to arc lengths. We also recall that a map is harmonic if $trace \nabla F_* = 0$. Geometrically, a harmonic map $F$ is a critical point of the energy functional $E$.

**Theorem 114.** *[255] Let $F$ be a generic Riemannian map from a Kähler manifold $(M, J, g_M)$ to a Riemannian manifold $(N, g_N)$. Then $F$ is totally geodesic if and only if the following conditions are satisfied:*

(1) $g_M(\mathcal{A}_X\phi U + \mathcal{H}\nabla_X\omega U, CY) = -g_M(\mathcal{A}_X\omega U + \mathcal{V}\nabla_X\phi U, BY)$,

(2) $g_M(\hat{\mathcal{V}}_U BX + \mathcal{T}_U CX, \phi V) = -g_M(\mathcal{T}_U BX + \mathcal{H}\nabla_U CX, \omega V)$, *and*

(3) $(\nabla F_*)(X, \omega P_2 Y + C\omega Y)$ *has no components in* $(F_*(TM))^\perp$
*for $X, Y \in \Gamma((\ker F_*)^\perp)$ and $U, V \in \Gamma(\ker F_*)$.*

*Proof.* First of all, from Lemma 45, we have $g_M((\nabla F_*)(X, Y), F_*(Z)) = 0$ for $X, Y, Z \in \Gamma((\ker F_*)^\perp)$. For $X, Y \in \Gamma((\ker F_*)^\perp)$ and $U \in \Gamma(\ker F_*)$, using (1.71) and (3.5), we get

$$g_N((\nabla F_*)(U, X), F_*(Y)) = g_M(\nabla_X JU, JY).$$

Then from (5.49) and (5.48), we derive

$$g_N((\nabla F_*)(U, X), F_*(Y)) = g_M(\nabla_X\phi U + \omega U, BY + CY).$$

Using (4.15) and (4.14), we have

$$\begin{aligned} g_N((\nabla F_*)(U, X), F_*(Y)) &= g_M(\mathcal{A}_X\phi U + \mathcal{H}\nabla_X\omega U, CY) \\ &+ g_M(\mathcal{A}_X\omega U + \mathcal{V}\nabla_X\phi U, BY), \end{aligned}$$

which gives us (1). In a similar way, we obtain (2). Now, for $X, Y \in \Gamma((\ker F_*)^\perp)$ and $\xi \in \Gamma((F_*(TM))^\perp)$, we have

$$g_N((\nabla F_*)(X, Y), \xi) = -g_N(F_*(Y), \nabla_X^F\xi).$$

Using (4.16) and (3.3) we get

$$g_N((\nabla F_*)(X, Y), \xi) = -g_N(F_*(J^2 Y), S_\xi F_*(X)).$$

Thus from (5.49) and (5.48), we obtain

$$g_N((\nabla F_*)(X, Y), \xi) = -g_N(F_*(\omega P_2 Y), S_\xi F_*(X)) - g_N(F_*(C\omega Y), S_\xi F_*(X)).$$

Now using (4.17), we arrive at

$$g_N((\nabla F_*)(X, Y), \xi) = -g_N((\nabla F_*)(X, \omega P_2 Y) + (\nabla F_*)(X, C\omega Y), \xi)$$

which gives us (3).                                                            $\square$

**Theorem 115.** *[255] Let $F$ be a generic Riemannian map from a Kähler manifold*

$(M, J, g_M)$ to a Riemannian manifold $(N, g_N)$. Then $F$ is harmonic if and only if

$$trace \mid_{(ker\, F_*)} F_*(C\mathcal{T}_{(.)}\phi(.) + C\mathcal{H}\nabla_{(.)}\omega(.) + \omega\hat{\nabla}_{(.)}\phi(.) + \omega\mathcal{T}_{(.)}\omega(.))$$
$$= trace \mid_{(ker\, F_*)^{\perp}} (\nabla^F_{(.)}F_*(\omega\mathcal{B}(.) + C^2(.)) - F_*(C\mathcal{A}_{(.)}\mathcal{B}(.)$$
$$+ C\mathcal{H}\nabla_{(.)}C(.) + \omega V\nabla_{(.)}\mathcal{B}(.) + \omega\mathcal{A}_{(.)}C(.))).$$

*Proof.* For $X \in \Gamma((ker\, F_*)^{\perp})$ and $U \in \Gamma(ker\, F_*)$, from (3.5), (1.71), (5.48), and (5.49), we have

$$(\nabla F_*)(X, X) \quad + \quad (\nabla F_*)(U, U) = -\nabla^F_X F_*(\omega\mathcal{B}X + C^2 X)$$
$$+ F_*(J(\nabla_X \mathcal{B}X + CX)) + F_*(J(\nabla_U \phi U + \omega U))$$

Using (5.48), (5.49), and (4.12)-(4.15), we have

$$(\nabla F_*)(X, X) + (\nabla F_*)(U, U) = -\nabla^F_X F_*(\omega\mathcal{B}X + C^2 X)$$
$$+ F_*(C\mathcal{T}_U \phi U + C\mathcal{H}\nabla_U \omega(U) + \omega\hat{\nabla}_U \phi U + \omega\mathcal{T}_U \omega U$$
$$+ F_*(C\mathcal{A}_X \mathcal{B}X + C\mathcal{H}\nabla_X CX + \omega V\nabla_X \mathcal{B}X + \omega\mathcal{A}_X CX),$$

which gives us our assertion.                                                                       □

## 5. Slant Riemannian maps from almost Hermitian manifolds

In this section, as a generalization of almost Hermitian submersions, slant submersions, and anti-invariant Riemannian submersions, we introduce slant Riemannian maps from an almost Hermitian manifold to a Riemannian manifold. We first focus on the existence of such maps by giving some examples. Then we investigate the effect of slant Riemannian maps on the geometry of the total manifold, the base manifold, and themselves. More precisely, we investigate the geometry of leaves of distributions on the total manifold arising from such maps. We also obtain necessary and sufficient conditions for slant Riemannian maps to be harmonic and totally geodesic. We first present the following definition.

**Definition 56.** [250] Let $F$ be a Riemannian map from an almost Hermitian manifold $(M_1, g_1, J_1)$ to a Riemannian manifold $(M_2, g_2)$. If for any nonzero vector $X \in \Gamma(ker F_*)$, the angle $\theta(X)$ between $JX$ and the space $ker\, F_*$ is a constant, i.e., it is independent of the choice of the point $p \in M_1$ and choice of the tangent vector $X$ in $ker\, F_*$, then we say that $F$ is a *slant Riemannian map*. In this case, the angle $\theta$ is called the *slant angle* of the slant Riemannian map.

Since $F$ is a subimmersion, it follows that the rank of $F$ is constant on $M_1$, then the rank theorem for functions implies that $ker\, F_*$ is an integrable subbundle of $TM_1$. Thus it follows from the above definition that the leaves of the distribution $ker\, F_*$ of a slant Riemannian map are slant submanifolds of $M_1$. We give some examples of slant Riemannian maps.

**Example 43.** Every Hermitian submersion from an almost Hermitian manifold onto an almost Hermitian manifold is a slant Riemannian map with $\theta = 0$ and $(rangeF_*)^\perp = \{0\}$.

**Example 44.** Every anti-invariant Riemannian submersion from an almost Hermitian manifold onto a Riemannian manifold is a slant Riemannian map with $\theta = \frac{\pi}{2}$ and $(rangeF_*)^\perp = \{0\}$.

**Example 45.** Every proper slant submersion with the slant angle $\theta$ is a slant Riemannian map with $(rangeF_*)^\perp = \{0\}$.

We now denote the Euclidean $2m-$ space with the standard metric by $\mathbb{R}^{2m}$. An almost complex structure $J$ on $\mathbb{R}^{2m}$ is said to be compatible if $(\mathbb{R}^{2m}, J)$ is complex analytically isometric to the complex number space $C^m$ with the standard flat Kählerian metric. Recall that the compatible almost complex structure $J$ on $\mathbb{R}^{2m}$ is defined by

$$J(a^1, ..., a^{2m}) = (-a^{m+1}, -a^{m+2}, ..., -a^{2m}, a^1, a^2, ..., a^m).$$

A slant Riemannian map is said to be proper if it is not a submersion. Here is an example of proper slant Riemannian maps.

**Example 46.** [250] Consider the following Riemannian map given by

$$F: \quad \mathbb{R}^4 \quad \longrightarrow \quad \mathbb{R}^4$$
$$(x_1, x_2, x_3, x_4) \quad (0, \tfrac{x_2 \sin \alpha + x_3 + x_4 \cos \alpha}{\sqrt{2}}, 0, x_2 \cos \alpha - x_4 \sin \alpha).$$

Then for any $0 < \alpha < \frac{\pi}{2}$, $F$ is a slant Riemannian map with respect to the compatible almost complex structure $J$ on $\mathbb{R}^4$ with slant angle $\frac{\pi}{4}$.

Let $F$ be a Riemannian map from a Kähler manifold $(M_1, g_1, J)$ to a Riemannian manifold $(M_2, g_2)$. Then for $X \in \Gamma(ker F_*)$, we write

$$JX = \phi X + \omega X, \tag{5.57}$$

where $\phi X$ and $\omega X$ are vertical and horizontal parts of $JX$. Also for $V \in \Gamma((ker F_*)^\perp)$, we have

$$JZ = \mathcal{B}Z + CZ, \tag{5.58}$$

where $\mathcal{B}Z$ and $CZ$ are vertical and horizontal components of $JZ$. Using (4.12), (4.13),

(5.57), and (5.59), we obtain

$$(\nabla_X \omega)Y = C\mathcal{T}_X Y - \mathcal{T}_X \phi Y \tag{5.59}$$

$$(\nabla_X \phi)Y = \mathcal{B}\mathcal{T}_X Y - \mathcal{T}_X \omega Y, \tag{5.60}$$

where $\nabla$ is the Levi-Civita connection on $M_1$ and

$$(\nabla_X \omega)Y = \mathcal{H}\nabla_X \omega Y - \omega\hat{\nabla}_X Y$$

$$(\nabla_X \phi)Y = \hat{\nabla}_X \phi Y - \phi\hat{\nabla}_X Y$$

for $X, Y \in \Gamma(ker F_*)$. Let $F$ be a slant Riemannian map from an almost Hermitian manifold $(M_1, g_1, J_1)$ to a Riemannian manifold $(M_2, g_2)$, then we say that $\omega$ is parallel with respect to the Levi-Civita connection $\nabla$ on $ker F_*$ if its covariant derivative with respect to $\nabla$ vanishes, i.e., we have

$$(\nabla_X \omega)Y = \nabla_X \omega Y - \omega(\nabla_X Y) = 0$$

for $X, Y \in \Gamma(ker F_*)$. Let $F$ be a slant Riemannian map from a complex $m$-dimensional Hermitian manifold $(M, g_1, J)$ to a Riemannian manifold $(N, g_2)$. Then, $\omega(ker F_*)$ is a subspace of $(ker F_*)^\perp$. Thus it follows that $ker F_{*p} \oplus \omega(ker F_{*p})$ is invariant with respect to $J$. Then for every $p \in M$, there exists an invariant subspace $\mu_p$ of $(ker F_{*p})^\perp$ such that

$$T_p M = ker F_{*p} \oplus \omega(ker F_{*p}) \oplus \mu_p.$$

The proof of the following result is exactly the same as that for slant immersions (see [70] or [53] and [48] for the Sasakian case), therefore we omit its proof.

**Theorem 116.** *[250] Let $F$ be a Riemannian map from an almost Hermitian manifold $(M_1, g_1, J)$ to a Riemannian manifold $(M_2, g_2)$. Then $F$ is a slant Riemannian map if and only if there exists a constant $\lambda \in [-1, 0]$ such that*

$$\phi^2 X = \lambda X$$

*for $X \in \Gamma(ker F_*)$. If $F$ is a slant Riemannian map, then $\lambda = -\cos^2 \theta$.*

By using the above theorem, it is easy to see that

$$g_1(\phi X, \phi Y) = \cos^2 \theta g_1(X, Y) \tag{5.61}$$

$$g_1(\omega X, \omega Y) = \sin^2 \theta g_1(X, Y) \tag{5.62}$$

for any $X, Y \in \Gamma(ker F_*)$. Also by using (5.61) we can easily conclude that

$$\{e_1, \sec \theta \phi e_1, e_2, \sec \theta \phi e_2, ..., e_n, \sec \theta \phi e_n\}$$

is an orthonormal frame for $\Gamma(ker F_*)$. On the other hand, by using (5.62), we can see that

$$\{\csc \theta \omega e_1, \csc \theta \omega e_2, ..., \csc \theta \omega e_n\}$$

is an orthonormal frame for $\Gamma(\omega(ker\, F_*))$. As in slant immersions, we call the frame

$$\{e_1, \sec \theta \phi e_1, e_2, \sec \theta \phi e_2, ..., e_n, \sec \theta \phi e_n, \csc \theta \omega e_1, \csc \theta \omega e_2, ..., \csc \theta \omega e_n\}$$

an adapted frame for slant Riemannian maps.

We note that since the distribution $ker\, F_*$ is integrable, it follows that $\mathcal{T}_X Y = \mathcal{T}_Y X$ for $X, Y \in \Gamma(ker\, F_*)$. Then the following lemma can be obtained by using Theorem 116.

**Lemma 65.** *[250] Let F be a slant Riemannian map from a Kähler manifold to a Riemannian manifold. If $\omega$ is parallel with respect to $\nabla$ on $ker\, F_*$, then*

$$\mathcal{T}_{\phi X} \phi X = - \cos^2 \theta \mathcal{T}_X X \tag{5.63}$$

*for $X \in \Gamma(ker\, F_*)$.*

In fact, the proof of the above lemma is exactly the sameas that for the Lemma 23 given in Chapter 3.

We now give necessary and sufficient conditions for $F$ to be harmonic.

**Theorem 117.** *[250] Let F be a slant Riemannian map from a Kähler manifold to a Riemannian manifold. Then F is harmonic if and only if*

$$\mathcal{T}_{\phi e_i} \phi e_i = -\cos^2 \theta \mathcal{T}_{e_i} e_i, \tag{5.64}$$

$$trace \mid_{\omega(ker\, F_*)} {}^*F_*(\mathcal{S}_{E_j} F_*(.)) \in \Gamma(\mu), \tag{5.65}$$

*and*

$$trace \mid_{\mu} {}^*F_*(\mathcal{S}_{E_j} F_*(.)) \in \Gamma(\omega(ker F_*)), \tag{5.66}$$

*where $\{e_1, \sec \theta \phi e_1, e_2, \sec \theta \phi e_2, ..., e_n, \sec \theta \phi e_n\}$ is an orthonormal frame for $\Gamma(ker\, F_*)$ and $\{E_k\}$ is an orthonormal frame of $\Gamma((range F_*)^\perp)$.*

*Proof.* We choose a canonical orthonormal frame

$$e_1, \sec \theta \phi e_1 ..., e_p, \sec \theta \phi e_p, \omega \csc \theta e_1, ..., \omega \csc \theta e_{2p}, \bar{e}_1, ..., \bar{e}_n$$

such that $\{e_1, \sec \theta \phi e_1 ..., e_p, \sec \theta \phi e_p\}$ is an orthonormal basis of $ker F_*$ and $\{\bar{e}_1, .., \bar{e}_n\}$

of $\mu$, where $\theta$ is the slant angle. Then $F$ is harmonic if and only if

$$\sum_{i=1}^{p}(\nabla F_*)(e_i, e_i) \quad + \quad \sec^2 \theta(\nabla F_*)(\phi e_i, \phi e_i) + \csc^2 \theta \sum_{i=1}^{2p}(\nabla F_*)(\omega e_i, \omega e_i)$$

$$+ \sum_{j=1}^{m}(\nabla F_*)(\bar{e}_j, \bar{e}_j) = 0. \tag{5.67}$$

By using (1.71) and (4.12), we have

$$\sum_{i=1}^{p}((\nabla F_*)(e_i, e_i) + \sec^2 \theta(\nabla F_*)(\phi e_i, \phi e_i) = -F_*(\mathcal{T}_{e_i}e_i + \sec^2 \theta \mathcal{T}_{\phi e_i}\phi e_i). \tag{5.68}$$

On the other hand, from Lemma 45, we know that

$$\csc^2 \theta \sum_{i=1}^{2p}(\nabla F_*)(\omega e_i, \omega e_i) + \sum_{j=1}^{m}(\nabla F_*)(\bar{e}_j, \bar{e}_j) \in \Gamma((rangeF_*)^\perp).$$

Thus we can write

$$\csc^2 \theta \sum_{i=1}^{2p}(\nabla F_*)(\omega e_i, \omega e_i) + \sum_{j=1}^{m}(\nabla F_*)(\bar{e}_j, \bar{e}_j) = \csc^2 \theta$$

$$\sum_{i=1}^{2p}\sum_{k=1}^{s} g_2((\nabla F_*)(\omega e_i, \omega e_i), E_k)E_k$$

$$+ \sum_{j=1}^{m}\sum_{k=1}^{s} g_2((\nabla F_*)(\bar{e}_j, \bar{e}_j), E_k)E_k,$$

where$\{E_k\}$ is an orthonormal basis of $\Gamma((rangeF_*)^\perp)$. Then, using (4.17), we have

$$\csc^2 \theta \sum_{i=1}^{2p}(\nabla F_*)(\omega e_i, \omega e_i) + \sum_{j=1}^{m}(\nabla F_*)(\bar{e}_j, \bar{e}_j) = \csc^2 \theta$$

$$\sum_{i=1}^{2p}\sum_{k=1}^{s} g_2(S_{E_k}F_*(\omega e_i), F_*(\omega e_i))E_k$$

$$+ \sum_{j=1}^{m}\sum_{k=1}^{s} g_2(S_{E_k}F_*(\bar{e}_j), F_*(\bar{e}_j))E_k. \tag{5.69}$$

Then, the proof comes from the adjoint of $F_*$ and (5.68) and (5.69). $\qquad\square$

**Example 47.** Consider the slant Riemannian map given in Example 46, then we have

$$(ker F_*) = S pan\{Z_1 = \frac{\partial}{\partial x_1}, Z_2 = \sin \alpha \frac{\partial}{\partial x_2} - \frac{\partial}{\partial x_3} + \cos \alpha \frac{\partial}{\partial x_4}\}$$

and

$$(ker F_*)^\perp = S pan\{Z_3 = \frac{\sin \alpha}{\sqrt{2}} \frac{\partial}{\partial x_2} + \frac{1}{\sqrt{2}} \frac{\partial}{\partial x_3} + \frac{\cos \alpha}{\sqrt{2}} \frac{\partial}{\partial x_4},$$

$$Z_4 = -\cos \alpha \frac{\partial}{\partial x_2} + \sin \alpha \frac{\partial}{\partial x_4}\}.$$

By direct computations, we have

$$JZ_1 = -\frac{1}{2}Z_2 + \frac{1}{\sqrt{2}}Z_3, JZ_2 = Z_1 + Z_4,$$

which imply that

$$\phi Z_1 = -\frac{1}{2}Z_2, \phi Z_2 = Z_1.$$

Then it is easy to see that

$$\phi^2 Z_i = -\cos^2 \frac{\pi}{4} Z_i = -\frac{1}{2}Z_i, i = 1, 2,$$

which is the statement of Theorem 116. On the other hand, since $\mathcal{T}$ and $\mathcal{S}$ vanish for this slant Riemannian map, it satisfies the claim of Theorem 117.

By using (4.12) and (5.59), we can notice that the equality (5.64) is satisfied in terms of the tensor field $\omega$. More precisely, we have the following.

**Lemma 66.** *[250] Let F be a slant Riemannian map from a Kähler manifold $(M_1, g_1, J)$ to a Riemannian manifold $(M_2, g_2)$. If $\omega$ is parallel, then (5.64) is satisfied.*

**Remark 18.** We note that the equality (5.63) (as a result of above lemma, parallel $\omega$) is enough for a slant submersion to be harmonic; however, for a slant Riemannian map, this case is not valid anymore.

We now investigate necessary and sufficient conditions for a slant Riemannian map $F$ to be totally geodesic.

**Theorem 118.** *[250] Let F be a slant Riemannian map from a Kähler manifold $(M_1, g_1, J)$ to a Riemannian manifold $(M_2, g_2)$. Then F is totally geodesic if and only if*

$$g_1(\mathcal{T}_U \omega V, \mathcal{B}X) = -g_2((\nabla F_*)(U, \omega \phi V), F_*(X)) + g_2((\nabla F_*)(U, \omega V), F_*(CX))$$

$$g_1(\mathcal{A}_X \omega U, \mathcal{B}Y) = g_2(\nabla^F_X F_*(\omega \phi U), F_*(Y)) - g_2(\nabla^F_X F_*(\omega U), F_*(CY))$$

*and*

$$\nabla^F_X F_*(Y) + F_*(C(\mathcal{A}_X \mathcal{B}Y + \mathcal{H}\nabla^1_X CY) + \omega(V\nabla^1_X \mathcal{B}Y + \mathcal{A}_X CY)) \in \Gamma(range F_*)$$

*for $X, Y \in \Gamma((\ker F_*)^\perp)$ and $U, V \in \Gamma(\ker F_*)$, where $\nabla^1$ is the Levi-Civita connection of $M_1$.*

*Proof.* From the decomposition of the total manifold of a slant Riemannian map, it follows that $F$ is totally geodesic if and only if

$$g_2((\nabla F_*)(U, V), F_*(X)) = 0, \, g_2((\nabla F_*)(X, U), F_*(Y)) = 0$$

and $(\nabla F_*)(X, Y) = 0$ for $X, Y \in \Gamma((\ker F_*)^\perp)$ and $U, V \in \Gamma(\ker F_*)$. First, since $F$ is a Riemannian map, from (1.71) we obtain

$$g_2((\nabla F_*)(U, V), F_*(X)) = -g_1(\nabla^1_U V, X).$$

Since $M_1$ is a Kähler manifold, using (5.57) and (5.58) we have

$$\begin{aligned}
g_2((\nabla F_*)(U, V), F_*(X)) &= -\cos^2 \theta \, g_1(\nabla^1_U V, X) + g_1(\nabla^1_U \omega \phi V, X) \\
&\quad - g_1(\nabla^1_U \omega V, \mathcal{B}X) - g_1(\nabla^1_U \omega V, CX).
\end{aligned}$$

Taking into account that $F$ is a Riemannian map, again using (1.71) and (4.13) we get

$$\begin{aligned}
g_2((\nabla F_*)(U, V), F_*(X)) &= \sec^2 \theta \{-g_1(\mathcal{T}_U \omega V, \mathcal{B}X) - g_2((\nabla F_*)(U, \omega \phi V), F_*(X)) \\
&\quad + g_2((\nabla F_*)(U, \omega V), F_*(CX))\}. \tag{5.70}
\end{aligned}$$

In a similar way, we also have

$$\begin{aligned}
g_2((\nabla F_*)(X, U), F_*(Y)) &= \sec^2 \theta \{-g_1(\mathcal{A}_X \omega U, \mathcal{B}Y) - g_2(\nabla^F_X F_*(\omega U), F_*(CY)) \\
&\quad + g_2(\nabla^F_X F_*(\omega \phi U), F_*(Y))\}. \tag{5.71}
\end{aligned}$$

On the other hand, by using (1.71) and (4.16), we derive

$$(\nabla F_*)(X, Y) = \nabla^F_X F_*(Y) + F_*(J\nabla^1_X JY)$$

for $X, Y \in \Gamma((\ker F_*)^\perp)$. Then, using (5.57), (5.58,) and (4.12)-(4.15), we obtain

$$\begin{aligned}
(\nabla F_*)(X, Y) &= \nabla^F_X F_*(Y) + F_*(\mathcal{B}\mathcal{A}_X \mathcal{B}Y \\
&\quad + C\mathcal{A}_X \mathcal{B}Y + \phi V\nabla^1_X \mathcal{B}Y + \omega V\nabla^1_X \mathcal{B}Y \\
&\quad + \mathcal{B}\mathcal{H}\nabla^1_X CY + C\mathcal{H}\nabla^1_X CY \\
&\quad + \phi \mathcal{A}_X CY + \omega \mathcal{A}_X CY).
\end{aligned}$$

Since

$$\mathcal{B}\mathcal{A}_X \mathcal{B}Y + \phi V\nabla^1_X \mathcal{B}Y + \mathcal{B}\mathcal{H}\nabla^1_X CY + \phi \mathcal{A}_X CY \in \Gamma(\ker F_*),$$

we have

$$
\begin{aligned}
(\nabla F_*)(X, Y) &= \nabla_X^F F_*(Y) + F_*(C\mathscr{A}_X \mathscr{B} Y \\
&+ \omega \mathscr{V} \nabla_X^1 \mathscr{B} Y + C \mathscr{H} \nabla_X^1 C Y \\
&+ \omega \mathscr{A}_X C Y).
\end{aligned}
\tag{5.72}
$$

Then, the proof comes from (5.70), (5.71), and (5.72).    □

**Remark 19.** We observe that the conditions for a slant Riemannian map to be totally geodesic are different from the conditions for a slant submersion to be totally geodesic. For a Riemannian submersion, the second fundamental form satisfies $(\nabla F_*)(X, Y) = 0$, $X, Y \in \Gamma((\ker F_*)^\perp)$. However, for a slant Riemannian map, there is no guarantee that $(\nabla F_*)(X, Y) = 0$, $X, Y \in \Gamma((\ker F_*)^\perp)$. From Lemma 45, we only know that $(\nabla F_*)(X, Y)$ is $\Gamma((range F_*)^\perp)$– valued. Due to the above reason, it is necessary to use extra geometric conditions to investigate the geometry of slant Riemannian maps.

In the rest of this section, we are going to obtain necessary and sufficient conditions for the total manifold of a slant Riemannian map to be a locally product Riemannian manifold. Let $g$ be a Riemannian metric tensor on the manifold $M = B \times F$ and assume that the canonical foliations $D$ and $\bar{D}$ intersect perpendicularly everywhere. Then from de Rham's theorem [91] (see also Theorem 6 (iv)), we know that $g$ is the metric tensor of a usual product Riemannian manifold if and only if $D$ and $\bar{D}$ are totally geodesic foliations.

**Theorem 119.** *[250] Let F be a slant Riemannian map from a Kähler manifold* $(M_1, g_1, J)$ *to a Riemannian manifold* $(M_2, g_2)$*. Then* $(M_1, g_1)$ *is a locally product Riemannian manifold if and only if*

$$
g_1(\mathcal{T}_U \omega V, \mathscr{B} X) = -g_2((\nabla F_*)(U, \omega \phi V), F_*(X)) + g_2((\nabla F_*)(U, \omega V), F_*(CX))
$$

*and*

$$
g_2((\nabla F_*)(X, \mathscr{B} Y), F_*(\omega U)) = g_2(F_*(Y), \nabla_X^F F_*(\omega \phi U)) - g_2(F_*(CY), \nabla_X^F F_*(\omega U))
$$

*for* $X, Y \in \Gamma((\ker F_*)^\perp)$ *and* $U, V \in \Gamma(\ker F_*)$*.*

*Proof.* For $X, Y \in \Gamma((\ker F_*)^\perp)$ and $U \in \Gamma(\ker F_*)$, from (4.16), (5.57), and (5.58), we have

$$
\begin{aligned}
g_1(\nabla_X^1 Y, U) &= -\cos^2 \theta \, g_1(Y, \nabla_X^1 U) + g_1(Y, \nabla_X^1 \omega \phi U) \\
&- g_1(\mathscr{B} Y, \nabla_X^1 \omega U) - g_1(CY, \nabla_X^1 \omega U).
\end{aligned}
$$

Taking into account that $F$ is a Riemannian map and using (1.71) we obtain

$$
\begin{aligned}
g_1(\nabla_X^1 Y, U) &= \sec^2 \theta \{ -g_2(F_*(Y), (\nabla F_*)(X, \omega\phi U)) + g_2(F_*(Y), \nabla_X^F F_*(\omega\phi U)) \\
&\quad - g_2((\nabla F_*)(X, \mathcal{B}Y), F_*(\omega U)) + g_2((\nabla F_*)(X, \omega U), F_*(CY)) \\
&\quad - g_2(F_*(CY), \nabla_X^F F_*(\omega U)) \}.
\end{aligned}
$$

Then Lemma 2.1 implies that

$$
\begin{aligned}
g_1(\nabla_X^1 Y, U) &= \sec^2 \theta \{ g_2(F_*(Y), \nabla_X^F F_*(\omega\phi U)) - g_2((\nabla F_*)(X, \mathcal{B}Y), F_*(\omega U)) \\
&\quad - g_2(F_*(CY), \nabla_X^F F_*(\omega U)).
\end{aligned}
\tag{5.73}
$$

Thus, the proof follows from (5.70) and (5.73).    □

## 6. Semi-slant Riemannian maps from almost Hermitian manifolds

In this section, as a generalization of slant Riemannian maps and semi-invariant Riemannian maps, we introduce and study semi-slant Riemannian maps from an almost Hermitian manifold to a Riemannian manifold.

**Definition 57.** [217] Let $(M, g_M, J)$ be an almost Hermitian manifold and $(N, g_N)$ a Riemannian manifold. A Riemannian map $F : (M, g_M, J) \mapsto (N, g_N)$ is called a *semi-slant Riemannian map* if there is a distribution $\mathcal{D}_1 \subset \ker F_*$ such that

$$
\ker F_* = \mathcal{D}_1 \oplus \mathcal{D}_2, \quad J(\mathcal{D}_1) = \mathcal{D}_1,
$$

and the angle $\theta = \theta(X)$ between $JX$ and the space $(\mathcal{D}_2)_p$ is constant for nonzero $X \in (\mathcal{D}_2)_p$ and $p \in M$, where $\mathcal{D}_2$ is the orthogonal complement of $\mathcal{D}_1$ in $\ker F_*$.

The angle $\theta$ is called a *semi-slant angle*. Let $F : (M, g_M, J) \mapsto (N, g_N)$ be a semi-slant Riemannian map. Then there is a distribution $\mathcal{D}_1 \subset \ker F_*$ such that

$$
\ker F_* = \mathcal{D}_1 \oplus \mathcal{D}_2, \quad J(\mathcal{D}_1) = \mathcal{D}_1,
$$

and the angle $\theta = \theta(X)$ between $JX$ and the space $(\mathcal{D}_2)_p$ is constant for nonzero $X \in (\mathcal{D}_2)_p$ and $p \in M$, where $\mathcal{D}_2$ is the orthogonal complement of $\mathcal{D}_1$ in $\ker F_*$.

Since $F$ is a subimmersion, it follows that the rank of $F$ is constant on $M_1$, then the rank theorem for functions implies that $\ker F_*$ is an integrable subbundle of $TM_1$. Thus, it follows from the above definition that the leaves of the distribution $\ker F_*$ of a semi-slant Riemannian map are semi-slant submanifolds of $M_1$.

We now give examples of semi-slant Riemannian maps. We note that throughout this section, we consider the canonical almost complex structure $J$ on $\mathbb{R}^{2n}$ as follows:

$$J(a_1\frac{\partial}{\partial x_1} + a_2\frac{\partial}{\partial x_2} + \cdots + a_{2n-1}\frac{\partial}{\partial x_{2n-1}} + a_{2n}\frac{\partial}{\partial x_{2n}}) = -a_2\frac{\partial}{\partial x_1} + a_1\frac{\partial}{\partial x_2} + \cdots$$

$$-a_{2n}\frac{\partial}{\partial x_{2n-1}} + a_{2n-1}\frac{\partial}{\partial x_{2n}},$$

where $a_1, \cdots, a_{2n} \in \mathbb{R}$.

**Example 48.** Let $F$ be an almost Hermitian submersion from an almost Hermitian manifold $(M, g_M, J_M)$ onto an almost Hermitian manifold $(N, g_N, J_N)$. Then the map $F$ is a semi-slant Riemannian map with $\mathcal{D}_1 = \ker F_*$.

**Example 49.** Let $F$ be a slant submersion from an almost Hermitian manifold $(M, g_M, J)$ onto a Riemannian manifold $(N, g_N)$ with the slant angle $\theta$. Then the map $F$ is a semi-slant Riemannian map such that $\mathcal{D}_2 = \ker F_*$ and the semi-slant angle $\theta$.

**Example 50.** Let $F$ be an anti-invariant submersion from an almost Hermitian manifold $(M, g_M, J)$ onto a Riemannian manifold $(N, g_N)$. Then the map $F$ is a semi-slant Riemannian map such that $\mathcal{D}_2 = \ker F_*$ and the semi-slant angle $\theta = \frac{\pi}{2}$.

**Example 51.** Let $F$ be a semi-invariant submersion from an almost Hermitian manifold $(M, g_M, J)$ onto a Riemannian manifold $(N, g_N)$. Then the map $F$ is a semi-slant Riemannian map with the semi-slant angle $\theta = \frac{\pi}{2}$.

**Example 52.** Let $F$ be a semi-slant submersion from an almost Hermitian manifold $(M, g_M, J)$ onto a Riemannian manifold $(N, g_N)$ with the semi-slant angle $\theta$. Then the map $F$ is a semi-slant Riemannian map with the semi-slant angle $\theta$.

**Example 53.** [217] Let $(M, g_M, J)$ be a $2m$-dimensional almost Hermitian manifold and $(N, g_N)$ a $(2m - 1)$-dimensional Riemannian manifold. Let $F$ be a Riemannian map from an almost Hermitian manifold $(M, g_M, J)$ to a Riemannian manifold $(N, g_N)$ with $rankF = 2m - 1$. Then the map $F$ is a semi-slant Riemannian map such that $\mathcal{D}_2 = \ker F_*$ and the semi-slant angle $\theta = \frac{\pi}{2}$.

**Example 54.** [217] Define a map $F : \mathbb{R}^8 \mapsto \mathbb{R}^5$ by

$$F(x_1, x_2, \cdots, x_8) = (x_2, x_1, \frac{x_5 \cos\alpha + x_6 \sin\alpha + x_4}{\sqrt{2}}, 0, x_5 \sin\alpha - x_6 \cos\alpha),$$

with $\alpha \in (0, \frac{\pi}{2})$. Then the map $F$ is a semi-slant Riemannian map such that

$$\mathcal{D}_1 = \{\frac{\partial}{\partial x_7}, \frac{\partial}{\partial x_8}\} \text{ and } \mathcal{D}_2 = \{\frac{\partial}{\partial x_3}, \cos \alpha \frac{\partial}{\partial x_5} + \sin \alpha \frac{\partial}{\partial x_6} - \frac{\partial}{\partial x_4}\}$$

with the semi-slant angle $\theta = \frac{\pi}{4}$.

**Example 55.** [217] Define a map $F : \mathbb{R}^6 \mapsto \mathbb{R}^3$ by

$$F(x_1, x_2, \cdots, x_6) = (x_1 \cos \alpha - x_3 \sin \alpha, c, x_4),$$

where $\alpha \in (0, \frac{\pi}{2})$ and $c \in \mathbb{R}$. Then the map $F$ is a semi-slant Riemannian map such that

$$\mathcal{D}_1 = \{\frac{\partial}{\partial x_5}, \frac{\partial}{\partial x_6}\} \text{ and } \mathcal{D}_2 = \{\frac{\partial}{\partial x_2}, \sin \alpha \frac{\partial}{\partial x_1} + \cos \alpha \frac{\partial}{\partial x_3}\}$$

with the semi-slant angle $\theta = \alpha$.

**Example 56.** [217] Define a map $F : \mathbb{R}^{10} \mapsto \mathbb{R}^7$ by

$$F(x_1, x_2, \cdots, x_{10}) = (x_4, 0, x_3, \frac{x_5 - x_6}{\sqrt{2}}, 0, \frac{x_7 + x_9}{\sqrt{2}}, \frac{x_8 + x_{10}}{\sqrt{2}}).$$

Then the map $F$ is a semi-slant Riemannian map such that

$$\mathcal{D}_1 = \{\frac{\partial}{\partial x_1}, \frac{\partial}{\partial x_2}, -\frac{\partial}{\partial x_7} + \frac{\partial}{\partial x_9}, -\frac{\partial}{\partial x_8} + \frac{\partial}{\partial x_{10}}\} \text{ and } \mathcal{D}_2 = \{\frac{\partial}{\partial x_5} + \frac{\partial}{\partial x_6}\},$$

with the semi-slant angle $\theta = \frac{\pi}{2}$.

**Example 57.** [217] Define a map $F : \mathbb{R}^{10} \mapsto \mathbb{R}^5$ by

$$F(x_1, x_2, \cdots, x_{10}) = (\frac{x_3 + x_5}{\sqrt{2}}, 2012, x_6, \frac{x_7 + x_9}{\sqrt{2}}, x_8).$$

Then the map $F$ is a semi-slant Riemannian map such that

$$\mathcal{D}_1 = \{\frac{\partial}{\partial x_1}, \frac{\partial}{\partial x_2}\} \text{ and } \mathcal{D}_2 = \{\frac{\partial}{\partial x_3} - \frac{\partial}{\partial x_5}, \frac{\partial}{\partial x_7} - \frac{\partial}{\partial x_9}, \frac{\partial}{\partial x_4}, \frac{\partial}{\partial x_{10}}\},$$

with the semi-slant angle $\theta = \frac{\pi}{4}$.

**Example 58.** [217] Define a map $F : \mathbb{R}^8 \mapsto \mathbb{R}^5$ by

$$F(x_1, x_2, \cdots, x_8) = (x_8, x_7, \gamma, x_3 \cos \alpha - x_5 \sin \alpha, x_4 \sin \beta - x_6 \cos \beta),$$

where $\alpha, \beta, \gamma$ are constant. Then the map $F$ is a semi-slant Riemannian map such that

$$\mathcal{D}_1 = \{\frac{\partial}{\partial x_1}, \frac{\partial}{\partial x_2}\} \text{ and } \mathcal{D}_2 = \{\sin \alpha \frac{\partial}{\partial x_3} + \cos \alpha \frac{\partial}{\partial x_5}, \cos \beta \frac{\partial}{\partial x_4} + \sin \beta \frac{\partial}{\partial x_6}\}$$

with the semi-slant angle $\theta$ with $\cos\theta = |\sin(\alpha + \beta)|$.

For $X \in \Gamma(\ker F_*)$, we write

$$X = PX + QX, \tag{5.74}$$

where $PX \in \Gamma(\mathcal{D}_1)$ and $QX \in \Gamma(\mathcal{D}_2)$. For $X \in \Gamma(\ker F_*)$, we write

$$JX = \phi X + \omega X, \tag{5.75}$$

where $\phi X \in \Gamma(\ker F_*)$ and $\omega X \in \Gamma((\ker F_*)^\perp)$. For $Z \in \Gamma((\ker F_*)^\perp)$, we have

$$JZ = BZ + CZ, \tag{5.76}$$

where $BZ \in \Gamma(\ker F_*)$ and $CZ \in \Gamma((\ker F_*)^\perp)$. For $U \in \Gamma(TM)$, we obtain

$$U = \mathcal{V}U + \mathcal{H}U, \tag{5.77}$$

where $\mathcal{V}U \in \Gamma(\ker F_*)$ and $\mathcal{H}U \in \Gamma((\ker F_*)^\perp)$. For $W \in \Gamma(F^{-1}TN)$, we write

$$W = \bar{P}W + \bar{Q}W, \tag{5.78}$$

where $\bar{P}W \in \Gamma(rangeF_*)$ and $\bar{Q}W \in \Gamma((rangeF_*)^\perp)$. Then

$$(\ker F_*)^\perp = \omega \mathcal{D}_2 \oplus \mu, \tag{5.79}$$

where $\mu$ is the orthogonal complement of $\omega\mathcal{D}_2$ in $(\ker F_*)^\perp$ and is invariant under $J$. Furthermore, we have the following facts

$$\phi\mathcal{D}_1 = \mathcal{D}_1, \omega\mathcal{D}_1 = 0, \phi\mathcal{D}_2 \subset \mathcal{D}_2, B((\ker F_*)^\perp) = \mathcal{D}_2$$
$$\phi^2 + B\omega = -id, C^2 + \omega B = -id, \omega\phi + C\omega = 0, BC + \phi B = 0.$$

For $X, Y \in \Gamma(\ker F_*)$, define

$$\widehat{\nabla}_X Y := \mathcal{V}\nabla_X Y \tag{5.80}$$

$$(\nabla_X \phi)Y \quad := \quad \widehat{\nabla}_X \phi Y - \phi\widehat{\nabla}_X Y \tag{5.81}$$
$$(\nabla_X \omega)Y \quad := \quad \mathcal{H}\nabla_X \omega Y - \omega\widehat{\nabla}_X Y. \tag{5.82}$$

Then from (5.80), (5.81), (5.82), (4.9), and (4.10), we have the following result.

**Lemma 67.** *[217] Let $(M, g_M, J)$ be a Kähler manifold and $(N, g_N)$ a Riemannian manifold. Let $F : (M, g_M, J) \mapsto (N, g_N)$ be a semi-slant Riemannian map. Then we get:*
**(1)**

$$\widehat{\nabla}_X \phi Y + \mathcal{T}_X \omega Y = \phi\widehat{\nabla}_X Y + B\mathcal{T}_X Y$$
$$\mathcal{T}_X \phi Y + \mathcal{H}\nabla_X \omega Y = \omega\widehat{\nabla}_X Y + C\mathcal{T}_X Y,$$

*for* $X, Y \in \Gamma(\ker F_*)$.

**(2)**

$$\mathcal{V}\nabla_Z BW + \mathcal{A}_Z CW = \phi\mathcal{A}_Z W + B\mathcal{H}\nabla_Z W$$

$$\mathcal{A}_Z BW + \mathcal{H}\nabla_Z CW = \omega\mathcal{A}_Z W + C\mathcal{H}\nabla_Z W, and$$

*for* $Z, W \in \Gamma((\ker F_*)^\perp)$.

**(3)**

$$\widehat{\nabla}_X BZ + \mathcal{T}_X CZ = \phi\mathcal{T}_X Z + B\mathcal{H}\nabla_X Z$$

$$\mathcal{T}_X BZ + \mathcal{H}\nabla_X CZ = \omega\mathcal{T}_X Z + C\mathcal{H}\nabla_X Z$$

$$\mathcal{V}\nabla_Z \phi X + \mathcal{A}_Z \omega X = \phi\mathcal{V}\nabla_Z X + B\mathcal{A}_Z X$$

$$\mathcal{A}_Z \phi X + \mathcal{H}\nabla_Z \omega X = \omega\mathcal{V}\nabla_Z X + C\mathcal{A}_Z X$$

*for* $X \in \Gamma(\ker F_*)$ *and* $Z \in \Gamma((\ker F_*)^\perp)$.

Let $F$ be a slant Riemannian map from an almost Hermitian manifold $(M, g_M, J)$ to a Riemannian manifold $(N, g_N)$ with the slant angle $\theta$. Then, given the non-vanishing $X \in \Gamma(\ker F_*)$, we have

$$\cos\theta = \frac{|\phi X|}{|JX|} \quad \text{and} \quad \cos\theta = \frac{g_M(JX, \phi X)}{|JX| \cdot |\phi X|} = \frac{-g_M(X, \phi^2 X)}{|X| \cdot |\phi X|}.$$

Thus we find

$$\cos^2\theta = \frac{-g_M(X, \phi^2 X)}{|X|^2},$$

which means

$$\phi^2 X = -\cos^2\theta \cdot X. \tag{5.83}$$

Furthermore, if $(M, g_M, J)$ is Kähler, then it is easy to get

$$(\nabla_X \omega)Y = C\mathcal{T}_X Y - \mathcal{T}_X \phi Y \tag{5.84}$$

$$(\nabla_X \phi)Y = B\mathcal{T}_X Y - \mathcal{T}_X \omega Y \tag{5.85}$$

for $X, Y \in \Gamma(\ker F_*)$. Assume that the tensor $\omega$ is parallel. Then we get

$$C\mathcal{T}_X Y = \mathcal{T}_X \phi Y \quad \text{for } X, Y \in \Gamma(\ker F_*).$$

Interchanging the roles of $X$ and $Y$, we derive

$$C\mathcal{T}_Y X = \mathcal{T}_Y \phi X \quad \text{for } X, Y \in \Gamma(\ker F_*).$$

Hence, we obtain

$$\mathcal{T}_X \phi Y = \mathcal{T}_Y \phi X \quad \text{for } X, Y \in \Gamma(\ker F_*).$$

Substituting $Y$ by $\phi X$ and using (5.83), we arrive at

$$\mathcal{T}_{\phi X}\phi X = -\cos^2\theta \cdot \mathcal{T}_X X \quad \text{for } X \in \Gamma(\ker F_*). \tag{5.86}$$

In a similar way, we have the following theorem.

**Theorem 120.** *[217] Let F be a semi-slant Riemannian map from an almost Hermitian manifold $(M, g_M, J)$ to a Riemannian manifold $(N, g_N)$ with the semi-slant angle $\theta$. Then we obtain*

$$\phi^2 X = -\cos^2\theta \cdot X \quad \text{for } X \in \Gamma(\mathcal{D}_2). \tag{5.87}$$

It is easy to see that

$$\begin{aligned} g_M(\phi X, \phi Y) &= \cos^2\theta g_M(X, Y) \\ g_M(\omega X, \omega Y) &= \sin^2\theta g_M(X, Y) \end{aligned}$$

are valid for $X, Y \in \Gamma(\mathcal{D}_2)$.

From (5.86) we have the following lemma.

**Lemma 68.** *[217] Let F be a semi-slant Riemannian map from a Kähler manifold $(M, g_M, J)$ to a Riemannian manifold $(N, g_N)$ with the semi-slant angle $\theta$. If the tensor $\omega$ is parallel, then we get*

$$\mathcal{T}_{\phi X}\phi X = -\cos^2\theta \cdot \mathcal{T}_X X \quad \text{for } X \in \Gamma(\mathcal{D}_2). \tag{5.88}$$

We now investigate the integrability of distributions. The proofs of the following Theorems are the same as those of Theorem 49 and Theorem 50.

**Theorem 121.** *[217] Let F be a semi-slant Riemannian map from an almost Hermitian manifold $(M, g_M, J)$ to a Riemannian manifold $(N, g_N)$. Then the complex distribution $\mathcal{D}_1$ is integrable if and only if we have*

$$\omega(\widehat{\nabla}_X Y - \widehat{\nabla}_Y X) = 0 \quad \text{for } X, Y \in \Gamma(\mathcal{D}_1).$$

Similarly, we get the following theorem.

**Theorem 122.** *[217] Let F be a semi-slant Riemannian map from an almost Hermitian manifold $(M, g_M, J)$ to a Riemannian manifold $(N, g_N)$. Then the slant distribution $\mathcal{D}_2$ is integrable if and only if we obtain*

$$P(\phi(\widehat{\nabla}_X Y - \widehat{\nabla}_Y X)) = 0 \quad \text{for } X, Y \in \Gamma(\mathcal{D}_2).$$

Given a semi-slant Riemannian map $F$ from an almost Hermitian manifold $(M, g_M, J)$ to a Riemannian manifold $(N, g_N)$ with the semi-slant angle $\theta \in [0, \frac{\pi}{2})$, we define an endomorphism $\widehat{J}$ of ker $F_*$ by

$$\widehat{J} := JP + \sec \theta \phi Q.$$

Then

$$\widehat{J}^2 = -id \quad \text{on ker } F_*. \tag{5.89}$$

Note that the distribution ker $F_*$ is integrable and does not need to be invariant under the almost complex structure $J$. Furthermore, its dimension may be odd. But with the endomorphism $\widehat{J}$, we have the following result.

**Theorem 123.** *[217] Let $F$ be a semi-slant Riemannian map from an almost Hermitian manifold $(M, g_M, J)$ to a Riemannian manifold $(N, g_N)$ with the semi-slant angle $\theta \in [0, \frac{\pi}{2})$. Then the fibers $(F^{-1}(x), \widehat{J})$ are almost complex manifolds for $x \in M$.*

We will now check the harmonicity of semi-slant Riemannian map $F$.

**Theorem 124.** *[217] Let $F$ be a semi-slant Riemannian map from a Kähler manifold $(M, g_M, J)$ to a Riemannian manifold $(N, g_N)$ such that $\mathcal{D}_1$ is integrable. Then $F$ is harmonic if and only if $trace(\nabla F_*) = 0$ on $\mathcal{D}_2$ and $\widetilde{H} = 0$, where $\widetilde{H}$ denotes the mean curvature vector field of range$F_*$.*

*Proof.* Using Lemma 45, we have $trace\nabla F_*|_{\text{ker } F_*} \in \Gamma(range F_*)$ and $trace\nabla F_*|_{(\text{ker } F_*)^\perp} \in \Gamma((range F_*)^\perp)$ so that

$$trace(\nabla F_*) = 0 \quad \Leftrightarrow \quad trace\nabla F_*|_{\text{ker } F_*} = 0 \text{ and } trace\nabla F_*|_{(\text{ker } F_*)^\perp} = 0.$$

Since $\mathcal{D}_1$ is invariant under $J$, we can choose locally an orthonormal frame $\{e_1, Je_1, \cdots, e_k, Je_k\}$ of $\mathcal{D}_1$. Using the integrability of the distribution $\mathcal{D}_1$,

$$
\begin{aligned}
(\nabla F_*)(Je_i, Je_i) &= -F_* \nabla_{Je_i} Je_i = -F_* J(\nabla_{e_i} Je_i + [Je_i, e_i]) \\
&= F_* \nabla_{e_i} e_i = -(\nabla F_*)(e_i, e_i) \quad \text{for } 1 \le i \le k.
\end{aligned}
$$

Hence,

$$trace\nabla F_*|_{\text{ker } F_*} = 0 \quad \Leftrightarrow \quad trace\nabla F_*|_{\mathcal{D}_2} = 0.$$

Moreover, it is easy to get that

$$trace\nabla F_*|_{(\text{ker } F_*)^\perp} = l\widetilde{H} \quad \text{for } l := \dim(\text{ker } F_*)^\perp$$

so that

$$trace \nabla F_*|_{(\ker F_*)^\perp} = 0 \quad \Leftrightarrow \quad \widetilde{H} = 0.$$

Therefore, we obtain the result. □

Using Lemma 77, we also have the following result.

**Corollary 53.** *[217] Let F be a semi-slant Riemannian map from a Kähler manifold $(M, g_M, J)$ to a Riemannian manifold $(N, g_N)$ such that $\mathcal{D}_1$ is integrable and the semi-slant angle $\theta \in [0, \frac{\pi}{2})$. Assume that the tensor $\omega$ is parallel. Then F is harmonic if and only if $\widetilde{H} = 0$.*

We will now study the condition for such a map $F$ to be totally geodesic.

**Theorem 125.** *[217] Let F be a semi-slant Riemannian map from a Kähler manifold $(M, g_M, J)$ to a Riemannian manifold $(N, g_N)$. Then F is a totally geodesic map if and only if*

$$\omega(\widehat{\nabla}_X \phi Y + \mathcal{T}_X \omega Y) + C(\mathcal{T}_X \phi Y + \mathcal{H} \nabla_X \omega Y) = 0$$
$$\omega(\widehat{\nabla}_X BZ + \mathcal{T}_X CZ) + C(\mathcal{T}_X BZ + \mathcal{H} \nabla_X CZ) = 0$$
$$\bar{Q}(\nabla^F_{Z_1} F_* Z_2) = 0$$

*for $X, Y \in \Gamma(\ker F_*)$ and $Z, Z_1, Z_2 \in \Gamma((\ker F_*)^\perp)$.*

*Proof.* If $Z_1, Z_2 \in \Gamma((\ker F_*)^\perp)$, then by Lemma 45, we have

$$(\nabla F_*)(Z_1, Z_2) = 0 \quad \Leftrightarrow \quad \bar{Q}((\nabla F_*)(Z_1, Z_2)) = \bar{Q}(\nabla^F_{Z_1} F_* Z_2) = 0.$$

Given $X, Y \in \Gamma(\ker F_*)$, we get

$$(\nabla F_*)(X, Y) = -F_*(\nabla_X Y) = F_*(J\nabla_X(\phi Y + \omega Y))$$
$$= F_*(\phi \widehat{\nabla}_X \phi Y + \omega \widehat{\nabla}_X \phi Y + B\mathcal{T}_X \phi Y + C\mathcal{T}_X \phi Y + \phi \mathcal{T}_X \omega Y + \omega \mathcal{T}_X \omega Y$$
$$+ B\mathcal{H} \nabla_X \omega Y + C\mathcal{H} \nabla_X \omega Y).$$

Hence, we obtain

$$(\nabla F_*)(X, Y) = 0 \Leftrightarrow \omega(\widehat{\nabla}_X \phi Y + \mathcal{T}_X \omega Y) + C(\mathcal{T}_X \phi Y + \mathcal{H} \nabla_X \omega Y) = 0.$$

If $X \in \Gamma(\ker F_*)$ and $Z \in \Gamma((\ker F_*)^\perp)$, then since the tensor $\nabla F_*$ is symmetric, we only need to consider the following:

$$(\nabla F_*)(X, Z) = -F_*(\nabla_X Z) = F_*(J\nabla_X(BZ + CZ))$$
$$= F_*(\phi\widehat{\nabla}_X BZ + \omega\widehat{\nabla}_X BZ + B\mathcal{T}_X BZ + C\mathcal{T}_X BZ + \phi\mathcal{T}_X CZ + \omega\mathcal{T}_X CZ$$
$$+ B\mathcal{H}\nabla_X CZ + C\mathcal{H}\nabla_X CZ).$$

Thus, we derive

$$(\nabla F_*)(X, Z) = 0 \Leftrightarrow \omega(\widehat{\nabla}_X BZ + \mathcal{T}_X CZ) + C(\mathcal{T}_X BZ + \mathcal{H}\nabla_X CZ) = 0.$$

Therefore, we obtain the result.      □

In the rest of this section, we obtain decomposition theorems by using semi-slant Riemannian maps.

**Theorem 126.** *[217] Let F be a semi-slant Riemannian map from a Kähler manifold* $(M, g_M, J)$ *to a Riemannian manifold* $(N, g_N)$. *Then* $(M, g_M, J)$ *is locally a Riemannian product manifold of the leaves of* ker $F_*$ *and* (ker $F_*)^\perp$ *if and only if*

$$\omega(\widehat{\nabla}_X \phi Y + \mathcal{T}_X \omega Y) + C(\mathcal{T}_X \phi Y + \mathcal{H}\nabla_X \omega Y) = 0 \quad \textit{for } X, Y \in \Gamma(\ker F_*)$$

*and*

$$\phi(\mathcal{V}\nabla_Z BW + \mathcal{A}_Z CW) + B(\mathcal{A}_Z BW + \mathcal{H}\nabla_Z CW) = 0 \quad \textit{for } Z, W \in \Gamma((\ker F_*)^\perp).$$

*Proof.* For $X, Y \in \Gamma(\ker F_*)$, using (3.5), we have

$$\nabla_X Y = -J\nabla_X JY = -J(\widehat{\nabla}_X \phi Y + \mathcal{T}_X \phi Y + \mathcal{T}_X \omega Y + \mathcal{H}\nabla_X \omega Y)$$
$$= -(\phi\widehat{\nabla}_X \phi Y + \omega\widehat{\nabla}_X \phi Y + B\mathcal{T}_X \phi Y + C\mathcal{T}_X \phi Y + \phi\mathcal{T}_X \omega Y + \omega\mathcal{T}_X \omega Y$$
$$+ B\mathcal{H}\nabla_X \omega Y + C\mathcal{H}\nabla_X \omega Y).$$

Thus, we obtain

$$\nabla_X Y \in \Gamma(\ker F_*) \Leftrightarrow \omega(\widehat{\nabla}_X \phi Y + \mathcal{T}_X \omega Y) + C(\mathcal{T}_X \phi Y + \mathcal{H}\nabla_X \omega Y) = 0.$$

In a similar way, given $Z, W \in \Gamma((\ker F_*)^\perp)$, we have

$$\nabla_Z W = -J\nabla_Z JW = -J(\mathcal{V}\nabla_Z BW + \mathcal{A}_Z BW + \mathcal{A}_Z CW + \mathcal{H}\nabla_Z CW)$$
$$= -(\phi\mathcal{V}\nabla_Z BW + \omega\mathcal{V}\nabla_Z BW + B\mathcal{A}_Z BW + C\mathcal{A}_Z BW + \phi\mathcal{A}_Z CW$$
$$+ \omega\mathcal{A}_Z CW + B\mathcal{H}\nabla_Z CW + C\mathcal{H}\nabla_Z CW).$$

Hence, we derive

$$\nabla_Z W \in \Gamma((\ker F_*)^\perp) \Leftrightarrow \phi(\mathcal{V}\nabla_Z BW + \mathcal{A}_Z CW) + B(\mathcal{A}_Z BW + \mathcal{H}\nabla_Z CW) = 0.$$

Therefore, the result follows.      □

**Theorem 127.** *[217] Let F be a semi-slant Riemannian map from a Kähler manifold* $(M, g_M, J)$ *to a Riemannian manifold* $(N, g_N)$. *Then the fibers of F are locally Riemannian product manifolds of the leaves of* $\mathcal{D}_1$ *and* $\mathcal{D}_2$ *if and only if*

$$Q(\phi\widehat{\nabla}_U\phi V + BT_U\phi V) = 0 \text{ and } \omega\widehat{\nabla}_U\phi V + CT_U\phi V = 0 \quad \text{for } U, V \in \Gamma(\mathcal{D}_1)$$

*and*

$$P(\phi(\widehat{\nabla}_X\phi Y + T_X\omega Y) + B(T_X\phi Y + \mathcal{H}\nabla_X\omega Y)) = 0$$

$$\omega(\widehat{\nabla}_X\phi Y + T_X\omega Y) + C(T_X\phi Y + \mathcal{H}\nabla_X\omega Y) = 0$$

*for* $X, Y \in \Gamma(\mathcal{D}_2)$.

*Proof.* Given $U, V \in \Gamma(\mathcal{D}_1)$, using (3.5) we get

$$\nabla_U V = -J\nabla_U JV = -J(\widehat{\nabla}_U\phi V + T_U\phi V)$$

$$= -(\phi\widehat{\nabla}_U\phi V + \omega\widehat{\nabla}_U\phi V + BT_U\phi V + CT_U\phi V).$$

Hence, we derive

$$\nabla_U V \in \Gamma(\mathcal{D}_1) \Leftrightarrow Q(\phi\widehat{\nabla}_U\phi V + BT_U\phi V) = 0 \text{ and } \omega\widehat{\nabla}_U\phi V + CT_U\phi V = 0.$$

For $X, Y \in \Gamma(\mathcal{D}_2)$, we obtain

$$\nabla_X Y = -J\nabla_X JY = -J(\widehat{\nabla}_X\phi Y + T_X\phi Y + T_X\omega Y + \mathcal{H}\nabla_X\omega Y)$$

$$= -(\phi\widehat{\nabla}_X\phi Y + \omega\widehat{\nabla}_X\phi Y + BT_X\phi Y + CT_X\phi Y + \phi T_X\omega Y$$

$$+ \omega T_X\omega Y + B\mathcal{H}\nabla_X\omega Y + C\mathcal{H}\nabla_X\omega Y).$$

Thus, we have $\nabla_X Y \in \Gamma(\mathcal{D}_2) \Leftrightarrow$

$$\begin{cases} P(\phi(\widehat{\nabla}_X\phi Y + T_X\omega Y) + B(T_X\phi Y + \mathcal{H}\nabla_X\omega Y)) = 0, \\ \omega(\widehat{\nabla}_X\phi Y + T_X\omega Y) + C(T_X\phi Y + \mathcal{H}\nabla_X\omega Y) = 0. \end{cases}$$

Therefore, we have the result.                                               $\square$

## 7. Hemi-slant Riemannian maps from almost Hermitian manifolds

In this section, we are going to introduce hemi-slant Riemannian maps from almost Hermitian manifolds to Riemannian manifolds as a generalization of hemi-slant submersions, semi-invariant Riemannian maps, and slant Riemannian maps. Of course, since invariant submersions, anti-invariant submersions, slant submersions, invariant Riemannian maps and anti-invariant Riemannian maps are included in the class of slant Riemannian maps, this new class includes all these maps. We give examples, obtain integrability conditions for distributions, and investigate the geometry of leaves. We also find necessary and sufficient conditions for a hemi-slant Riemannian map to be a totally geodesic map and a harmonic map, respectively.

**Definition 58.** [259] Let $M$ be a $2m$-dimensional almost Hermitian manifold with Hermitian metric $g$ and almost complex structure $J$, and $N$ be a Riemannian manifold with Riemannian metric $g_N$. A Riemannian map $F : (M, g, J) \to (N, g_N)$ is called a *hemi-slant Riemannian map* if the vertical distribution $kerF_*$ of $F$ admits two orthogonal complementary distributions $\mathcal{D}^\theta$ and $\mathcal{D}^\perp$ such that $\mathcal{D}^\theta$ is slant and $\mathcal{D}^\perp$ is anti-invariant, i.e., we have

$$kerF_* = \mathcal{D}^\theta \oplus \mathcal{D}^\perp. \tag{5.90}$$

In this case, the angle $\theta$ is called the hemi-slant angle of the Riemannian map.

Since $F$ is a subimmersion, it follows that the rank of $F$ is constant on $M_1$, then the rank theorem for functions implies that $ker F_*$ is an integrable subbundle of $TM_1$. Thus it follows from the above definition that the leaves of the distribution $ker F_*$ of a hemi-slant Riemannian map are hemi-slant submanifolds of $M_1$.

We next give some examples of hemi-slant Riemannian maps.

**Example 59.** Every slant submersion from almost an Hermitian manifold to a Riemannian manifold is a hemi-slant Riemannian map with $(rangeF_*)^\perp = \{0\}$ and $\mathcal{D}^\perp = \{0\}$.

**Example 60.** Every hemi-slant submersion from an almost Hermitian manifold to a Riemannian manifold is a hemi-slant Riemannian map with $(rangeF_*)^\perp = \{0\}$.

**Example 61.** Every slant Riemannian map from an almost Hermitian manifold to a Riemannian manifold is a hemi-slant Riemannian map with $\mathcal{D}^\perp = \{0\}$.

**Example 62.** Every semi-invariant Riemannian map from an almost Hermitian manifold to a Riemannian manifold is a hemi-slant Riemannian map with $\theta = \frac{\pi}{2}$.

We say that the hemi-slant Riemannian map $F : (M, g, J) \to (N, g_N)$ is *proper* if $\mathcal{D}^\perp \neq \{0\}$ and $\theta \neq 0, \frac{\pi}{2}$.

For any $V \in \Gamma(kerF_*)$, we put

$$V = \mathcal{P}V + \mathcal{Q}V, \tag{5.91}$$

where $\mathcal{P}V \in \Gamma(\mathcal{D}^\theta)$ and $\mathcal{Q}V \in \Gamma(\mathcal{D}^\perp)$ and put

$$JV = \phi V + \omega V, \tag{5.92}$$

where $\phi V \in \Gamma(kerF_*)$ and $\omega V \in \Gamma((kerF_*)^\perp)$. Also for any $\xi \in (kerF_*)^\perp$, we have

$$J\xi = \mathcal{B}\xi + \mathcal{C}\xi, \tag{5.93}$$

where $\mathcal{B}\xi \in \Gamma(kerF_*)$ and $\mathcal{C}\xi \in (kerF_*)^\perp$.

As for hemi-slant submersions, the proof of the following theorem is exactly the same as that for hemi-slant submanifolds; see Theorem 3.8. of [286]. we therefore omit it.

**Theorem 128.** *[259] Let F be a Riemannian submersion from an almost Hermitian manifold $(M, g, J)$ onto a Riemannian manifold. Then F is a hemi-slant Riemannian map if and only if there exists a constant $\lambda \in [-1, 0]$ and a distribution $\mathcal{D}$ on $kerF_*$ such that*
**(a)** $\mathcal{D} = \{V \in kerF_* \mid \phi^2 V = \lambda V\}$, *and*
**(b)** *for any $V \in kerF_*$ orthogonal to $\mathcal{D}$, we have $\phi V = 0$.*
*Moreover, in this case $\lambda = -\cos^2\theta$, where $\theta$ is the slant angle of F.*

It is easy to see that the following expressions are valid:

$$g(\phi Z, \phi W) = \cos^2\theta g(Z, W) \tag{5.94}$$
$$g(\omega Z, \omega W) = \sin^2\theta g(Z, W) \tag{5.95}$$

for any $Z, W \in \mathcal{D}^\theta$. We now present an example of a proper hemi-slant Riemannian map.

**Example 63.** [259] Define a map $F : \mathbb{R}^8 \to \mathbb{R}^7$ by

$$F(x_1, ..., x_8) = \left( \frac{x_1 - x_3}{\sqrt{2}}, 0, x_4, x_5, x_6, \frac{x_7 + x_8}{\sqrt{2}}, 0 \right).$$

Then the map $F$ is a proper hemi-slant Riemannian map such that

$$\mathcal{D}^\theta = span\{\frac{1}{\sqrt{2}}(\partial x_1 + \partial x_3), \partial x_2\},$$

with the slant angle $\theta = \frac{\pi}{4}$ and

$$\mathcal{D}^\perp = span\{\frac{1}{2}(\partial x_7 - \partial x_8)\}.$$

Moreover, $kerF_*^\perp = span\{\frac{1}{\sqrt{2}}(\partial x_1 - \partial x_3), \partial x_4, \partial x_5, \partial x_5, \frac{1}{2}(\partial x_7 + \partial x_8)\}, \partial_i = \frac{\partial}{\partial x_i}$. Furthermore, we have

$$(rangeF_*) = span\{\partial y_1, \partial y_3, \partial y_4, \partial y_5, \partial y_6\} \text{ and } (rangeF_*)^\perp = span\{\partial y_2, \partial y_7\},$$

where $\partial y_i = \frac{\partial}{\partial y_i}$, and $x_1, ..., x_8, y_1, ..., y_7$ are the local coordinates on $\mathbb{R}^8$, $\mathbb{R}^7$, respectively.

We now investigate the geometry of the total manifold by assuming the existence of a hemi-slant Riemannian map. For the distribution $\mathcal{D}^\perp$, we have the following theorem.

**Theorem 129.** *[259] Let F be a hemi-slant Riemannian map from a Kähler manifold $(M, g_M, J)$ to a Riemannian manifold $(N, g_N)$. Then the distribution $\mathcal{D}^\perp$ is integrable.*

*Proof.* For $U_1, V_1 \in \Gamma(\mathcal{D}^\perp)$ and $U_2 \in \Gamma(\mathcal{D}^\theta)$, we have

$$3d\Omega(U_1, V_1, U_2) = U_1\Omega(V_1, U_2) + V_1\Omega(U_2, U_1) + U_2\Omega(U_1, V_1)$$
$$-\Omega([V_1, U_2], U_1) - \Omega([U_1, V_1], U_2) - \Omega([U_2, U_1], V_1),$$

where $\Omega(X, Y) = g_M(X, JY)$ is the fundamental 2–form of $M$, which vanishes for the Kähler case. Thus we get

$$0 = -g_M([U_1, V_1], \phi U_2),$$

which gives us the assertion.    □

For the distribution $\mathcal{D}^\theta$, we have the following integrability condition.

**Theorem 130.** *[259] Let F be a hemi-slant Riemannian map from a Kähler manifold $(M, g_M, J)$ to a Riemannian manifold $(N, g_N)$. Then the distribution $\mathcal{D}^\theta$ is integrable if and only if*

$$g_M(\mathcal{T}_{V_2}\omega\phi U_2 - \mathcal{T}_{U_2}\omega\phi V_2, U_1) = g_N((\nabla F_*)(U_2, \omega V_2) - (\nabla F_*)(V_2, \omega U_2), F_*(JU_1))$$

*for $U_2, V_2 \in \Gamma(\mathcal{D}^\theta)$ and $U_1 \in \Gamma(\mathcal{D}^\perp)$.*

*Proof.* From (4.7), (5.92), and (5.93), we have

$$g_M([U_2, V_2], U_1) = -g_M(\nabla_{U_2}J\phi V_2, U_1) + g_M(\nabla_{U_2}\omega V_2, JU_1)$$
$$+g_M(\nabla_{V_2}J\phi U_2, U_1) - g_M(\nabla_{V_2}\omega U_2, JU_1).$$

Using Theorem 128 and (4.13), we derive

$$\sin^2\theta g_M([U_2, V_2], U_1) = -g_M(\mathcal{T}_{V_2}\omega\phi U_2 - \mathcal{T}_{U_2}\omega\phi V_2, U_1)$$
$$+g_M(\mathcal{H}\nabla_{U_2}\omega V_2 - \mathcal{H}\nabla_{V_2}\omega U_2, JU_1).$$

Thus (1.71) completes the proof.    □

For the leaf of the distribution $\mathcal{D}^\theta$, we have the following result.

**Theorem 131.** *[259] Let F be a hemi-slant Riemannian map from a Kähler manifold $(M, g_M, J)$ to a Riemannian manifold $(N, g_N)$. Then the distribution $\mathcal{D}^\theta$ defines a totally geodesic foliation on M if and only if*

$$g_M(\mathcal{T}_{V_2}\omega\phi U_2, U_1) = g_N((\nabla F_*)(V_2, \omega U_2), F_*(JU_1))$$

*for $U_2, V_2 \in \Gamma(\mathcal{D}^\theta)$ and $U_1 \in \Gamma(\mathcal{D}^\perp)$.*

In a similar way, we have the following condition for the leaf of the distribution $\mathcal{D}^\perp$.

**Theorem 132.** *[259] Let F be a hemi-slant Riemannian map from a Kähler manifold $(M, g_M, J)$ to a Riemannian manifold $(N, g_N)$. Then the distribution $\mathcal{D}^\perp$ defines a totally geodesic foliation on M if and only if*

$$g_M(\mathcal{T}_{U_1}\omega\phi U_2, V_1) = g_N((\nabla F_*)(U_1, \omega U_2), F_*(JU_1))$$

*for $U_2 \in \Gamma(\mathcal{D}^\theta)$ and $U_1, V_1 \in \Gamma(\mathcal{D}^\perp)$.*

From Theorem 131 and Theorem 132, we have the following decomposition theorem for the fibers.

**Theorem 133.** *[259] Let F be a hemi-slant Riemannian map from a Kähler manifold $(M, g_M, J)$ to a Riemannian manifold $(N, g_N)$. Then the fiber of F is a locally product Riemannian manifold of the form $M_{\mathcal{D}^\perp} \times M_{\mathcal{D}^\theta}$, where $M_{\mathcal{D}^\perp}$ and $M_{\mathcal{D}^\theta}$ are leaves of $\mathcal{D}^\perp$ and $\mathcal{D}^\theta$, respectively, if and only if*

$$g_M(\mathcal{T}_X\omega\phi U_2, V_1) = g_N((\nabla F_*)(X, \omega U_2), F_*(JU_1))$$

*for $U_2, V_2 \in \Gamma(\mathcal{D}^\theta)$ and $X \in \Gamma(\ker F_*)$, $V_1 \in \Gamma(\mathcal{D}^\perp)$.*

We now find necessary and sufficient conditions for the distributions $(\ker F_*)$ and $(\ker F_*)^\perp$ define totally geodesic foliations on the total manifold $M$.

**Theorem 134.** *[259] Let F be a hemi-slant Riemannian map from a Kähler manifold $(M, g_M, J)$ to a Riemannian manifold $(N, g_N)$. Then the distribution $(\ker F_*)$ defines a totally geodesic foliation on M if and only if*

$$g_N((\nabla F_*)(U, JV_1), CZ) = g_M(\mathcal{T}_U JV_1, \mathcal{B}Z)$$

*and*

$$g_N(\nabla^F_{\omega\phi V_2} F_*(CZ), F_*(\omega U)) + g_N(\nabla^F_{\omega V_2} F_*(\omega U), F_*(C^2 Z)) = g_M(\mathcal{B}Z, \mathcal{A}_{\omega\phi V_2}\omega U)$$
$$-g_M(\mathcal{V}\nabla_{\omega\phi V_2}\mathcal{B}Z + \mathcal{A}_{\omega\phi V_2}CZ, \phi U)) + g_M(\phi U, \mathcal{A}_{\omega V_2}C^2 Z) + g_M(\omega V_2, \mathcal{T}_U \omega\mathcal{B}Z)$$
$$-g_M(\mathcal{V}\nabla_{\omega V_2}\phi U + \mathcal{A}_{\omega V_2}\omega U, \mathcal{B}CZ)$$

*for* $U \in \Gamma(\ker F_*)$, $V_1 \in \Gamma(\mathcal{D}^\perp)$, $V_2 \in \Gamma(\mathcal{D}^\theta)$ *and* $Z \in \Gamma((\ker F_*)^\perp)$.

*Proof.* For $U \in \Gamma(\ker F_*)$, $V_1 \in \mathcal{D}^\perp$ and $Z \in \Gamma((\ker F_*)^\perp)$, using (4.7), (4.12), and (5.93), we obtain

$$g_M(\nabla_U V_1, Z) = g_M(\mathcal{H}\nabla_U J V_1, CZ) + g_M(\mathcal{T}_U J V_1, \mathcal{B}Z).$$

Then from (1.71) and (4.7), we get

$$g_M(\nabla_U V_1, Z) = -g_N((\nabla F_*)(U, J V_1), CZ) + g_M(\mathcal{T}_U J V_1, \mathcal{B}Z). \tag{5.96}$$

For $V_2 \in \Gamma(\mathcal{D}^\theta)$, using (4.7), (5.92), (5.93), (4.14) and Theorem 128, we get

$$\sin^2\theta g_M(\nabla_U V_2, Z) = -g_M(\mathcal{H}\nabla_U \omega\phi V_2, Z) + g_M(\mathcal{H}\nabla_U \omega V_2, CZ) \tag{5.97}$$
$$+ g_M(\mathcal{T}_U \omega V_2, \mathcal{B}Z).$$

Since $[X, U] \in \Gamma(\ker F_*)$ for $X \in \Gamma((\ker F_*)^\perp)$ and $U \in \Gamma(\ker F_*)$, we have

$$g_M(\mathcal{H}\nabla_U \omega\phi V_2, Z) = -g_M(\nabla_{\omega\phi V_2} Z, U).$$

Using (4.7), (5.92), (5.93), (4.14), and (4.15) we obtain

$$g_M(\mathcal{H}\nabla_U \omega\phi V_2, Z) = -g_M(\mathcal{A}_{\omega\phi V_2}\mathcal{B}Z, \omega U) - g_M(\mathcal{V}\nabla_{\omega\phi V_2}\mathcal{B}Z, \phi U)$$
$$- g_M(\mathcal{H}\nabla_{\omega\phi V_2}CZ, \omega U) - g_M(\mathcal{A}_{\omega\phi V_2}CZ, \phi U).$$

Then from (1.71), we have

$$g_M(\mathcal{H}\nabla_U \omega\phi V_2, Z) = -g_M(\mathcal{A}_{\omega\phi V_2}\mathcal{B}Z, \omega U) - g_M(\mathcal{V}\nabla_{\omega\phi V_2}\mathcal{B}Z + \mathcal{A}_{\omega\phi V_2}CZ, \phi U))$$
$$- g_N(\nabla^F_{\omega\phi V_2} F_*(CZ), F_*(\omega U)). \tag{5.98}$$

In a similar way, we get

$$g_M(\mathcal{H}\nabla_U \omega V_2, CZ) = g_M(\mathcal{A}_{\omega V_2}\phi U, C^2 Z) + g_M(\mathcal{V}\nabla_{\omega V_2}\phi U + \mathcal{A}_{\omega V_2}\omega U, \mathcal{B}CZ)$$
$$+ g_N(\nabla^F_{\omega V_2} F_*(\omega U), F_*(C^2 Z)). \tag{5.99}$$

Thus putting (5.98) and (5.99) in (5.97), we arrive at

$$
\begin{aligned}
\sin^2\theta g_M(\nabla_U V_2, Z) \;=\; & g_M(\mathcal{A}_{\omega\phi V_2}\mathcal{B}Z, \omega U) + g_M(\mathcal{V}\nabla_{\omega\phi V_2}\mathcal{B}Z + \mathcal{A}_{\omega\phi V_2}CZ, \phi U)) \\
& - g_N(\nabla^F_{\omega\phi V_2}F_*(CZ), F_*(\omega U)) + g_M(\mathcal{A}_{\omega V_2}\phi U, C^2 Z) \\
& + g_M(\mathcal{V}\nabla_{\omega V_2}\phi U + \mathcal{A}_{\omega V_2}\omega U, \mathcal{B}CZ) \qquad\qquad (5.100) \\
& + g_N(\nabla^F_{\omega V_2}F_*(\omega U), F_*(C^2 Z)) + g_M(\mathcal{T}_U \omega V_2, \mathcal{B}Z).
\end{aligned}
$$

Thus, the proof comes from (5.96) and (5.100).                                   □

We now find necessary and sufficient conditions for the distribution $(ker\, F_*)^\perp$ to define a totally geodesic foliation on $M$.

**Theorem 135.** *[259] Let $F$ be a hemi-slant Riemannian map from a Kähler manifold $(M, g_M, J)$ to a Riemannian manifold $(N, g_N)$. Then the distribution $(ker\, F_*)^\perp$ defines a totally geodesic foliation on $M$ if and only if*

$$
g_N(\nabla^F_{Z_1}F_*(CZ_2), F_*(JV_1)) = g_M(\mathcal{B}Z_2, \mathcal{A}_{Z_1}JV_1)
$$

*and*

$$
g_M(\omega V, \mathcal{A}_{Z_1}\mathcal{B}Z_2) = g_N(F_*(CZ_2), \nabla^F_{Z_1}F_*(\omega V)) - g_N(F_*(Z_2), \nabla^F_{Z_1}F_*(\omega\phi V))
$$

*for $Z_1, Z_2 \in \Gamma((ker\, F_*)^\perp)$ and $V \in \Gamma(\mathcal{D}^\theta)$.*

*Proof.* For $Z_1, Z_2 \in \Gamma((ker\, F_*)^\perp)$ and $V_1 \in \Gamma(\mathcal{D}^\perp)$, using (3.5), (4.14), and (4.15), we get

$$
g_M(\nabla_{Z_1}Z_2, V_1) = g_M(\mathcal{H}\nabla_{Z_1}CZ_2 + \mathcal{A}_{Z_1}\mathcal{B}Z_2, JV_1).
$$

Then, using (1.71), we obtain

$$
g_M(\nabla_{Z_1}Z_2, V_1) = g_N(\nabla^F_{Z_1}F_*(CZ_2), F_*(JV_1)) + g_M(\mathcal{A}_{Z_1}\mathcal{B}Z_2, JV_1). \qquad (5.101)
$$

Now, for $V \in \Gamma(\mathcal{D}^\theta)$, from Theorem 128 we have

$$
\sin^2\theta g_M(\nabla_{Z_1}Z_2, V) = g_M(Z_2, \mathcal{H}\nabla_{Z_1}\omega\phi V) - g_M(CZ_2, \mathcal{H}\nabla_{Z_1}\omega V) - g_M(\mathcal{B}Z_2, \mathcal{A}_{Z_1}\omega V).
$$

Using (4.7) and (1.71), we derive

$$
\begin{aligned}
\sin^2\theta g_M(\nabla_{Z_1}Z_2, V) \;=\; & g_N(F_*(Z_2), \nabla^F_{Z_1}F_*(\omega\phi V)) - g_N(F_*(CZ_2), \nabla^F_{Z_1}F_*(\omega V)) \\
& - g_M(\mathcal{B}Z_2, \mathcal{A}_{Z_1}\omega V). \qquad\qquad (5.102)
\end{aligned}
$$

Then (5.101) and (5.102) complete the proof.                                   □

We now find necessary and sufficient conditions for $F$ to be harmonic and totally geodesic, respectively.

**Theorem 136.** *[259] Let F be a hemi-slant Riemannian map from a Kähler manifold $(M, g_M, J)$ to a Riemannian manifold $(N, g_N)$. Then F is harmonic if and only if*

$$trace \mid_{\mathcal{D}^\perp} \{F_*(\mathcal{T}_{(.)}(.) - C\mathcal{A}_{J(.)}(.) - \omega V \nabla_{J(.)}(.)) + \nabla^F_{J(.)} F_* J(.)\} + trace \mid_{\mathcal{D}^\theta}$$
$$\{F_*(C\mathcal{T}_{(.)}\phi(.) + \omega V \nabla_{(.)}\phi(.) + C\mathcal{H}\nabla_{(.)}\omega(.) + J\mathcal{T}_{(.)}\omega(.) - \sec^2\theta\mathcal{T}_{\phi(.)}\phi(.)$$
$$- \csc^2\theta(C\mathcal{A}_{\omega(.)}(.) + \omega V \nabla_{\omega(.)}(.) + \theta\mathcal{A}_{\omega(.)}\phi(.))) + \csc^2\theta\nabla^F_{\omega(.)} F_*(\omega(.)))\} = 0$$

*Proof.* First of all, it is easy to see that $(ker F_*)$ has an orthonormal frame $\{e_1, ..., e_{r_1}, \tilde{e}_1, ..., \tilde{e}_{r_2}, \sec\theta\phi\tilde{e}_1, ..., \sec\theta\phi\tilde{e}_{r_2}\}$ such that $\{e_1, ..., e_{r_1}\}$ is an orthonormal frame of $\mathcal{D}^\perp$ and $\{\tilde{e}_1, ..., \tilde{e}_{r_2}, \sec\theta\phi\tilde{e}_1, ..., \sec\theta\phi\tilde{e}_{r_2}\}$ is an orthonormal frame of $\mathcal{D}^\theta$. Hence it follows that $\{Je_1, ..., Je_r, \csc\theta\omega\tilde{e}_1, ..., \csc\theta\omega\tilde{e}_{r_2}\}$ is an orthonormal frame of $(ker F_*)^\perp$. Now for $X \in \Gamma(\mathcal{D}^\perp)$ and $Y \in \Gamma(\mathcal{D}^\theta)$, define $\delta(X, Y)$ as

$$\delta(X, Y) = (\nabla F_*)(X, X) + (\nabla F_*)(JX, JX) + (\nabla F_*)(Y, Y)$$
$$+ \sec^2\theta(\nabla F_*)(\phi Y, \phi Y) + \csc^2\theta(\nabla F_*)(\omega Y, \omega Y).$$

Thus, using (3.5) and (1.71), we have

$$\delta(X, Y) = -F_*(\nabla_X X) + \nabla^F_{JX} F_* JX - F_*(J\nabla_{JX} X) + F_*(J\nabla_Y JY)$$
$$- \sec^2\theta F_*(\nabla_{\phi Y}\phi Y) + \csc^2\theta\nabla^F_{\omega Y} F_*(\omega Y) - \csc^2\theta F_*(\nabla_{\omega Y}\omega Y).$$

From (5.92), (5.93), and (4.12)-(4.14), we get

$$\delta(X, Y) = -F_*(\mathcal{T}_X X) + \nabla^F_{JX} F_* JX - F_*(C\mathcal{A}_{JX} X) - F_*(\omega V \nabla_{JX} X)$$
$$+ F_*(C\mathcal{T}_Y\phi Y) + F_*(\omega V \nabla_Y\phi Y) + F_*(C\mathcal{H}\nabla_Y\omega Y) + F_*(J\mathcal{T}_Y\omega Y)$$
$$- \sec^2\theta F_*(\mathcal{T}_{\phi Y}\phi Y) + \csc^2\theta\nabla^F_{\omega Y} F_*(\omega Y) - \csc^2\theta F_*(C\mathcal{A}_{\omega Y} Y)$$
$$- \csc^2\theta F_*(\omega V \nabla_{\omega Y} Y) + \csc^2\theta F_*(\mathcal{A}_{\omega Y}\phi Y).$$

which gives us the assertion. □

For the total geodesicity of $F$, we have the following result.

**Theorem 137.** *[259] Let F be a hemi-slant Riemannian map from a Kähler manifold $(M, g_M, J)$ to a Riemannian manifold $(N, g_N)$. Then F is totally geodesic if and only if*

$$\omega\mathcal{T}_U JV + C\mathcal{H}\nabla_U JV = 0$$

$$\mathcal{H}\nabla_\xi\omega\phi Z + C\mathcal{H}\nabla_\xi\omega Z + \omega\mathcal{T}_\xi\omega Z = 0$$

$$\mathcal{H}\nabla_X\omega\phi Z + C\mathcal{H}\nabla_X\omega Z + \omega\mathcal{A}_X\omega Z = 0$$

*and*

$$\nabla_X^F F_*(Y) = -F_*(\mathcal{A}_X \phi BY + \mathcal{H}\nabla_X \omega BY) + C\mathcal{H}\nabla_X CY + \omega \mathcal{A}_X CY)$$

*for $\xi \in \Gamma(ker\, F_*)$, $U, V \in \Gamma(\mathcal{D}^{\perp})$, $Z \in \Gamma(\mathcal{D}^{\theta})$, and $X, Y \in \Gamma((ker\, F_*)^{\perp})$.*

*Proof.* For $U, V \in \Gamma(\mathcal{D}^{\perp})$, from (3.5) and (1.71) we have

$$(\nabla F_*)(U, V) = F_*(J\nabla_U JV).$$

Using (4.13), (5.92), and (5.93), we get

$$(\nabla F_*)(U, V) = F_*(\omega \mathcal{T}_U JV + C\mathcal{H}\nabla_U JV). \tag{5.103}$$

For $\xi \in \Gamma(ker\, F_*)$ and $Z \in \mathcal{D}^{\theta})$, (5.92), (3.5), and (1.71) imply

$$(\nabla F_*)(U, Z) = F_*(\nabla_U \phi^2 Z + \nabla_U \omega \phi Z + J\nabla_U \omega Z).$$

Then, using Theorem 128, and (4.13) and (5.93) we derive

$$\sin^2 \theta (\nabla F_*)(U, Z) = F_*(\mathcal{H}\nabla_U \omega \phi Z + C\mathcal{H}\nabla_U \omega Z + \omega \mathcal{T}_U \omega Z). \tag{5.104}$$

In a similar way, for $X \in \Gamma((ker\, F_*)^{\perp})$ and $Z \in \mathcal{D}^{\theta})$, we obtain

$$\sin^2 \theta (\nabla F_*)(X, Z) = F_*(\mathcal{H}\nabla_X \omega \phi Z + C\mathcal{H}\nabla_X \omega Z + \omega \mathcal{A}_X \omega Z). \tag{5.105}$$

For $X, Y \in \Gamma((ker\, F_*)^{\perp})$, from (5.92), (3.5), and (1.71), we have

$$(\nabla F_*)(X, Y) = \nabla_X^F F_*(Y) + F_*(\nabla_X JBY) + F_*(J\nabla_X CY).$$

Then using (5.92), (5.93), (4.14), and (4.15), we find

$$(\nabla F_*)(X, Y) = \nabla_X^F F_*(Y) + F_*(\mathcal{A}_X \phi BY + \mathcal{H}\nabla_X \omega BY) + C\mathcal{H}\nabla_X CY + \omega \mathcal{A}_X CY). \tag{5.106}$$

Thus proof is complete due to (5.103)-(5.106). □

**Remark 20.** We conclude this chapter by adding that the concept of Riemannian maps has been also considered from a few total bases. This implies that there are many new problems to study in this research area.

(i)   Riemannian maps from almost quaternion Kähler manifolds (see Park [218]).

(ii)  Riemannian maps from cosymplectic manifolds (see Panday, Jaiswal, and Ojha [213]).

(iii) Riemannian maps from Sasakian manifolds (see Jaiswal [148]).

(iv)  Slant Riemannian maps from almost contact manifolds (see Prasad, Pandey [228]).

# CHAPTER 6

# Riemannian Maps To Almost Hermitian Manifolds

## Contents

## Abstract

In this chapter, we study Riemannian maps from Riemannian manifolds to almost Hermitian manifolds. In section 1, we study invariant Riemannian maps, that is, the image of derivative map is invariant under the almost complex structure of the base manifold. We give examples, investigate the geometry of foliations, and find new conditions for the harmonicity of such maps. In section 2, we study anti-invariant Riemannian maps from Riemannian manifolds to almost Hermitian manifolds. We give several examples, and obtain a method to find anti-invariant Riemannian maps. We also obtain a characterization for umbilical anti-invariant Riemannian maps. In section 3, we study semi-invariant Riemannian maps from Riemannian manifolds to almost Hermitian manifolds as a generalization of CR-submanifolds. We give examples, obtain the totally geodesicity of such maps, and relate it with PHWC maps. In section 4, we investigate generic Riemannian maps from Riemannian manifolds to almost Hermitian manifolds as a generalization of generic submanifolds. We give examples and obtain necessary and sufficient conditions for such maps to be totally geodesic and harmonic. In section 5, we introduce slant Riemannian maps from Riemannian manifolds to almost Hermitian manifolds as a generalization of slant submanifolds. We obtain characterizations for such maps and investigate the geometry of maps. In section 6, we study semi-slant Riemannian maps as a generalization of semi-slant submanifolds, give examples, and obtain new characterizations. In section 7, we define hemi-slant Riemannian maps from Riemannian manifolds to almost Hermitian manifolds as a generalization of hemi-slant submanifolds and slant Riemannian maps to almost Hermitian manifolds. We give examples, obtain a decomposition theorem and find necessary and sufficient conditions to be a totally geodesic map.

**Keywords:** Kähler manifold, Riemannian maps, invariant Riemannian maps, anti-invariant Riemannian map, semi-invariant Riemannian map, generic Riemannian map, slant Riemannian map, semi-slant Riemannian map, hemi-slant Riemannian map, holomorphic

Reimannian Submersions, Reimannian Maps in Hermitian Geometry, and their Applications
http://dx.doi.org/10.1016/B978-0-12-804391-2.50006-3

submanifold, totally real submanifold, CR-submanifold, slant submanifold, semi-slant submanifold, hemi-slant submanifold

*If we all worked on the assumption that what is accepted as true is really true, there would be little hope of advance.*

*Orville Wright*

## 1. Invariant Riemannian maps to almost Hermitian manifolds

In this section, we introduce invariant and anti-invariant Riemannian maps between Riemannian manifolds and almost Hermitian manifolds as a generalization of invariant immersions and totally real immersions, respectively. Then we give examples, present a characterization, and obtain a geometric characterization of harmonic invariant Riemannian maps in terms of the distributions that are involved in the definition of such maps. We also give a decomposition theorem by using the existence of invariant Riemannian maps to Kähler manifolds. Moreover, we study anti-invariant Riemannian maps, give examples, and obtain a classification theorem for umbilical anti-invariant Riemannian maps.

First recall that if we write the conditions of a submanifold of a Hermitian manifold to be invariant and anti-invariant in terms of the inclusion map $i$, we get $Ji_*(T_xM) \subset i_*(T_xM)$ and $Ji_*(T_xM) \subset (i_*(T_xM))^\perp$, respectively. For arbitrary Riemannian maps between Riemannian manifolds and Hermitian manifolds, we present the following definitions.

**Definition 59.** [241] Let $F : (M, g_M) \to (N, g_N)$ be a proper Riemannian map between Riemannian manifold $M$ and a Hermitian manifold $N$ with an almost complex structure $J_N$. We say that $F$ is an *invariant Riemannian map* at $p$ if $J_N(rangeF_{*p}) = rangeF_{*p}$. If $F$ is an invariant Riemannian map for every $p \in M$, then $F$ is called an invariant Riemannian map. In a similar way, we say that $F$ is an *anti-invariant Riemannian map* at $p \in M$ if $J_N(rangeF_{*p}) \subseteq (rangeF_*p)^\perp$. If $F$ is an anti-invariant Riemannian map for every $p \in M$, then $F$ is called an anti-invariant Riemannian map.

First of all, we have the following trivial example of invariant Riemannian maps.

**Example 64.** [241] Let $(M, g_M)$ be a real manifold and $(N, g_N)$ a Kähler manifold with complex structure $J_N$. Suppose that $F : (M, g_M) \to (N, g_N)$ is an isometric immersion.

If $M$ is a complex submanifold of $N$, then the above isometric immersion $F$ is an invariant Riemannian map with $\ker F_* = \{0\}$.

From the above definition, we have the following result.

**Proposition 53.** *Let $F : (M, g_M) \to (N, g_N, J_N)$ be an invariant Riemannian map between a Riemannian manifold $M$ and a Hermitian manifold $N$ with integrable distribution $(rangeF_*)$. Then the leaves of the distribution $(rangeF_*)$ of an invariant Riemannian map are holomorphic submanifolds of $N$.*

The following result can be derived from the notation of invariant (holomorphic) submanifolds, and it gives a useful way to obtain examples of invariant Riemannian maps.

**Proposition 54.** *[241] Let $F_1$ be a Riemannian submersion from Riemannian manifold $(M, g_M)$ to a Riemannian manifold $(N, g_N)$. Let also $F_2$ be a holomorphic immersion from $(N, g_N)$ to a Hermitian manifold $(P, g_P, J_P)$, where $J_P$ is the complex structure of $P$. Then the composite map $F_2 \circ F_1$ is an invariant Riemannian map.*

**Remark 21.** In the above proposition the existence of holomorphic immersion is crucial. So the question is, are there such immersions. Here we present two examples: (1) A $n$-dimensional complex projective space $CP^n(c)$ of constant holomorphic sectional curvature $c$ can be imbedded into $CP^{n+p}(c)$ as a totally geodesic Kähler submanifold; (2) an $n$-dimensional complex hyperbolic space $CH^n(c)$ of constant holomorphic sectional curvature $c$ can be imbedded into $CH^{n+p}(c)$ as a totally geodesic Kähler submanifold [156]. Thus the above proposition gives a useful method to find examples of invariant Riemannian maps to Kähler manifolds.

Note that the above example and proposition justify the name of invariant Riemannian maps given in Definition 59. We denote by $\mathbb{R}^{2m}, m \geq 1$ the Euclidean $2m$ space with the standard metric and recall that the canonical complex structure of $\mathbb{R}^{2m}$ is defined by

$$J(x_1, y_1 ..., x_m, y_m) = (-y_1, x_1, ..., -y_m, x_m).$$

**Example 65.** [241] Consider the following map:

$$F : \quad \mathbb{R}^3 \quad \longrightarrow \quad \mathbb{R}^4$$
$$(x_1, x_2, x_3) \quad \longrightarrow \quad (\tfrac{x_1+x_2}{\sqrt{2}}, \tfrac{-x_1+x_2}{\sqrt{2}}, 0, 0),$$

where $\mathbb{R}^3$ and $\mathbb{R}^4$ are Euclidean spaces with usual inner products $g_{\mathbb{R}^3}$ and $g_{\mathbb{R}^4}$, respectively. We denote the Cartesian coordinates of $\mathbb{R}^3$ and $\mathbb{R}^4$ by $x_1, x_2, x_3$ and $y_1, y_2, y_3, y_4$, respectively. It is easy to see that $rank F = 2$. Then $ker F_* = span\{X_1 = \partial x_3\}$ and $(ker F_*)^\perp = \{X_2 = \partial x_2, X_3 = \partial x_1\}$, where $\{\partial x_1, \partial x_2, \partial x_3\}$ is a canonical basis of $(\mathbb{R}^3, g_{\mathbb{R}^3})$. By direct computations, we have

$$range F_* = span\{F_*(X_2) = \frac{1}{\sqrt{2}}(\partial y_1 + \partial y_2), F_*(X_3) = \frac{1}{\sqrt{2}}(\partial y_1 - \partial y_2)\},$$

where $\{\partial y_1, \partial y_2, \partial y_3, \partial y_4\}$ is a canonical basis of $(\mathbb{R}^4, g_{\mathbb{R}^4})$. It is also easy to check that

$$g_{\mathbb{R}^3}(X_2, X_2) = g_{\mathbb{R}^4}(F_*(X_2), F_*(X_2))$$

and

$$g_{\mathbb{R}^3}(X_3, X_3) = g_{\mathbb{R}^4}(F_*(X_3), F_*(X_3)),$$

which show that $F$ is a Riemannian map. We denote the canonical complex structure of $\mathbb{R}^4$ by $J$. Then it follows that $J(F_*(X_2)) = -F_*(X_3)$, so $range F_*$ is an almost complex distribution. Thus $F$ is an invariant Riemannian map.

As in the theory of submanifolds [235], we give the following definition.

**Definition 60.** [241] Let $F : (M, g_M) \longrightarrow (N, g_N)$ be an invariant Riemannian map between a Riemannian manifold $(M, g_M)$ and a Kähler manifold $(N, g_N)$. Then $F$ is called a *curvature invariant Riemannian map* with respect to $R^N$ if for any $x \in M$ and any $F_*(X), F_*(Y) \in T_{F(x)}N$ the $range F_{*x}$ is invariant under the curvature transformation

$$R^N_{F(x)}(F_*(X), F_*(Y)) : T_{F(x)}N \longrightarrow T_{F(x)}N,$$

that is,

$$R^N_{F(x)}(range F_{*x}) \subseteq (range F_{*x}). \tag{6.1}$$

**Theorem 138.** *[241] Let $F : (M, g_M) \longrightarrow (N, g_N)$ be a curvature invariant Riemannian map from a Riemannian manifold $(M, g_M)$ to a complex space form $(N(c), g_N)$. Then $F$ is an invariant Riemannian map or an anti-invariant Riemannian map.*

*Proof.* Let $F$ be a curvature invariant Riemannian map, then for $X, Y \in \Gamma((ker F_*)^\perp)$ and $F_*(X), F_*(Y) \in \Gamma(range F_*)$, using (3.16), we get

$$
\begin{aligned}
R^N(F_*(X), F_*(Y))F_*(X) = {} & \frac{c}{4}\{g_N(F_*(Y), F_*(X))F_*(X) - g_N(F_*(X), F_*(X))F_*(Y) \\
& + 3g_N(F_*(X), J_N F_*(Y))J_N F_*(X)\}.
\end{aligned}
$$

Then, since $F$ is curvature invariant, we have $R^N(F_*(X), F_*(Y))F_*(X) \in \Gamma(rangeF_*)$, from the above equation either $g_N(F_*(X), J_N F_*(Y)) = 0$ or $J_N F_*(X) \in \Gamma(rangeF_*)$. Then the first equation implies that $F$ is anti-invariant and the other tells us that $F$ is invariant. Thus the proof is complete. $\qquad\square$

We now investigate the harmonicity of an invariant Riemannian map. To do this, we need the following preparatory lemmas.

**Lemma 69.** *[241] Let $F : (M, g_M) \longrightarrow (N, g_N)$ be an invariant Riemannian map between a Riemannian manifold $(M, g_M)$ and a Kähler manifold $(N, g_N)$. If $rangeF_*$ is integrable, then*

$$B^R(\tilde{X}, \tilde{Y}) = -B^R(J_N \tilde{X}, J_N \tilde{Y}) \qquad (6.2)$$

*for $\tilde{X}, \tilde{Y} \in \Gamma(rangeF_*)$, where $B^R$ is the second fundamental form of $rangeF_*$.*

*Proof.* For $V \in \Gamma((rangeF_*)^\perp)$, we have $g_N(B^R(\tilde{X}, \tilde{Y}), V) = g_N(\nabla^N_{\tilde{X}} \tilde{Y}, V)$, where $\nabla^N$ is the Levi-Civita connection of $N$. Since $N$ is a Kähler manifold, we get $g_N(B^R(\tilde{X}, \tilde{Y}), V) = g_N(\nabla^N_{\tilde{X}} J_N \tilde{Y}, J_N V)$. Hence we derive

$$g_N(B^R(\tilde{X}, \tilde{Y}), V) = g_N([\tilde{X}, J_N \tilde{Y}] + \nabla^N_{J_N \tilde{Y}} \tilde{X}, J_N V).$$

Since $F$ is invariant and $rangeF_*$ is integrable, we have $[\tilde{X}, J_N \tilde{Y}] \in \Gamma(rangeF_*)$ and $J_N V \in \Gamma((rangeF_*)^\perp)$. Hence we arrive at $g_N(B^R(\tilde{X}, \tilde{Y}), V) = g_N(\nabla^N_{J_N \tilde{Y}} \tilde{X}, J_N V)$. Using again (3.2), we get

$$g_N(B^R(\tilde{X}, \tilde{Y}), V) = -g_N(\nabla^N_{J_N \tilde{Y}} J_N \tilde{X}, V)$$

which gives (6.2). $\qquad\square$

**Lemma 70.** *[241] Let $F : (M, g_M) \longrightarrow (N, g_N)$ be an invariant Riemannian map from a Riemannian manifold $(M, g_M)$ to a Kähler manifold $(N, g_N)$ such that $rangeF_*$ is integrable. Then the distribution $rangeF_*$ is minimal.*

*Proof.* First note, since $rangeF_*$ is invariant with respect to $J_N$, that we can choose an orthonormal basis $\{\tilde{e}_1, \tilde{e}_2, ..., \tilde{e}_s, \tilde{e}_1^*, ..., \tilde{e}_s^*\}$ such that $\tilde{e}_i^* = J_N \tilde{e}_i$. From (4.39) we have

$$g_N(\mu^R, V) = \frac{1}{n_1} \sum_{i=1}^s \{g_N(B^R(\tilde{e}_i, \tilde{e}_i) + B^R(\tilde{e}_i^*, \tilde{e}_i^*), V)\}.$$

where $n_1 = 2s = dim(rangeF_*)$, $\mu^R$ is the mean curvature vector field of $rangeF_*$ and $V \in \Gamma((rangeF_*)^\perp)$. Using (6.2) in the above equation, we obtain $g_N(\mu^R, V) = 0$, which shows that $rangeF_*$ is minimal. □

We also have the following result for harmonicity of $F$.

**Theorem 139.** *[241] Let $F : (M^m, g_M) \longrightarrow (N^n, g_N)$ be an invariant Riemannian map from a Riemannian manifold $(M, g_M)$ to a Kähler manifold $(N, g_N)$ such that $rangeF_*$ is integrable. Then $F$ is a harmonic map if and only if the distribution $\ker F_*$ is minimal.*

*Proof.* From Lemma 45, we have

$$\mathcal{R}(\nabla F_*)(X, Y) = 0 \tag{6.3}$$

for $X, Y \in \Gamma((\ker F_*)^{\perp})$, where $\mathcal{R}$ denotes the orthogonal projection onto distribution $rangeF_*$. Now let $\{e_1, ..., e_{n_1}\}$ be a local orthonormal basis for the $(\ker F_*)^{\perp}$, the trace (restricted to $(\ker F_*)^{\perp} \times (\ker F_*)^{\perp}$) of the second fundamental form is given by

$$
\begin{aligned}
trace^{\mathcal{H}} \nabla F_* &= \sum_{i=1}^{n_1} (\nabla F_*)(e_i, e_i) \\
&= \sum_{i,j=1}^{n_1} g_N((\nabla F_*)(e_i, e_i), \tilde{e}_j) \tilde{e}_j + \sum_{\alpha=1}^{n_2} \sum_{i=1}^{n_1} g_N((\nabla F_*)(e_i, e_i), \bar{e}_\alpha) \bar{e}_\alpha,
\end{aligned}
$$

where $\{\bar{e}_1, ..., \bar{e}_{n_2}\}$ is an orthonormal basis of $(rangeF_*)^{\perp}$. Using (6.3), we derive

$$trace^{\mathcal{H}} \nabla F_* = \sum_{\alpha=1}^{n_2} \sum_{i=1}^{n_1} g_N((\nabla F_*)(e_i, e_i), \bar{e}_\alpha) \bar{e}_\alpha.$$

Hence, we have

$$trace^{\mathcal{H}} \nabla F_* = \sum_{\alpha=1}^{n_2} \sum_{i=1}^{n_1} g_N(\nabla_{e_i}^F F_*(e_i), \bar{e}_\alpha) \bar{e}_\alpha.$$

Thus, from (4.39) we obtain

$$trace^{\mathcal{H}} \nabla F_* = n_1 \sum_{\alpha=1}^{n_2} g_N(\mu^{\mathcal{R}}, \bar{e}_\alpha) \bar{e}_\alpha.$$

Then minimal $rangeF_*$ implies that $\mu^{\mathcal{R}} = 0$, i.e.,

$$trace^{\mathcal{H}} \nabla F_* = 0. \tag{6.4}$$

Now let $\{e_{n_1+1}, ..., e_m\}$ be an orthonormal basis of $kerF_*$. Then we have

$$trace^{\mathcal{V}}\nabla F_* = \sum_{r=n_1+1}^{m}(\nabla F_*)(e_r, e_r) = -\sum_{r=n_1+1}^{m} F_*(\nabla_{e_r} e_r),$$

where $\mathcal{V}$ denotes the orthogonal projection onto $ker\, F_*$. Hence, we derive

$$trace^{\mathcal{V}}\nabla F_* = -(m-n_1)F_*(\mu^{\mathcal{V}}), \qquad (6.5)$$

where $\mu^{\mathcal{V}}$ is the mean curvature vector field of $ker\, F_*$. Thus from (6.4) and (6.5) we obtain

$$\tau = -(m-n_1)F_*(\mu^{\mathcal{V}}). \qquad (6.6)$$

Since a Riemannian map is an isometry between $range F_*$ and $(ker F_*)^{\perp}$, we get the assertion of the theorem. $\qquad\qquad\square$

**Remark 22.** It is known that an invariant submanifold of a Kähler manifold is minimal [311]. The above theorem is a generalization of this result. Indeed, if $m = n_1$ which means that $F$ is an invariant isometric immersion, then it follows that $\tau(F) = 0$. For an isometric immersion, $\tau(F)$ is exactly the mean curvature vector field of the immersion; this tells us that the invariant isometric immersion $F$ is minimal.

**Remark 23.** We note that besides complex and invariant submanifolds, the notion of holomorphic submanifolds is also used in the theory of submanifolds of Kähler manifolds. Because holomorphic Riemannian maps are different from the notion of invariant maps of this chapter, comparing Definition 52, Lemma 53, and Definition 59, one can conclude that if $M$ is an almost Hermitian manifold and $N$ is a Kähler manifold, then it follows that every holomorphic (Riemannian) map is an invariant (Riemannian) map, but the converse is not true. If $M$ is not an almost complex manifold, there is no relation between holomorphic maps and invariant Riemannian maps.

## 2. Anti-invariant Riemannian maps to almost Hermitian manifolds

In this section, we introduce anti-invariant Riemannian maps, give examples and investigate umbilical anti-invariant Riemannian maps. First, we give examples.

**Example 66.** [241] Let

$$\begin{aligned} F: \quad \mathbb{R}^2 \quad &\longrightarrow \quad \mathbb{R}^4 \\ (x_1, x_2) \quad &\longrightarrow \quad (0, \tfrac{x_1+x_2}{\sqrt{2}}, 0, 0). \end{aligned}$$

be a map. Then it is easy to see that $rank F = 1$. We can also see that $F$ is an anti-invariant Riemannian map.

Motivated from the above example, we have the following result.

**Proposition 55.** *[241] Every Riemannian map from a Riemannian manifold to an almost Hermitian manifold with rank one is an anti-invariant Riemannian map.*

**Example 67.** [241] Consider the following map:

$$F: \quad \mathbb{R}^3 \quad \longrightarrow \quad \mathbb{R}^6$$
$$(x_1, x_2, x_3) \quad \longrightarrow \quad (\sin x_1, \cos x_1, 0, 0 \sin x_2, \cos x_2).$$

we can see that $F$ is an anti-invariant Riemannian map with $rankF = 2$. At any point $x \in \mathbb{R}^3$, the vertical distribution is spanned by $Z_1 = \partial x_3$, the horizontal distribution is spanned by $Z_2 = \partial x_2, Z_3 = \partial x_1$, and $rangeF_*$ is spanned by $N_1 = \cos x_1 \partial x_1 - \sin x_2 \partial x_2$, $N_3 = \cos x_2 \partial x_5 - \sin x_2 \partial x_6$.

From Definition 59, we have the following result.

**Proposition 56.** *Let $F : (M, g_M) \to (N, g_N, J_N)$ be an anti-invariant Riemannian map between Riemannian manifold M and a Hermitian manifold N with integrable distribution $(rangeF_*)$. Then the leaves of the distribution $(rangeF_*)$ of an anti-invariant Riemannian map are totally real submanifolds of N.*

**Proposition 57.** *Let S be a Riemannian submersion from a Riemannian manifold $(M, g_M)$ to a Riemannian manifold $(N, g_N)$ and I be a totally real submanifold from Riemannian manifold $(N, g_N)$ to an almost Hermitian manifold $(P, g_P, J_P)$. Then the Riemannian map $I \circ S$ is an anti-invariant Riemannian map.*

Proposition 57 is useful to obtain examples of anti-invariant Riemannian maps.

**Example 68.** [241] Consider the orthogonal projection $S : \mathbb{R}^{n+k} \longrightarrow \mathbb{R}^n$; then it is obviously a Riemannian submersion. Also let $I : \mathbb{R}^n \longrightarrow \mathbb{C}^n$ be an isometric immersion. It is known that $\mathbb{R}^n$ is a totally geodesic anti-invariant submanifold $\mathbb{C}^n$. Thus $I \circ S$ is an anti-invariant Riemannian map.

In the rest of this section, we study umbilical anti-invariant Riemannian maps. We say that an anti-invariant Riemannian map $F:(M, g_M) \longrightarrow (N, g_N)$ is *Lagrangian Riemannian map*, if $rankF = dim(rangeF_*)^{\perp}$. This definition is a generalization of Lagrangian immersions. Using the above method given in Proposition 57, we have the following result which gives the harmonicity of Lagrangian Riemannian maps.

**Proposition 58.** *Let S be a Riemannian submersion from a Riemannian manifold* $(M, g_M)$ *to a Riemannian manifold* $(N, g_N)$ *with minimal fibers and I be a totally umbilical Lagrangian submanifold from Riemannian manifold* $(N, g_N)$ *to a Kähler manifold* $(P, g_P, J_P)$. *Then the Lagrangian Riemannian map* $I \circ S$ *is harmonic.*

*Proof.* From Theorem 90, we see that it is enough to show that the immersion $I$ is totally geodesic. From a result in [311, Proposition 2.2, page 206], it follows that a totally umbilical Lagrangian submanifold in a Kähler manifold is totally geodesic. This proves our assertion.                                                                □

We now give an example of umbilical Riemannian maps.

**Example 69.** [241] Let $M_1 \times_f M_2$ be a warped product manifold. Then the projection $\pi$: $M_1 \times_f M_2 \longrightarrow M_1$ is a Riemannian submersion whose vertical and horizontal spaces at any point $(p, q)$ are, respectively, identified with $T_{p_2} M_2$ and $T_{p_1} M_1$. Any fiber $\pi$ which is identified with $M_2$ is totally umbilical. Moreover, the horizontal distribution is integrable, and the invariant $A$ associated with $\pi$ vanishes [111]. Let $\mathbf{RP}^n(\frac{c}{4})$ be a real projective $n$–space and $N$ be a totally umbilical hypersurface of $RP^n(\frac{c}{4})$. Then it is known that ([83]) $N$ can be imbedded in a complex projective $2n$–space $\mathbf{CP}^{2n}(c)$ as a totally real and totally umbilical submanifold. Now consider the composition of the Riemannian submersion $\pi : N \times_f M_2 \longrightarrow N \subset \mathbf{RP}^n$ followed by the isometric immersion $I : N \subset \mathbf{RP}^n \longrightarrow \mathbf{CP}^{2n}$; then, from Proposition 58, we conclude that $I \circ \pi$ is an umbilical anti-invariant Riemannian map.

Finally, we give a classification result for umbilical Riemannian maps.

**Theorem 140.** *Let F be an umbilical anti-invariant Riemannian map from a Riemannian manifold* $(M^m, g_M)$ *to a Kähler manifold* $(N^n, g_N, J_N)$. *Then either the distribution* $(ker F_*)^\perp$ *is one-dimensional or* $H_2 = 0$.

*Proof.* From (1.71) and (4.49), we have $\nabla_X^F F_*(Y) - F_*(\nabla_X Y) = g_M(X, Y) H_2$ for $X, Y \in \Gamma((ker F_*)^\perp)$. Then, for $Z \in \Gamma(ker F_*)^\perp$, we get

$$g_N(\nabla_X^F F_*(Y), J_N F_*(Z)) = g_M(X, Y) g_N(J_N F_*(Z), H_2).$$

Hence, we obtain

$$g_M(X, Y) g_N(J_N F_*(Z), H_2) = X g_N(F_*(Y), J_N F_*(Z))$$
$$- g_N(F_*(Y), \nabla_X^F J_N F_*(Z)).$$

Since $N$ is Kähler, we derive

$$g_M(X, Y)g_N(J_N F_*(Z), H_2) = -g_N(F_*(Y), J_N \nabla^F_X F_*(Z)).$$

Using again (1.71), we get

$$g_M(X, Y)g_N(J_N F_*(Z), H_2) = -g_N(F_*(Y), J_N (\nabla F_*)(X, Z))$$
$$-g_N(F_*(Y), J_N F_*(\nabla_X Z)).$$

Thus, we have

$$g_M(X, Y)g_N(J_N F_*(Z), H_2) = g_N(J_N F_*(Y), (\nabla F_*)(X, Z)).$$

Then, using (4.49) for $X = Y$, we obtain

$$g_M(X, X)g_N(J_N F_*(Z), H_2) = g_M(X, Z)g_N(J_N F_*(X), H_2). \tag{6.7}$$

Interchanging the role of $X$ and $Z$, we get

$$g_M(Z, Z)g_N(J_N F_*(X), H_2) = g_M(Z, X)g_N(J_N F_*(Z), H_2). \tag{6.8}$$

Thus, from (6.7) and (6.8), we obtain

$$g_N(J_N F_*(Z), H_2) = \frac{g_M(X, Z)^2}{g_M(X, X)g_M(Z, Z)} = g_N(H_2, J_N F_*(Z)). \tag{6.9}$$

Since $H_2$ belongs to $(range F_*)^\perp$, then (6.9) implies that either $X$ and $Z$ are linearly dependent or $H_2 = 0$, which shows that either $(ker F_*)^\perp$ is one-dimensional or $H_2 = 0$. $\qquad\qquad\square$

## 3. Semi-invariant Riemannian maps to almost Hermitian manifolds

In this section, we define semi-invariant Riemannian maps, give examples, and investigate the geometry of leaves of the distributions which are defined on $M_1$ and $M_2$, respectively. We also give a decomposition theorem and obtain necessary and sufficient conditions for such Riemannian maps to be totally geodesic.

**Definition 61.** [244] Let $F$ be a Riemannian map from a Riemannian manifold $(M_1, g_1)$ to an almost Hermitian manifold $(M_2, g_2, J_2)$. Then we say that $F$ is a semi-invariant Riemannian map at $p \in M_1$ if there are subbundles $\mathcal{D}_1$ and $\mathcal{D}_2$ in $range F_*$ such that

$$J_2(\mathcal{D}_1) = \mathcal{D}_1, J_2(\mathcal{D}_2) \subseteq (range F_*)^\perp. \tag{6.10}$$

If $F$ is a semi-invariant Riemannian map at every point $p \in M_1$, then we say that $F$ is a semi-invariant Riemannian map between $M_1$ and $M_2$.

From the above definition, we have the following result.

**Proposition 59.** *Let $F : (M, g_M) \rightarrow (N, g_N, J_N)$ be a semi-invariant Riemannian map between Riemannian manifold M and a Hermitian manifold N with integrable distribution ($rangeF_*$). Then the leaves of the distribution ($rangeF_*$) of a semi-invariant Riemannian map are CR-submanifolds of N.*

We present the following examples which show that semi-invariant Riemannian maps from Riemannian manifolds to almost Hermitian manifolds are generalizations of holomorphic submanifolds, totally real submanifolds, and CR-submanifolds of almost Hermitian manifolds. They are also a generalization of holomorphic submersions.

**Example 70.** Every CR-submanifold of an almost Hermitian manifold is a semi-invariant Riemannian map with $kerF_* = \{0\}$.

**Example 71.** Every holomorphic submersion between almost Hermitian manifolds is a semi-invariant Riemannian map with $(rangeF_*)^{\perp} = \{0\}$, so $\mathcal{D}_2 = \{0\}$.

**Example 72.** Every invariant Riemannian map from a Riemannian manifold to an almost Hermitian manifold is a semi-invariant Riemannian map with $\mathcal{D}_2 = \{0\}$.

**Example 73.** Every anti-invariant Riemannian map from a Riemannian manifold to an almost Hermitian manifold is a semi-invariant Riemannian map with $\mathcal{D}_1 = \{0\}$.

**Example 74.** [244] Consider the following map defined by

$$F : \quad \begin{matrix} \mathbb{R}^5 \\ (x_1, x_2, x_3, x_4, x_5) \end{matrix} \quad \longrightarrow \quad \begin{matrix} \mathbb{R}^4 \\ (\frac{x_1+x_2}{\sqrt{2}}, \frac{x_3+x_4}{\sqrt{2}}, x_5). \end{matrix}$$

Then we have

$$ker\, F_* = span\{Z_1 = \frac{\partial}{\partial x_1} - \frac{\partial}{\partial x_2}, Z_2 = \frac{\partial}{\partial x_3} - \frac{\partial}{\partial x_4}\}$$

and

$$(ker\, F_*)^{\perp} = span\{Z_3 = \frac{\partial}{\partial x_1} + \frac{\partial}{\partial x_2}, Z_4 = \frac{\partial}{\partial x_3} + \frac{\partial}{\partial x_4}, Z_5 = \frac{\partial}{\partial x_5}\}.$$

Hence it is easy to see that

$$g_{\mathbb{R}^4}(F_*(Z_i), F_*(Z_i)) = g_{\mathbb{R}^4}(Z_i, Z_i) = 2, g_{\mathbb{R}^4}(F_*(Z_5), F_*(Z_5)) = g_{\mathbb{R}^4}(Z_5, Z_5) = 1$$

and

$$g_{\mathbb{R}^4}(F_*(Z_i), F_*(Z_j)) = g_{\mathbb{R}^4}(Z_i, Z_j) = 0,$$

$i \neq j$, for $i, j = 3, 4$. Thus $F$ is a Riemannian map. On the other hand, we have $JF_*(Z_3) = F_*(Z_4)$ and $JF_*(Z_5) = V \in \Gamma((range F_*)^\perp)$, where $J$ is the complex structure of $\mathbb{R}^4$. Thus $F$ is a semi-invariant Riemannian map.

We now study the integrability of distributions which are the outcome of the above definition. First note that for $F_*(X) \in \Gamma(\mathcal{D}_1)$ and $F_*(Z) \in \Gamma(\mathcal{D}_2)$, we have $g_2(F_*(X), F_*(Z)) = 0$. Then Riemannian map $F$ implies that $g_1(X, Z) = 0$. Thus we have two orthogonal distributions $\bar{D}_1$ and $\bar{D}_2$ such that

$$(ker F_*)^\perp = \bar{D}_1 \oplus \bar{D}_2.$$

The next proposition shows that the base manifold $M_2$ is foliated by the distribution $\mathcal{D}_2$.

**Proposition 60.** *[244] Let $F$ be a semi-invariant Riemannian map from a Riemannian manifold $(M_1, g_1)$ to a Kähler manifold $(M_2, g_2)$. Then the distribution $\mathcal{D}_2$ is integrable.*

*Proof.* Since $M_2$ is Kähler manifold, we have $d\Omega_2 = 0$, where $\Omega_2$ is the fundamental 2–form of $M_2$. Hence we get

$$d\Omega_2(F_*(X), F_*(Z_1), F_*(Z_2)) = -\Omega_2([F_*(Z_1), F_*(Z_2)], F_*(X)) = 0$$

for $F_*(X) \in \Gamma(\mathcal{D}_1)$ and $F_*(Z_1), F_*(Z_2) \in \Gamma(\mathcal{D}_2)$. Thus we have $[F_*(Z_1), F_*(Z_2)] \in \Gamma(\mathcal{D}_2)$, which proves the assertion. □

For the distribution $(ker F_*)^\perp$, we have the following result.

**Proposition 61.** *[244] Let $F$ be a semi-invariant Riemannian map from a Riemannian manifold $(M_1, g_1)$ to a Kähler manifold $(M_2, g_2, J_2)$. Then $(ker F_*)^\perp$ is integrable if and only if*

$$g_2((\nabla F_*)(X, Z), F_*(Y)) = g_2((\nabla F_*)(Y, Z), F_*(X))$$

*for $X, Y \in \Gamma((ker F_*)^\perp)$ and $Z \in \Gamma(ker F_*)$.*

*Proof.* Since $F$ is a Riemannian map, we get

$$g_1([X, Y], Z) = -g_2(F_*(\nabla_X^1 Z), F_*(Y)) + g_2(F_*(X), F_*(\nabla_Y^1 Z))$$

for $X, Y \in \Gamma((ker\, F_*)^\perp)$ and $Z \in \Gamma(ker\, F_*)$, where $\nabla^1$ is the Levi-Civita connection of $M_1$. Then, using (1.71,) we have

$$g_1([X, Y], Z) = g_2((\nabla F_*)(X, Z), F_*(Y)) - g_2((\nabla F_*)(Y, Z), F_*(X)),$$

which proves the assertion. $\qquad\qquad\square$

For the leaves of distributions of $\bar{D}_1$ and $\bar{D}_2$, we first have the following.

**Proposition 62.** *[244] Let F be a Riemannian map from a Riemannian manifold $(M_1, g_1)$ to a Kähler manifold $(M_2, g_2, J_2)$. Then the distribution $\bar{D}_1$ defines a totally geodesic foliation if and only if*
**(a)** *$(\nabla F_*)(X_1, Z)$ has no components in $F_*(\bar{D}_1)$, and*
**(b)** *$(\nabla F_*)(X_1, Y_1')$ has no components in $J_2 \mathcal{D}_2$,*
*where $Z \in \Gamma(ker\, F_*)$, $X_1, Y_1 \in \Gamma(\bar{D}_1)$ and $F_*(Y_1') = J_2 F_*(Y_1)$.*

*Proof.* We note that $\bar{D}_1$ defines a totally geodesic foliation if and only if $g_1(\nabla^1_{X_1} Y_1, Z) = g_1(\nabla^1_{X_1} Y_1, X_2) = 0$ for $Z \in \Gamma(ker\, F_*)$, $X_1, Y_1 \in \Gamma(\bar{D}_1)$ and $X_2 \in \Gamma(\bar{D}_2)$. Using (1.71), we have

$$g_1(\nabla^1_{X_1} Y_1, Z) = g_1((\nabla F_*)(X_1, Z), F_*(Y_1)). \qquad (6.11)$$

In a similar way, from (1.71) we get

$$g_1(\nabla^1_{X_1} Y_1, X_2) = -g_2((\nabla F_*)(X_1, Y_1), F_*(X_2)) + g_2(\nabla^F_{X_1} F_*(Y_1), F_*(X_2)).$$

Then from Proposition 3.1, we obtain

$$g_1(\nabla^1_{X_1} Y_1, X_2) = g_2(\nabla^F_{X_1} F_*(Y_1), F_*(X_2)).$$

Since $M_2$ is a Kähler manifold, we arrive at

$$g_1(\nabla^1_{X_1} Y_1, X_2) = g_2(\nabla^F_{X_1} J_2 F_*(Y_1), J F_*(X_2)).$$

Hence, using (1.71), we have

$$g_1(\nabla^1_{X_1} Y_1, X_2) = g_2((\nabla F_*)(X_1, Y_1'), J F_*(X_2)), F_*(Y_1') = J_2 F_*(Y_1). \qquad (6.12)$$

Thus proof follows from (6.11) and (6.12). $\qquad\qquad\square$

The following result can be obtained similar to the above proposition.

**Proposition 63.** *[244] Let F be a semi-invariant Riemannian map from a Riemannian manifold $(M_1, g_1)$ to a Kähler manifold $(M_2, g_2, J_2)$. Then $\bar{D}_2$ defines a totally geodesic foliation if and only if*
**(a)** *$(\nabla F_*)(X_2, Z)$ has no components in $F_*(\bar{D}_2)$ for $X_2 \in \Gamma(\bar{D}_2)$ and $Z \in \Gamma(ker\, F_*)$, and*

**(b)** $(\nabla F_*)(X_2, X_1')$ *has no components in* $J\mathcal{D}_2$ *for* $X_1' \in \Gamma(\bar{D}_1)$ *such that* $F_*(X_1') = J_2 F_*(X_1)$.

Thus from Proposition 62 and Proposition 63, we have the following decomposition theorem.

**Theorem 141.** *[244] Let F be a semi-invariant Riemannian map from a Riemannian manifold* $(M_1, g_1)$ *to a Kähler manifold* $(M_2, J, g_2)$. *Then the integral manifold of* $(ker F_*)^\perp$ *is a locally product manifold if and only if*
**(a)** $(\nabla F_*)(X_2, Z)$ *has no components in* $F_*(\bar{D}_2)$ *for* $X_2 \in \Gamma(\bar{D}_2)$ *and* $Z \in \Gamma(ker F_*)$,
**(b)** $(\nabla F_*)(X_2, X_1')$ *has no components in* $J\mathcal{D}_2$ *for* $X_1' \in \Gamma(\bar{D}_1)$ *such that* $F_*(X_1') = J_2 F_*(X_1)$.
**(c)** $(\nabla F_*)(X_1, Z)$ *has no components in* $F_*(\bar{D}_1)$, *and*
**(d)** $(\nabla F_*)(X_1, Y_1')$ *has no components in* $J_2\mathcal{D}_2$
*where* $Z \in \Gamma(ker F_*)$, $X_1, Y_1 \in \Gamma(\bar{D}_1)$ *and* $F_*(Y_1') = J_2 F_*(Y_1)$.

For $X \in \Gamma(range F_*)$, we then write

$$JX = \phi X + \omega X, \tag{6.13}$$

where $\phi X \in \Gamma(\mathcal{D}_1)$ and $\omega X \in \Gamma(J\mathcal{D}_2)$. On the other hand, for $V \in \Gamma((range F_*)^\perp)$, we then have

$$JV = \mathcal{B}V + CV, \tag{6.14}$$

where $\mathcal{B}V \in \Gamma(\mathcal{D}_1)$ and $CV \in \Gamma(\mu)$. Here $\mu$ is the complementary orthogonal distribution to $\omega(\mathcal{D}_2)$ in $(range F_*)^\perp$. It is easy to see that $\mu$ is invariant with respect to $J$.

**Theorem 142.** *[244] Let F be a Riemannian manifold* $(M_1, g_1)$ *to a Kähler manifold* $(M_2, J, g_2)$. *Then the base manifold is locally a product manifold* $M_{(range F_*)} \times M_{(range)^\perp}$ *if and only if*

$$g_2(-A_{CV}X + F_*(\nabla_X Z'), \mathcal{B}V) = -g_2([V, F_*(X)], W) - g_2((\nabla F_*)(X, Z') + \nabla_X^{F\perp} CV, CW) \tag{6.15}$$

*and*

$$g_2(-(\nabla F_*)(X, Y') + \nabla_X^{F\perp} \omega F_*(Y), CV_1) = g_2(-F_*(\nabla_X Y') + A_{\omega F_*(Y)}X, \mathcal{B}V_1) \tag{6.16}$$

*for* $V, W, V_1 \in \Gamma((range F_*)^\perp)$, $X, Y, X', Y' \in \Gamma((ker F_*)^\perp)$ *such that* $F_*(X') = \phi F_*(X)$, $F_*(Y') = \phi F_*(Y)$ *and* $F_*(Z') = \mathcal{B}V$.

*Proof.* For $X, Y \in \Gamma((rangeF_*))$ and $V \in \Gamma((rangeF_*)^\perp)$, using (3.2), we have

$$g_2(\nabla^2_X F_*(Y), V) = g_2(\nabla^2_X J F_*(Y), J_2 V).$$

Thus from (6.13) and (6.14), we obtain

$$g_2(\nabla^2_X F_*(Y), V) = g_2(\nabla^2_X \phi Y + \omega Y, \mathcal{B}V + \mathcal{C}V).$$

Then, since $(\nabla F_*)(X, Y')$ has no components in $(rangeF_*)$, where $F_*(Y') = \phi F_*(Y)$ for $Y' \in \Gamma((ker F_*)^\perp)$, by using (1.71) and (4.16), we get

$$\begin{aligned} g_2(\nabla^2_X F_*(Y), V) &= g_2((\nabla F_*)(X, Y') + \nabla^{F\perp}_X \omega F_*(Y), \mathcal{C}V) \\ &+ g_2(-A_{\omega F_*(Y)}X + F_*(\nabla_X Y), \mathcal{B}V). \end{aligned}$$

In a similar way, one can obtain that the distribution $(rangeF_*)$ defines a totally geodesic foliation on $M_2$ if and only if Equation (6.16) is satisfied. Thus proof is complete.    □

We now give necessary and sufficient conditions for a semi-invariant Riemannian map to be totally geodesic.

**Theorem 143.** *[244] Let $F$ be a semi-invariant Riemannian map from a Riemannian manifold $(M_1, g_1)$ to a Kähler manifold $(M_2, g_2, J)$. Then $F$ is totally geodesic if and only if*

**(a)** *$ker F_*$ is totally geodesic,*
**(b)** *$(ker F_*)^\perp$ is totally geodesic, and*
**(c)** *for $X, Y, Z \in \Gamma((ker F_*)^\perp)$ such that $F_*(Z) = \phi F_*(Y)$, we have*

$$C(\nabla F_*)(X, Z) - C\nabla^{F\perp}_X \omega F_*(Y) = -\omega F_*(\nabla_X Z) - \omega A_{\omega F_*(Y)}X.$$

*Proof.* We first note that by direct computations, we can obtain that (a) and (b) are satisfied if and only if $(\nabla F_*)(V, W) = 0$, $(\nabla F_*)(X, V) = 0$ for $X \in \Gamma((ker F_*)^\perp)$ and $V, W \in \Gamma(ker F_*)$. On the other hand, for $X, Y \in \Gamma((ker F_*)^\perp)$, using (1.71) and (3.5,) we have

$$(\nabla F_*)(X, Y) = -J\nabla^2_X J F_*(Y) - F_*(\nabla_X Y).$$

Then from (6.13) we get

$$\begin{aligned} (\nabla F_*)(X, Y) &= -J(\nabla^2_X F_*(Z) + \nabla^2_X \omega F_*(Y)) \\ &- F_*(\nabla_X Y), \end{aligned}$$

where $F_*(Z) = \phi F_*(Y)$, $Z \in \Gamma((ker F_*)^\perp)$. Then (1.71) and (4.16) imply that

$$
\begin{aligned}
(\nabla F_*)(X, Y) \;=\;& -J(\nabla F_*)(X, Z) - JF_*(\nabla_X Z) \\
&+\; J(-A_{\omega F_*(Y)}X + \nabla_X^{F\perp}\omega F_*(Y)) \\
&-\; F_*(\nabla_X Y).
\end{aligned}
$$

Since $(\nabla F_*)(X, Z) \in \Gamma((range F_*)^{\perp})$, using (6.14) we obtain

$$
\begin{aligned}
(\nabla F_*)(X, Y) \;=\;& -\mathcal{B}(\nabla F_*)(X, Z) - C(\nabla F_*)(X, Z) \\
&-\; \phi F_*(\nabla_X Z) - \omega F_*(\nabla_X Z) \\
&-\; \phi A_{\omega F_*(Y)}X - \omega A_{\omega F_*(Y)}X \\
&+\; \mathcal{B}\nabla_X^{F\perp}\omega F_*(Y)) + C\nabla_X^{F\perp}\omega F_*(Y)) \\
&-\; F_*(\nabla_X Y).
\end{aligned}
$$

Then, taking the $(range F_*)^{\perp}$ components, we arrive at

$$
\begin{aligned}
(\nabla F_*)(X, Y) \;=\;& -C(\nabla F_*)(X, Z) - \omega F_*(\nabla_X Z) \\
&-\; \omega A_{\omega F_*(Y)}X + C\nabla_X^{F\perp}\omega F_*(Y)),
\end{aligned}
$$

which completes the proof. $\qquad\qquad\square$

In the rest of this section, we consider umbilical semi-invariant Riemannian maps from Riemannian manifolds to Kähler manifolds. We mainly show that if $F$ is an umbilical semi-invariant Riemannian map from a Riemannian manifold to a Kähler manifold, then either $(range F_*)$ defines a totally geodesic foliation on the base manifold or $\bar{D}_2$ is one-dimensional. However, we first need the following lemma, which will be useful when we deal with the theorem of this section.

**Lemma 71.** *[244] Let F be an umbilical semi-invariant Riemannian map from a Riemannian manifold $(M_1, g_1)$ to a Kähler manifold $(M_2, g_2, J)$. Then $H_2 \in \Gamma(J\mathcal{D}_2)$.*

*Proof.* From (4.49) and (1.71) we have

$$
g_2(\nabla_X^F F_*(X), V) = g_1(X, X)g_2(H_2, V)
$$

for $X \in \Gamma(\bar{D}_1)$ and $V \in \Gamma(\mu)$. Hence we get

$$
g_2(\nabla_X^F F_*(Y'), JV) = g_1(X, X)g_2(H_2, V),
$$

where $Y' = {}^*F_*(\phi F_*(X))$. Thus, from (1.71) we have

$$
g_2((\nabla F_*)(X, Y'), JV) = g_1(X, X)g_2(H_2, V).
$$

Since $F$ is umbilical, it follows that

$$g_1(X, {}^*F_*(\phi F_*(X)))g_2(H_2, JV) = g_1(X, X)g_2(H_2, V).$$

Hence, we have

$$g_1(F_*X, \phi F_*(X))g_2(H_2, JV) = g_1(X, X)g_2(H_2, V).$$

Thus we derive $g_1(X, X)g_2(H_2, V) = 0$; since $g_1$ and $g_2$ are Riemannian metrics, we obtain $H_2 \in \Gamma(J\mathcal{D}_2)$. $\square$

**Theorem 144.** *[244] Let $F$ be an umbilical semi-invariant Riemannian map from a Riemannian manifold $(M_1, g_1)$ to a Kähler manifold $(M_2, g_2, J)$. Then either $(\text{range}F_*)$ defines a totally geodesic foliation or $\bar{D}_2$ is one-dimensional.*

*Proof.* For $Z_1, Z_2 \in \Gamma(\bar{D}_2)$, we have

$$g_2((\nabla F_*)(Z_1, Z_2), JF_*(Z_2)) = g_1(Z_1, Z_2)g_2(H_2, JF_*(Z_2)).$$

Using (1.71), we obtain

$$g_2(\nabla^F_{Z_2} F_*(Z_1), JF_*(Z_2)) = g_1(Z_1, Z_2)g_2(H_2, JF_*(Z_2)).$$

Then, from (3.2), (3.5), and (1.71), we get

$$g_2(JF_*(Z_1), \nabla^F_{Z_2} F_*(Z_2)) = g_1(Z_1, Z_2)g_2(H_2, JF_*(Z_2)).$$

Using again (1.71) and (4.49), we arrive at

$$g_2(JF_*(Z_1), H_2)g_1(Z_2, Z_2) = g_1(Z_1, Z_2)g_2(H_2, JF_*(Z_2)). \tag{6.17}$$

Interchanging the roles of $Z_2$ and $Z_1$, we obtain

$$g_2(JF_*(Z_2), H_2)g_1(Z_1, Z_1) = g_1(Z_1, Z_2)g_2(H_2, JF_*(Z_1)). \tag{6.18}$$

Then (6.17) and (6.18) imply that

$$g_2(F_*(Z_1), JH_2) = \frac{g_1(Z_1, Z_2)^2}{\| Z_1 \|^2 \| Z_2 \|^2} g_2(JH_2, F_*(Z_1)).$$

Thus the above equation tells us that either $H_2 = 0$, which implies that $(\text{range}F_*)$ is totally geodesic or $\bar{D}_2$ is one-dimensional. $\square$

## 4. Generic Riemannian maps to Kähler manifolds

In this section, we define generic Riemannian maps from Riemannian manifolds to almost Hermitian manifolds, provide examples, and investigate necessary and sufficient conditions for such maps to be totally geodesic and harmonic.

Let $F$ be a Riemannian map from an almost Hermitian manifold $(M_1, g_1)$ to an almost Hermitian manifold $(M_2, g_2, J)$. Define

$$\mathfrak{D}_p = (rangeF_{*p}) \cap J(rangeF_{*p})), p \in M$$

the complex subspace of the image subspace $F_*(T_pM)$.

**Definition 62.** [255] Let $F$ be a Riemannian map from a Riemannian manifold $(M_1, g_1)$ to an almost Hermitian manifold $(M_2, g_2, J)$. If the dimension $\mathfrak{D}_p$ is constant along $F$ and it defines a differentiable distribution on $M_2$, then we say that $F$ is a *generic Riemannian map*.

From the above definition, we have the following result.

**Proposition 64.** *Let $F : (M, g_M) \to (N, g_N, J_N)$ be a generic Riemannian map between Riemannian manifold $M$ and a Hermitian manifold $N$ with integrable distribution $(rangeF_*)$. Then the leaves of the distribution $(rangeF_*)$ of a generic Riemannian map are generic submanifolds of $N$.*

A generic Riemannian map is *purely real* (respectively, *complex*) if $\mathfrak{D}_p = \{0\}$ (respectively, $\mathfrak{D}_p = rangeF_{*p}$). For a generic Riemannian map, the orthogonal complementary distribution $\mathfrak{D}^\perp$, called purely real distribution, satisfies

$$rangeF_* = \mathfrak{D} \oplus \mathfrak{D}^\perp \qquad (6.19)$$

$$\mathfrak{D} \cap \mathfrak{D}^\perp = \{0\}. \qquad (6.20)$$

Let $F$ be a generic Riemannian map from a Riemannian manifold $(M_1, g_1)$ to an almost Hermitian manifold $(M_2, g_2, J)$. Then for $F_*(X) \in \Gamma(rangeF_*)$, $X \in \Gamma((ker\, F_*)^\perp)$ we write

$$JF_*(X) = \varphi F_*(X) + \varpi F_*(X), \qquad (6.21)$$

where $\varphi F_*(X) \in \Gamma(rangeF_*)$ and $\varpi F_*(X) \in \Gamma((rangeF_*)^\perp)$. Now we consider the complementary orthogonal distribution $\upsilon$ to $\varpi \mathfrak{D}^\perp$ in $(rangeF_*)^\perp$. Then it is obvious that

$$(rangeF_*)^\perp = \varpi \mathfrak{D}^\perp \oplus \upsilon, \varphi \mathfrak{D}^\perp \subseteq \mathfrak{D}^\perp.$$

Also for $V \in \Gamma((rangeF_*)^\perp)$, we write

$$JV = \mathfrak{B}V + \mathfrak{C}V, \qquad (6.22)$$

where $\mathfrak{B}V \in \Gamma(\mathfrak{D}^\perp)$ and $\mathfrak{C}V \in \Gamma(\upsilon)$. We also have

$$\mathfrak{B}(rangeF_*)^\perp = \mathfrak{D}^\perp. \qquad (6.23)$$

We now give some examples of generic Riemannian maps from Riemannian manifolds to almost Hermitian manifolds.

**Example 75.** Every CR-submanifold $F$ is a generic Riemannian map with $(ker F_*) = \{0\}$ and $\mathcal{D}^\perp$ is a totally real distribution.

**Example 76.** Every generic submanifold $F$ is a generic Riemannian map with $(ker F_*) = \{0\}$.

**Example 77.** Every semi-invariant Riemannian map $F$ from a Riemannian manifold to an almost Hermitian manifold is a generic Riemannian map such that $\mathcal{D}^\perp$ is a totally real distribution.

Since semi-invariant Riemannian maps include invariant Riemannian maps and anti-invariant Riemannian maps, such Riemannian maps are also examples of generic Riemannian maps.

**Example 78.** Every holomorphic Riemannian map is a generic Riemannian map between almost Hermitian manifolds with $\mathcal{D} = rangeF_*$.

We say that a generic Riemannian map is proper if $\mathcal{D}^\perp$ is neither complex nor purely real. We now present an example of a proper generic Riemannian map from a Riemannian manifold to a Kähler manifold.

**Example 79.** [255] Consider the following map defined by

$$F: \quad \begin{matrix} \mathbb{R}^9 \\ (x^1, ..., x^9) \end{matrix} \quad \longrightarrow \quad \begin{matrix} \mathbb{R}^6 \\ (x^1, x^9, x^3, \frac{x^4+x^5}{\sqrt{2}}, A\frac{x^4+x^5}{\sqrt{2}}, 0), A \neq 0. \end{matrix}$$

Then $ker F_*$ is spanned by

$$U_1 = \frac{\partial}{\partial x^2} \quad , \quad U_2 = \frac{\partial}{\partial x^4} - \frac{\partial}{\partial x^5}$$

$$U_3 = \frac{\partial}{\partial x^6} \quad , \quad U_4 = \frac{\partial}{\partial x^7}, U_4 = \frac{\partial}{\partial x^8}.$$

$(ker F_*)^\perp$ is spanned by

$$Z_1 = \frac{\partial}{\partial x^1} \quad , \quad Z_2 = \frac{\partial}{\partial x^9}$$

$$Z_3 = \frac{\partial}{\partial x^3} \quad , \quad Z_4 = \frac{1}{\sqrt{2}}(\frac{\partial}{\partial x^4} + \frac{\partial}{\partial x^5}). \tag{6.24}$$

Hence it is easy to see that

$$g_{\mathbb{R}^6}(F_*(Z_i), F_*(Z_i)) = g_{\mathbb{R}^9}(Z_i, Z_i), i = 1, 2, 3, 4$$

and

$$g_{\mathbb{R}^6}(F_*(Z_i), F_*(Z_j)) = g_{\mathbb{R}^9}(Z_i, Z_j) = 0,$$

$i \neq j$, Thus $F$ is a Riemannian map. Moreover, $(range F_*)^\perp$ is spanned by

$$V_1 = \frac{\partial}{\partial y^4} - \frac{\partial}{\partial y^5}, V_2 = \frac{\partial}{\partial y^6}.$$

On the other hand, we have $JF_*(Z_1) = F_*(Z_2)$ and $JF_*(Z_3) = \frac{1}{2}F_*(Z_4) + \frac{1}{2}V_1$ and $JF_*(Z_4) = -F_*(Z_3) + V_2$, where $J$ is the complex structure of $\mathbb{R}^6$. Thus $F$ is a generic Riemannian map with $\mathfrak{D} = span\{F_*(Z_1), F_*(Z_2)\}$, $\mathfrak{D}^\perp = span\{F_*(Z_3), F_*(Z_4)\}$, and $\upsilon = span\{V_1, V_2\}$.

We now find necessary and sufficient conditions for generic Riemannian maps from Riemannian manifolds to Kähler manifolds to be totally geodesic and harmonic. We first give the following result for totaly geodesicity.

**Theorem 145.** *[255] Let $F$ be a generic Riemannian map from a Riemannian manifold $(M_1, g_1)$ to a Kähler manifold $(M_2, g_2, J)$. Then $F$ is totally geodesic if and only if*
**(1)** *for $X, Y, \acute{Y} \in \Gamma((\ker F_*)^\perp)$,*

$$g_2(S_{\varpi F_*(Y)}F_*(X), \mathcal{B}V) = g_2((\nabla F_*)(X, \acute{Y}), \mathfrak{C}V) + g_1(\mathcal{H}\nabla_X\acute{Y},^* F_*\mathcal{B}V)$$
$$+ g_2(\nabla_X^{F^\perp} \varpi F_*(Y), \mathfrak{C}V)$$

*is satisfied, where $F_*(\acute{Y}) = \varphi F_*(Y)$,*
**(2)** *$\ker F_*$ is totally geodesic, and*
**1.** *$(\ker F_*)^\perp$ is totally geodesic.*

*Proof.* For $X, Y \in \Gamma((\ker F_*)^\perp)$ and $V \in \Gamma((range F_*)^\perp)$, using (3.5), (3.3), (6.22), and (6.21) we have

$$g_2((\nabla F_*)(X, Y), V) = g_2(\nabla_X^F \varphi F_*(Y) + \varpi F_*(Y), \mathcal{B} + V\mathfrak{C}V).$$

Then from (4.16) we get

$$g_2((\nabla F_*)(X, Y), V) = g_2(\nabla_X^F F_*(\acute{Y}), \mathcal{B}V + \mathfrak{C}V)$$
$$- g_2(S_{\varpi F_*(Y)}F_*(X), \mathcal{B}V) + g_2(\nabla_X^{F^\perp} \varpi F_*(Y), \mathfrak{C}V).$$

Then (1.71) and (4.15) imply

$$g_2((\nabla F_*)(X, Y), V) = g_2((\nabla F_*)(X, \acute{Y}), \mathfrak{C}V) + g_1(\mathcal{H}\nabla_X \acute{Y}, {}^* F_* \mathcal{B}V)$$
$$- g_2(S_{\varpi F_*(Y)} F_*(X), \mathcal{B}V) + g_2(\nabla_X^{F\perp} \varpi F_*(Y), \mathfrak{C}V).$$

which gives us (1). (2) and (3) can be obtained in a similar way.    □

For the harmonicity, we have the following result.

**Theorem 146.** *[255] Let F be a generic Riemannian map from a Riemannian manifold* $(M_1, g_1)$ *to a Kähler manifold* $(M_2, g_2, J)$. *Then F is harmonic if and only if*

$$trace \mid_{(ker F_*)^\perp} \{-\mathfrak{C}(\nabla F_*)(., {}^* F\varphi F_*(.)) - \varpi F_*(\mathcal{H}\nabla_{(.)} {}^* F\varphi F_*(.))$$
$$-\varpi S_{\varpi F_*(.)} F_*(.) + C\nabla_{(.)}^{F\perp} \varpi F_*(.)\} = 0 \qquad (6.25)$$

*and the fibers are minimal.*

*Proof.* From (1.71), (3.5), and (6.21) we have

$$(\nabla F_*)(X, X) = -J[\nabla_X^F \varphi F_*(X) + \varpi F_*(X)] - F_*(\nabla_X^1 X)$$

for $X \in \Gamma((ker F_*)^\perp)$. For $F_*(\acute{X}) = \varphi F_*(X)$, from (1.71), (4.15), (4.16), (6.21), and (6.22) we get

$$(\nabla F_*)(X, X) = -\mathcal{B}(\nabla F_*)(X, \acute{X}) - \mathfrak{C}(\nabla F_*)(X, \acute{X})$$
$$- \varphi F_*(\nabla_X \acute{X}) - \varpi F_*(\nabla_X \acute{X}) + \varphi S_{\varpi F_*(X)} F_*(\acute{X})$$
$$+ \varpi S_{\varpi F_*(X)} F_*(\acute{X}) + \mathcal{B}\nabla_X^{F\perp} \varpi F_*(X) + \mathfrak{C}\nabla_X^{F\perp} \varpi F_*(X)$$
$$- F_*(\nabla_X \acute{X}),$$

where $\nabla$ is the Levi-Civita connection on $M_1$. Now considering $(range F_*)$ parts of this equation and taking a trace on the resulting equation, we get (6.25). The second assertion comes from $(\nabla F_*)(U, U) = -F_*(\nabla_U U)$ for $U \in \Gamma(ker F_*)$.    □

## 5. Slant Riemannian maps to Kähler manifolds

In this section, we introduce slant Riemannian maps from Riemannian manifolds to almost Hermitian manifolds. We show that slant Riemannian maps include slant immersions (therefore holomorphic immersions and totally real immersions), invariant Riemannian maps, and anti-invariant Riemannian maps. We also obtain an example that is not included in immersions, or invariant Riemannian maps or anti-invariant Riemannian maps. We investigate the harmonicity of slant Riemannian maps and obtain necessary and sufficient conditions for such maps to be totally geodesic. We also show that every slant Riemannian map is a pseudo-horizontally weakly conformal map, then we obtain necessary and sufficient conditions for slant Riemannian maps to be a pseudo-homothetic map.

**Definition 63.** [251] Let $F$ be a Riemannian map from a Riemannian manifold $(M_1, g_1)$ to an almost Hermitian manifold $(M_2, g_2, J)$. If for any non-zero vector $X \in \Gamma(ker F_*)^\perp$, the angle $\theta(X)$ between $JF_*(X)$ and the space $range F_*$ is a constant, i.e., it is independent of the choice of the point $p \in M_1$ and choice of the tangent vector $F_*(X)$ in $range F_*$, then we say that $F$ is a *slant Riemannian map*. In this case, the angle $\theta$ is called the slant angle of the slant Riemannian map.

Since $F$ is a subimmersion, it follows that the rank of $F$ is constant on $M_1$, then the rank theorem for functions implies that $ker F_*$ is an integrable subbundle of $TM_1$. From the above definition, we have the following result.

**Proposition 65.** *Let $F : (M, g_M) \to (N, g_N, J_N)$ be a slant Riemannian map between Riemannian manifold $M$ and a Hermitian manifold $N$ with integrable distribution $(range F_*)$. Then the leaves of the distribution $(range F_*)$ of a slant Riemannian map are slant submanifolds of $N$.*

We first give some examples of slant Riemannian maps.

**Example 80.** Every slant immersion (slant submanifold) from a Riemannian manifold to an almost Hermitian manifold is a slant Riemannian map with $ker F_* = \{0\}$.

**Example 81.** Every invariant Riemannian map from a Riemannian manifold to an almost Hermitian manifold is a slant Riemannian map with $\theta = 0$.

**Example 82.** Every anti-invariant Riemannian map from a Riemannian manifold to an almost Hermitian manifold is a slant Riemannian map with $\theta = \frac{\pi}{2}$.

A slant Riemannian map is said to be proper if it is not an immersion and $\theta \neq 0, \frac{\pi}{2}$. Here is an example of proper slant Riemannian maps.

**Example 83.** [251] Consider the following Riemannian map given by

$$F : \quad \mathbb{R}^4 \quad \longrightarrow \quad \mathbb{R}^4$$
$$(x_1, x_2, x_3, x_4) \quad \quad (x_1, \tfrac{x_2+x_3}{\sqrt{3}}, \tfrac{x_2+x_3}{\sqrt{6}}, 0).$$

Then $rank F = 2$ and for any $0 < \alpha < \frac{\pi}{2}$, $F$ is a slant Riemannian map with slant angle $\cos^{-1}(\sqrt{\frac{2}{3}})$.

We also have the following result which is based on the fact that the composition of a Riemannian submersion $F_1$ from a Riemannian manifold $(M_1, g_1)$ onto a Riemannian manifold $(M_2, g_2)$ and an isometric immersion $F_2$ from $(M_2, g_2)$ to a Riemannian manifold $(M_3, g_3)$ is a Riemannian map.

**Proposition 66.** *[251] Let $F_1$ be a Riemannian submersion from a Riemannian manifold $(M_1, g_1)$ onto a Riemannian manifold $(M_2, g_2, J)$ and $F_2$ a slant immersion from $(M_2, g_2, J)$ to an almost Hermitian manifold $(M_3, g_3, J)$. Then $F_2 \circ F_1$ is a slant Riemannian map.*

Let $F$ be a Riemannian map from a Riemannian manifold $(M_1, g_1)$ to a Riemannian manifold an almost Hermitian manifold $(M_2, g_2, J)$. Then for $F_*(X) \in \Gamma(rangeF_*)$, $X \in \Gamma((ker F_*)^{\perp})$, we write

$$JF_*(X) = \phi F_*(X) + \omega F_*(X), \tag{6.26}$$

where $\phi F_*(X) \in \Gamma(rangeF_*)$ and $\omega F_*(X) \in \Gamma((rangeF_*)^{\perp})$. Also for $V \in \Gamma((rangeF_*)^{\perp})$, we have

$$JV = \mathcal{B}V + \mathcal{C}V, \tag{6.27}$$

where $\mathcal{B}V \in \Gamma(rangeF_*)$ and $\mathcal{C}V \in \Gamma((rangeF_*)^{\perp})$.

Let $F$ be a Riemannian map from a Riemannian manifold $(M_1, g_1)$ to a Riemannian manifold a Kähler manifold $(M_2, g_2, J)$, then from (6.26), (6.27), (4.7), and (1.71) we obtain

$$(\tilde{\nabla}_X \omega)F_*(Y) = C(\nabla F_*)(X, Y) - (\nabla F_*)(X, Y') \tag{6.28}$$

and

$$F_*(\nabla^1_X Y') - \phi F_*(\nabla^1_X Y) = \mathcal{B}(\nabla F_*)(X, Y) + \mathcal{S}_{\omega F_*(Y)} F_*(X) \tag{6.29}$$

for $X, Y \in \Gamma((ker F_*)^{\perp})$, where $(\tilde{\nabla}_X \omega)F_*(Y) = \nabla^{F^1}_X F_*(Y) - \omega F_*(\nabla^1_X Y)$, $\nabla^1$ is the Levi-Civita connection of $M_1$ and $\phi F_*(Y) = F_*(Y')$, $Y' \in \Gamma((kerF_*)^{\perp})$.

Let $F$ be a slant Riemannian map from a Riemannian manifold $(M_1, g_1)$ to an almost Hermitian manifold $(M_2, g_2, J)$, then we say that $\omega$ is parallel if $(\tilde{\nabla}_X \omega)F_*(Y) = 0$. We also say that $\phi$ is parallel if $F_*(\nabla^1_X Y') - \phi F_*(\nabla^1_X Y) = 0$ for $X, Y \in \Gamma((ker F_*)^{\perp})$ such that $\phi F_*(Y) = F_*(Y')$, $Y' \in \Gamma((kerF_*)^{\perp})$.

The proof of the following result is exactly same with slant immersions (see [70] or [48] for the Sasakian case), therefore we omit its proof.

**Theorem 147.** *[251] Let $F$ be a Riemannian map from a Riemannian manifold $(M_1, g_1)$ to an almost Hermitian manifold $(M_2, g_2, J)$. Then $F$ is a slant Riemannian map if and* *only if there exists a constant $\lambda \in [-1, 0]$ such that*

$$\phi^2 F_*(X) = \lambda F_*(X)$$

*for $X \in \Gamma((ker\, F_*)^\perp)$. If $F$ is a slant Riemannian map, then $\lambda = -\cos^2 \theta$.*

By using the above theorem, it is easy to see that

$$
\begin{align}
g_2(\phi F_*(X), \phi F_*(Y)) &= \cos^2 \theta g_1(X, Y) \tag{6.30} \\
g_2(\omega F_*(X), \omega F_*(Y)) &= \sin^2 \theta g_1(X, Y) \tag{6.31}
\end{align}
$$

for any $X, Y \in \Gamma((rangeF_*)^\perp)$.

We now recall here that considering $F_*^h$ at each $p_1 \in M_1$ as a linear transformation

$$F_{*p_1}^h : ((ker\, F_*)^\perp(p_1), g_{1\,p_1((ker\, F_*)^\perp(p_1))}) \rightarrow (rangeF_*(p_2), g_{2\,p_2(rangeF_*)(p_2))}),$$

we will denote the adjoint of $F_*^h$ by $^*F^h{}_{*p_1}$. Let $^*F_{*p_1}$ be the adjoint of

$$F_{*p_1} : (T_{p_1}M_1, g_{1\,p_2}) \longrightarrow (T_{p_2}M_2, g_{2\,p_2}).$$

Then the linear transformation

$$(^*F_{*p_1})^h : rangeF_*(p_2) \longrightarrow (ker\, F_*)^\perp(p_1),$$

defined by $(^*F_{*p_1})^h y = {}^*F_{*p_1}y$, where $y \in \Gamma(rangeF_{*p_1})$, $p_2 = F(p_1)$, is an isomorphism and $(F_{*p_1}^h)^{-1} = (^*F_{*p_1})^h = {}^*(F_{*p_1}^h)$.

Let $\{e_1, .., e_n\}$ be an orthonormal basis of $(ker\, F_*)^\perp$. Then $\{F_*(e_1), ..., F_*(e_n)\}$ is an orthonormal basis of $rangeF_*$. By using (6.30), we can easily conclude that

$$\{F_*(e_1), \sec \theta \phi F_*(e_1), F_*(e_2), \sec \theta \phi e_2, ..., e_p, \sec \theta \phi e_p\}$$

is an orthonormal frame for $\Gamma(rangeF_*)$, where $2p = n = rankF_*$. Then we have the following result.

**Lemma 72.** *[251] Let $F$ be a slant Riemannian map from a Riemannian manifold $(M_1, g_1)$ to an almost Hermitian manifold $(M_2, g_2, J)$ with $rankF_* = n$. Let*

$$\{e_1, e_2, .., e_p\}, p = \frac{n}{2}$$

*be a set of orthonormal vector fields in $(ker\, F_*)^\perp$. Then*

$$\{e_1, \sec \theta^* F_* \phi F_*(e_1), e_2, \sec \theta^* F_* \phi F_*(e_2), ..., e_p, \sec \theta^* F_* \phi F_*(e_p)\}$$

*is a local orthonormal basis of $(ker\, F_*)^\perp$.*

*Proof.* First, by direct computation, we have

$$g_1(e_i, \sec \theta^* F_* \phi F_*(e_i)) = \sec \theta g_2(F_*(e_i), \phi F_*(e_i)) = \sec \theta g_2(F_*(e_i), JF_*(e_i)) = 0.$$

In a similar way, we have

$$g_1(e_i, \sec \theta^* F_* \phi F_*(e_j)) = 0.$$

Since for a Riemannian map, we have $^*F_* \circ F_* = I$, (Identity map), by using (6.30), we get

$$g_1(\sec \theta^* F_* \phi F_*(e_i), \sec \theta^* F_* \phi F_*(e_j)) = \sec^2 \theta \cos^2 \theta g_1(e_i, e_j) = \delta_{ij},$$

which gives the assertion.    □

We now denote $^*F_* \phi F_*$ by $Q$, then we have the following characterization of slant Riemannian maps.

**Corollary 54.** *[251] Let F be a Riemannian map from a Riemannian manifold $(M_1, g_1)$ to an almost Hermitian manifold $(M_2, g_2, J)$. Then F is a slant Riemannian map if and only if there exists a constant $\mu \in [-1, 0]$ such that*

$$Q^2 X = \mu X$$

*for $X \in \Gamma((\ker F_*)^\perp)$. If F is a slant Riemannian map, then $\mu = -\cos^2 \theta$.*

*Proof.* By direct computation, we have

$$Q^2 X = {}^* F_* \phi^2 F_*(X).$$

Then the proof comes from Theorem 147.    □

In the sequel, we are going to show that the notion $\omega$ is useful to investigate the harmonicity of slant Riemannian map. To see this, we need the following lemma.

**Lemma 73.** *[251] Let F be a slant Riemannian map from a Riemannian manifold $(M_1, g_1)$ to a Kähler manifold $(M_2, g_2, J)$. If $\omega$ is parallel, then we have*

$$(\nabla F_*)(QX, QY) = -\cos^2 \theta (\nabla F_*)(X, Y) \tag{6.32}$$

*for $X, Y \in \Gamma((\ker F_*)^\perp)$.*

*Proof.* If $\omega$ is parallel, then from (6.28) we have

$$C(\nabla F_*)(X, Y) = (\nabla F_*)(X, QY).$$

Interchanging the role of $X$ and $Y$ and taking into account that the second fundamental form is symmetric, we obtain

$$(\nabla F_*)(QX, Y) = (\nabla F_*)(X, QY).$$

Hence we get

$$(\nabla F_*)(QX, QY) = (\nabla F_*)(X, Q^2 Y).$$

Then Corollary 54 implies (6.32). □

**Lemma 74.** *[251] Let F be a slant Riemannian map from a Riemannian manifold $(M_1^{m_1}, g_1)$ to a Kähler manifold $(M_2^{m_2}, g_2, J)$. If $\omega$ is parallel, then F is harmonic if and only if the distribution $(ker F_*)$ is minimal.*

*Proof.* We choose an orthonormal basis of $TM_1$ as

$$\{v_1, ..., v_{p_1}, e_1, \sec\theta Qe_1, e_2, \sec\theta Qe_2, ..., e_{p_2}, \sec\theta Qe_{p_2}\}, \quad p_1 + 2p_2 = m_1,$$

where $\{v_1, ..., v_{p_1}\}$ is an orthonormal basis of $kerF_*$ and $\{e_1, \sec\theta Qe_1, e_2, \sec\theta Qe_2, ..., e_{p_2}, \sec\theta Qe_{p_2}\}$ is an orthonormal basis of $(ker F_*)^\perp$. Since the second fundamental form is linear in every slot, we have

$$\tau = \sum_{i=1}^{p_1}(\nabla F_*)(v_i, v_i) + \sum_{j=1}^{p_2}(\nabla F_*)(e_j, e_j) + (\nabla F_*)(\sec\theta Qe_j, \sec\theta Qe_j).$$

Then, from (6.32), we obtain

$$\tau = \sum_{i=1}^{p_1}(\nabla F_*)(v_i, v_i) = -\sum_{i=1}^{p_1} F_*(\nabla_{v_i} v_i),$$

which proves the assertion. □

We now investigate necessary and sufficient conditions for a slant Riemannian map to be totally geodesic.

**Theorem 148.** *[251] Let F be a slant Riemannian map from a Riemannian manifold $(M_1^{m_1}, g_1)$ to a Kähler manifold $(M_2^{m_2}, g_2, J)$. Then F is totally geodesic if and only if*
**(i)** *the fibers are totally geodesic,*
**(ii)** *the horizontal distribution $(ker F_*)^\perp$ is totally geodesic, and*
**(iii)** *For $X, Y \in \Gamma((ker F_*)^\perp)$ and $V \in \Gamma((rangeF_*)^\perp)$ we have*

$$g_2(S_{\omega F_*(Y)}F_*(X), \mathcal{B}V) = g_2(\nabla_X^{F^\perp}\omega F_*((Y), CV) - g_2(\nabla_X^{F^\perp}\omega\phi F_*((Y), V).$$

*Proof.* For $X, Y \in \Gamma((ker\, F_*)^\perp)$ and $V \in \Gamma((rangeF_*)^\perp)$, we have

$$g_2((\nabla F_*)(X, Y), V) = g_2(\nabla_X^F F_*(Y), V).$$

Then, using (3.5), (6.26) and (6.27), we obtain

$$g_2((\nabla F_*)(X, Y), V) = -g_2(\nabla_X^F J\phi F_*(Y), V) + g_2(\nabla_X^F \omega F_*(Y), JV).$$

Again using (6.26), (6.27), (4.16) and Theorem 147, we get

$$\begin{aligned} g_2((\nabla F_*)(X, Y), V) &= \cos^2 \theta\, g_2(\nabla_X^F F_*(Y), V) - g_2(\nabla_X^{F\perp} \omega\phi F_*(Y), V) \\ &\quad - g_2(S_{\omega F_*(Y)} F_*(X), \mathcal{B}V) \\ &\quad + g_2(\nabla_X^{F\perp} \omega F_*(Y), CV). \end{aligned} \tag{6.33}$$

For $X, Y \in \Gamma((ker\, F_*)^\perp)$ and $W \in \Gamma(ker\, F_*)$, from (1.71) we derive

$$g_2((\nabla F_*)(X, W), F_*(Y)) = g_1(\nabla_X^1 Y, W). \tag{6.34}$$

In a similar way, for $U \in \Gamma(ker\, F_*)$, we obtain

$$g_2((\nabla F_*)(U, W), F_*(Y)) = -g_1(\nabla_U^1 W, Y). \tag{6.35}$$

Then, the proof follows from (4.7), (6.33), (6.34), and (6.35).   $\square$

## Slant Riemannian maps and PHWC maps

In this paragraph, we are going to show that every slant Riemannian map is a pseudo-horizontally weakly conformal (PHWC) map, then we investigate the conditions for slant Riemannian maps to be pseudo-horizontally homothetic map.

**Proposition 67.** *[251] Let F be a slant Riemannian map from a Riemannian manifold $(M_1, g_1)$ to an almost Hermitian manifold $(M_2, g_2, J)$. Then F is a PHWC map.*

*Proof.* We first note that $\tilde{J} = \sec \theta\, \phi$ is a complex structure on $(rangeF_*)$ and $rangeF_*$ is invariant with respect to $\tilde{J}$. Then we define $\hat{J} = \sec \theta\, Q = \sec \theta\, ^*F_* \phi F_*$; it is easy to see that $\hat{J}$ is a complex structure on $(kerF_*)^\perp$. Thus $((ker\, F_*)^\perp, \hat{J})$ is an almost complex distribution. We now consider $\hat{g} = g_1\,|_{(ker\, F_*)^\perp}$, then by direct computation, we obtain

$$\hat{g}(\hat{J}X, \hat{J}Y) = \hat{g}(X, Y)$$

for $X, Y \in \Gamma((ker\, F_*)^\perp)$. Thus $\hat{g}$ is $\hat{J}-$ Hermitian and $((ker\, F_*)^\perp, \hat{g}, \hat{J})$ is an almost Hermitian distribution. Thus from (3.11), $F$ is a PHWC map.   $\square$

From (6.28) and (1.71), we have the following.

**Lemma 75.** *[251] Let F be a slant Riemannian map from a Riemannian manifold*

$(M_1, g_1)$ *to a Kähler manifold* $(M_2, g_2, J)$. *Then* $\phi$ *is parallel if and only if*

$$(\nabla F_*)(X, QY) = \nabla^F_X \phi F_*(Y) - \phi F_*(\nabla^1_X Y)$$

*for* $X, Y \in \Gamma((ker F_*)^{\perp})$.

We now give necessary and sufficient conditions for a slant Riemannian map to be pseudo horizontally homothetic map.

**Theorem 149.** *[251] Let F be a slant Riemannian map from a Riemannian manifold* $(M_1, g_1)$ *to a Kähler manifold* $(M_2, g_2, J)$. *Then F is pseudo-horizontally homothetic map if and only if* $\phi$ *is parallel and*

$$(\nabla F_*)(X, U) = 0$$

*for* $X \in \Gamma((ker F_*)^{\perp})$ *and* $U \in \Gamma(ker F_*)$.

*Proof.* By direct computations, we have

$$(\nabla_X \hat{J})Y = \sec \theta \, \nabla^1_X QY - Q\nabla^1_X Y.$$

Hence we obtain

$$F_*(\nabla_X \hat{J})Y = \sec \theta \, F_*(\nabla^1_X QY - \phi F_*(\nabla^1_X Y).$$

Then, using (1.71), we get

$$F_*(\nabla_X \hat{J})Y = \sec \theta \, (-(\nabla F_*)(X, QY) + \nabla^F_X \phi F_*(Y) - \phi F_*(\nabla^1_X Y)). \tag{6.36}$$

On the other hand, since $QY$ and $U$ are orthogonal, we have

$$g_1((\nabla^1_X \hat{J})Y, U) = \sec \theta \, g_1(QY, \nabla^1_X U).$$

Then adjoint map $^*F_*$ and (1.71) imply

$$g_1((\nabla^1_X \hat{J})Y, U) = \sec \theta \, g_2(\phi F_*(Y), (\nabla F_*)(X, U)). \tag{6.37}$$

Then the proof comes from (6.36) and (6.37).    □

## 6. Semi-slant Riemannian maps to Kähler manifolds

In this section, we introduce semi-slant Riemannian maps from Riemannian manifolds to almost Hermitian manifolds and show that such Riemannian maps include semi-slant immersions, invariant Riemannian maps, anti-invariant Riemannian maps, and slant Riemannian maps. After we give many examples of such maps, we obtain characterizations, investigate the harmonicity of such maps, and find necessary and sufficient conditions for semi-slant Riemannian maps to be totally geodesic. Then we relate the notion of semi-slant Riemannian maps to the notion of pseudo-horizontally weakly conformal maps. In fact, we show that every semi-slant Riemannian map is also a pseudo-horizontally weakly conformal map. In this direction, we find necessary and sufficient conditions for such maps to be pseudo-homothetic maps.

We first present the following definition, which is a generalization of previous Riemannian maps.

**Definition 64.** [222] Let $(N, g_N)$ be a Riemannian manifold and $(M, g_M, J)$ an almost Hermitian manifold. A Riemannian map $F : (N, g_N) \mapsto (M, g_M, J)$ is called a *semi-slant Riemannian map* if there is a distribution $\mathcal{D}_1 \subset (ker F_*)^\perp$ such that

$$(ker F_*)^\perp = \mathcal{D}_1 \oplus \mathcal{D}_2, \quad J(F_*\mathcal{D}_1) = F_*\mathcal{D}_1,$$

and the angle $\theta = \theta(X)$ between $JF_*X$ and the space $F_*(\mathcal{D}_2)_p$ is constant for nonzero $X \in (\mathcal{D}_2)_p$ and $p \in N$, where $\mathcal{D}_2$ is the orthogonal complement of $\mathcal{D}_1$ in $(ker F_*)^\perp$. We call the angle $\theta$ a *semi-slant angle*.

From the above definition, we have the following result.

**Proposition 68.** *Let* $F : (M, g_M) \to (N, g_N, J_N)$ *be a semi-slant Riemannian map between Riemannian manifold M and a Hermitian manifold N with integrable distribution* $(range F_*)$. *Then the leaves of the distribution* $(range F_*)$ *of a semi-slant Riemannian map are semi-slant submanifolds of N.*

For a Euclidean space $\mathbb{R}^{2n}$ with coordinates $(y_1, y_2, \cdots, y_{2n})$, we canonically choose an almost complex structure $J$ on $\mathbb{R}^{2n}$ as follows:

$$J(a_1 \frac{\partial}{\partial y_1} + a_2 \frac{\partial}{\partial y_2} + \cdots + a_{2n-1} \frac{\partial}{\partial y_{2n-1}} + a_{2n} \frac{\partial}{\partial y_{2n}})$$
$$= -a_2 \frac{\partial}{\partial y_1} + a_1 \frac{\partial}{\partial y_2} + \cdots - a_{2n} \frac{\partial}{\partial y_{2n-1}} + a_{2n-1} \frac{\partial}{\partial y_{2n}},$$

where $a_1, \cdots, a_{2n} \in \mathbb{R}$. Throughout this section, we will use this notation.

**Example 84.** Let $F$ be an invariant Riemannian map from a Riemannian manifold $(M, g_M)$ to an almost Hermitian manifold $(N, g_N, J)$. Then the map $F$ is a semi-slant Riemannian map with $\mathcal{D}_1 = (ker F_*)^\perp$.

**Example 85.** Let $F$ be an anti-invariant Riemannian map from a Riemannian manifold $(M, g_M)$ to an almost Hermitian manifold $(N, g_N, J)$. Then the map $F$ is a semi-slant Riemannian map such that $\mathcal{D}_2 = (ker F_*)^\perp$ with the semi-slant angle $\theta = \frac{\pi}{2}$.

**Example 86.** Let $F$ be a semi-invariant Riemannian map from a Riemannian manifold

$(M, g_M)$ to an almost Hermitian manifold $(N, g_N, J)$. Then the map $F$ is a semi-slant Riemannian map with the semi-slant angle $\theta = \frac{\pi}{2}$.

**Example 87.** Let $F$ be a slant Riemannian map from a Riemannian manifold $(M, g_M)$ to an almost Hermitian manifold $(N, g_N, J)$ with the slant angle $\theta$. Then the map $F$ is a semi-slant Riemannian map such that $\mathcal{D}_2 = (ker F_*)^\perp$ and the semi-slant angle $\theta$.

**Example 88.** [222] Let $(M, g_M)$ be a $m$-dimensional Riemannian manifold and $(N, g_N, J)$ a $2n$-dimensional almost Hermitian manifold. Let $F$ be a Riemannian map from a Riemannian manifold $(M, g_M)$ to an almost Hermitian manifold $(N, g_N, J)$ with $rank F = 2n - 1$. Then the map $F$ is a semi-slant Riemannian map such that

$$F_*\mathcal{D}_2 = J((F_*[(ker F_*)^\perp])^\perp)$$

and the semi-slant angle $\theta = \frac{\pi}{2}$.

**Example 89.** [222] Define a map $F : \mathbb{R}^8 \mapsto \mathbb{R}^6$ by

$$F(x_1, x_2, \cdots, x_8) = (y_1, y_2, \cdots, y_6) = (x_3, \frac{x_4 - x_5}{\sqrt{6}}, \frac{x_4 - x_5}{\sqrt{3}}, c, x_2, x_1),$$

where $c$ is constant. Then the map $F$ is a semi-slant Riemannian map such that

$$ker F_* = span\{\frac{\partial}{\partial x_4} + \frac{\partial}{\partial x_5}, \frac{\partial}{\partial x_6}, \frac{\partial}{\partial x_7}, \frac{\partial}{\partial x_8}\},$$

$$\mathcal{D}_1 = span\{\frac{\partial}{\partial x_1}, \frac{\partial}{\partial x_2}\}, \mathcal{D}_2 = span\{\frac{\partial}{\partial x_3}, \frac{\partial}{\partial x_4} - \frac{\partial}{\partial x_5}\},$$

$$F_*\mathcal{D}_1 = span\{\frac{\partial}{\partial y_5}, \frac{\partial}{\partial y_6}\}, F_*\mathcal{D}_2 = span\{\frac{\partial}{\partial y_1}, \frac{\partial}{\partial y_2} + \sqrt{2}\frac{\partial}{\partial y_3}\},$$

the semi-slant angle $\theta$ with $\cos \theta = \frac{1}{\sqrt{3}}$.

**Example 90.** [222] Define a map $F : \mathbb{R}^9 \mapsto \mathbb{R}^6$ by

$$F(x_1, x_2, \cdots, x_9) = (y_1, y_2, \cdots, y_6) = (x_1, x_9, x_3, \frac{(x_4 + x_5)\cos \alpha}{\sqrt{2}}, \frac{(x_4 + x_5)\sin \alpha}{\sqrt{2}}, \beta),$$

where $\alpha$ and $\beta$ are constant with $\alpha \in (0, \frac{\pi}{2})$. Then the map $F$ is a semi-slant Riemannian map such that

$$ker\, F_* = span\{\frac{\partial}{\partial x_2}, \frac{\partial}{\partial x_4} - \frac{\partial}{\partial x_5}, \frac{\partial}{\partial x_6}, \frac{\partial}{\partial x_7}, \frac{\partial}{\partial x_8}\},$$

$$\mathcal{D}_1 = span\{\frac{\partial}{\partial x_1}, \frac{\partial}{\partial x_9}\}, \quad \mathcal{D}_2 = span\{\frac{\partial}{\partial x_3}, \frac{\partial}{\partial x_4} + \frac{\partial}{\partial x_5}\},$$

$$F_*\mathcal{D}_1 = span\{\frac{\partial}{\partial y_1}, \frac{\partial}{\partial y_2}\}, \quad F_*\mathcal{D}_2 = span\{\frac{\partial}{\partial y_3}, \sqrt{2}\cos\alpha\frac{\partial}{\partial y_4} + \sqrt{2}\sin\alpha\frac{\partial}{\partial y_5}\},$$

the semi-slant angle $\theta = \alpha$.

**Example 91.** [222] Define a map $F : \mathbb{R}^7 \mapsto \mathbb{R}^6$ by

$$F(x_1, x_2, \cdots, x_7) = (y_1, y_2, \cdots, y_6) = (x_2\sin\alpha, 0, x_3, x_5, x_2\cos\alpha, x_7),$$

where $\alpha \in (0, \frac{\pi}{2})$. Then the map $F$ is a semi-slant Riemannian map such that

$$ker\, F_* = span\{\frac{\partial}{\partial x_1}, \frac{\partial}{\partial x_4}, \frac{\partial}{\partial x_6}\},$$

$$\mathcal{D}_1 = span\{\frac{\partial}{\partial x_3}, \frac{\partial}{\partial x_5}\}, \quad \mathcal{D}_2 = span\{\frac{\partial}{\partial x_2}, \frac{\partial}{\partial x_7}\},$$

$$F_*\mathcal{D}_1 = span\{\frac{\partial}{\partial y_3}, \frac{\partial}{\partial y_4}\}, \quad F_*\mathcal{D}_2 = span\{\sin\alpha\frac{\partial}{\partial y_1} + \cos\alpha\frac{\partial}{\partial y_5}, \frac{\partial}{\partial y_6}\},$$

the semi-slant angle $\theta = \alpha$.

**Example 92.** [222] Define a map $F : \mathbb{R}^6 \mapsto \mathbb{R}^8$ by

$$F(x_1, x_2, \cdots, x_6) = (y_1, y_2, \cdots, y_8) = (x_1, \frac{x_2 + x_3}{2}, \frac{x_2 + x_3}{2}, 0, 0, 0, x_5, x_6).$$

Then the map $F$ is a semi-slant Riemannian map such that

$$ker\, F_* = span\{\frac{\partial}{\partial x_2} - \frac{\partial}{\partial x_3}, \frac{\partial}{\partial x_4}\},$$

$$\mathcal{D}_1 = span\{\frac{\partial}{\partial x_5}, \frac{\partial}{\partial x_6}\}, \quad \mathcal{D}_2 = span\{\frac{\partial}{\partial x_1}, \frac{\partial}{\partial x_2} + \frac{\partial}{\partial x_3}\},$$

$$F_*\mathcal{D}_1 = span\{\frac{\partial}{\partial y_7}, \frac{\partial}{\partial y_8}\}, \quad F_*\mathcal{D}_2 = span\{\frac{\partial}{\partial y_1}, \frac{\partial}{\partial y_2} + \frac{\partial}{\partial y_3}\},$$

the semi-slant angle $\theta = \frac{\pi}{4}$.

Let $F : (N, g_N) \mapsto (M, g_M, J)$ be a semi-slant Riemannian map. Then for $X \in \Gamma((ker\, F_*)^{\perp})$, we write

$$X = PX + QX, \tag{6.38}$$

where $PX \in \Gamma(\mathcal{D}_1)$ and $QX \in \Gamma(\mathcal{D}_2)$. For $U \in \Gamma(rangeF_*)$, we get

$$JU = \phi U + \omega U, \tag{6.39}$$

where $\phi U \in \Gamma(rangeF_*)$ and $\omega U \in \Gamma((rangeF_*)^\perp)$.

For $V \in \Gamma((rangeF_*)^\perp)$, we have

$$JV = BV + CV, \tag{6.40}$$

where $BV \in \Gamma(rangeF_*)$ and $CV \in \Gamma((rangeF_*)^\perp)$.

For $Y \in \Gamma(TN)$, we obtain

$$Y = \mathcal{V}Y + \mathcal{H}Y, \tag{6.41}$$

where $\mathcal{V}Y \in \Gamma(ker F_*)$ and $\mathcal{H}Y \in \Gamma((ker F_*)^\perp)$.

For $W \in \Gamma(F^{-1}TM)$, we write

$$W = \bar{P}W + \bar{Q}W, \tag{6.42}$$

where $\bar{P}W \in \Gamma(rangeF_*)$ and $\bar{Q}W \in \Gamma((rangeF_*)^\perp)$.

For $X, Y \in \Gamma((ker F_*)^\perp)$ and $V \in \Gamma((rangeF_*)^\perp)$, define

$$\widehat{\nabla}^F_X F_* Y := \bar{P}\nabla^F_X F_* Y, \tag{6.43}$$

$$S_V F_* Y := -\bar{P}\nabla^F_X V, \tag{6.44}$$

$$\nabla^{F\perp}_X V := \bar{Q}\nabla^F_X V. \tag{6.45}$$

Then

$$\nabla^F_X V = -S_V F_* Y + \nabla^{F\perp}_X V \tag{6.46}$$

and $\nabla^{F\perp}$ is a connection on $(rangeF_*)^\perp$ such that $\nabla^{F\perp}g_M = 0$.

For $X, Y \in \Gamma((ker F_*)^\perp)$, define

$$(\nabla^F_X \phi)F_* Y \quad := \quad \widehat{\nabla}^F_X \phi F_* Y - \phi \widehat{\nabla}^F_X F_* Y, \tag{6.47}$$

$$(\nabla^F_X \omega)F_* Y \quad := \quad \nabla^{F\perp}_X \omega F_* Y - \omega \widehat{\nabla}^F_X F_* Y. \tag{6.48}$$

Then we have

$$(\nabla^F_X \phi)F_* Y \quad = \quad S_{\omega F_* Y} F_* X + B(\nabla F_*)(X, Y), \tag{6.49}$$

$$(\nabla^F_X \omega)F_* Y \quad = \quad C(\nabla F_*)(X, Y) - (\nabla F_*)(X, Y') \tag{6.50}$$

for some $Y' \in \Gamma((ker F_*)^\perp)$ with $F_* Y' = \phi F_* Y$.

We call the tensor $\phi$ *parallel* if $\nabla^F \phi = 0$ and the tensor $\omega$ is said to be *parallel* if $\nabla^F \omega = 0$.

Then we easily obtain the following lemma.

**Lemma 76.** *[222] Let $(M, g_M, J)$ be a Kähler manifold and $(N, g_N)$ a Riemannian manifold. Let $F : (N, g_N) \mapsto (M, g_M, J)$ be a semi-slant Riemannian map.*
**(1)** *For $X, Y \in \Gamma(\mathcal{D}_1)$, we get*

$$\widehat{\nabla}_X^F \phi F_* Y = \phi \widehat{\nabla}_X^F F_* Y + B \bar{Q} \nabla_X^F F_* Y$$

$$\bar{Q} \nabla_X^F \phi F_* Y = \omega \widehat{\nabla}_X^F F_* Y + C \bar{Q} \nabla_X^F F_* Y.$$

**(2)** *For $X, Y \in \Gamma(\mathcal{D}_2)$, we have*

$$\widehat{\nabla}_X^F F_* Y' - \mathcal{S}_{\omega F_* Y} F_* X = \phi \widehat{\nabla}_X^F F_* Y + B \bar{Q} \nabla_X^F F_* Y$$

$$\bar{Q} \nabla_X^F F_* Y' + \nabla_X^{F\perp} \omega F_* Y = \omega \widehat{\nabla}_X^F F_* Y + C \bar{Q} \nabla_X^F F_* Y,$$

*where $Y' \in \Gamma((\ker F_*)^\perp)$ and $F_* Y' = \phi F_* Y$.*

In a similar way to Theorem 147, we have the following result.

**Theorem 150.** *[222] Let $F$ be a semi-slant Riemannian map from a Riemannian manifold $(N, g_N)$ to an almost Hermitian manifold $(M, g_M, J)$ with the semi-slant angle $\theta$. Then we obtain*

$$\phi^2 F_* X = -\cos^2 \theta \cdot F_* X \quad \text{for } X \in \Gamma(\mathcal{D}_2). \tag{6.51}$$

**Remark 24.** It is easy to check that the converse of Theorem 150 is also true. Furthermore, we get

$$g_M(\phi F_* X, \phi F_* Y) = \cos^2 \theta \, g_M(F_* X, F_* Y), \tag{6.52}$$

$$g_M(\omega F_* X, \omega F_* Y) = \sin^2 \theta \, g_M(F_* X, F_* Y) \tag{6.53}$$

for $X, Y \in \Gamma(\mathcal{D}_2)$ so that with $\theta \in [0, \frac{\pi}{2})$, there is locally an orthonormal frame

$$\{F_* e_1, \sec \theta \phi F_* e_1, \cdots, F_* e_k, \sec \theta \phi F_* e_k\}$$

of $F_* \mathcal{D}_2$ for some $\{e_1, \cdots, e_k\} \subset \Gamma(\mathcal{D}_2)$.

Let $F$ be a $C^\infty$-map from a Riemannian manifold $(N, g_N)$ into a Riemannian manifold $(M, g_M)$. Then the adjoint map $^*(F_*)_p$ of the differential $(F_*)_p$, $p \in N$, is given by

$$g_M((F_*)_p X, Z) = g_N(X, {}^*(F_*)_p Z) \quad \text{for } X \in T_p N \text{ and } Z \in T_{F(p)} M.$$

Moreover, if the map $F$ is a Riemannian map, then we easily have

$$(F_*)_p {}^*(F_*)_p Z = Z \quad \text{for } Z \in (range F_*)_{F(p)}$$

and

$$*(F_*)_p (F_*)_p X = X \quad \text{for } X \in (\ker(F_*)_p)^\perp$$

so that the linear map

$$*(F_*)_p : (range F_*)_{F(p)} \mapsto (\ker(F_*)_p)^\perp$$

is an isomorphism.

Define $Q := *(F_*)\phi(F_*)$. Using Theorem 150, we obtain the following corollary.

**Corollary 55.** *[222] Let F be a semi-slant Riemannian map from a Riemannian manifold $(N, g_N)$ to an almost Hermitian manifold $(M, g_M, J)$ with the semi-slant angle $\theta$. Then we obtain*

$$Q^2 X = -\cos^2\theta \cdot X \quad \text{for } X \in \Gamma(\mathcal{D}_2). \tag{6.54}$$

In the same way as Lemma 73, we have the following lemma.

**Lemma 77.** *[222] Let F be a semi-slant Riemannian map from a Riemannian manifold $(N, g_N)$ to a Kähler manifold $(M, g_M, J)$ with the semi-slant angle $\theta$. If the tensor $\omega$ is parallel, then we get*

$$(\nabla F_*)(QX, QY) = -\cos^2\theta \cdot (\nabla F_*)(X, Y) \quad \text{for } X, Y \in \Gamma(\mathcal{D}_2). \tag{6.55}$$

*Proof.* Assume that the tensor $\omega$ is parallel. Then by (6.50), we obtain

$$C(\nabla F_*)(X, Y) = (\nabla F_*)(X, QY) \quad \text{for } X, Y \in \Gamma(\mathcal{D}_2).$$

Interchanging the role of $X$ and $Y$ implies

$$C(\nabla F_*)(Y, X) = (\nabla F_*)(Y, QX).$$

Since the tensor $\nabla F_*$ is symmetric, we have

$$(\nabla F_*)(X, QY) = (\nabla F_*)(Y, QX)$$

so that

$$(\nabla F_*)(QX, QY) = (\nabla F_*)(X, Q^2 Y) = -\cos^2\theta \cdot (\nabla F_*)(X, Y).$$

$\square$

**Theorem 151.** *[222] Let F be a semi-slant Riemannian map from a Riemannian manifold $(N, g_N)$ to a Kähler manifold $(M, g_M, J)$ with the semi-slant angle $\theta \in [0, \frac{\pi}{2})$. If the tensor $\omega$ is parallel, then F is harmonic if and only if all the fibers $F^{-1}(y)$ are minimal submanifolds of N for $y \in M$.*

*Proof.* We know

$$TN = (ker\, F_*) \oplus (ker\, F_*)^\perp = (ker\, F_*) \oplus \mathcal{D}_1 \oplus \mathcal{D}_2.$$

Moreover, all the fibers $F^{-1}(y)$ are minimal submanifolds of $N$ for $y \in M$ if and only if $trace(\nabla F_*)|_{(ker\, F_*)} = 0$. Since $JF_*\mathcal{D}_1 = F_*\mathcal{D}_1$, there is locally an orthonormal frame $\{F_*v_1, JF_*v_1, \cdots, F_*v_l, JF_*v_l\}$ of $F_*\mathcal{D}_1$ so that $\{v_1, Qv_1, \cdots, v_l, Qv_l\}$ is locally an orthonormal frame of $\mathcal{D}_1$. We can also choose locally an orthonormal frame

$$\{e_1, \sec\theta Qe_1, \cdots, e_k, \sec\theta Qe_k\}$$

of $\mathcal{D}_2$. It is easy to get that $Q^2 v_i = -v_i$ for $1 \leq i \leq l$. Since $\omega$ is parallel, by using both (6.50) and the proof of Lemma 77, we get

$$
\begin{aligned}
trace(\nabla F_*)|_{\mathcal{D}_1} &= \sum_{i=1}^{l} \{(\nabla F_*)(v_i, v_i) + (\nabla F_*)(Qv_i, Qv_i)\} \\
&= \sum_{i=1}^{l} \{(\nabla F_*)(v_i, v_i) + (\nabla F_*)(v_i, Q^2 v_i)\} \\
&= \sum_{i=1}^{l} \{(\nabla F_*)(v_i, v_i) - (\nabla F_*)(v_i, v_i)\} = 0.
\end{aligned}
$$

Furthermore, by using Corollary 55,

$$
\begin{aligned}
trace(\nabla F_*)|_{\mathcal{D}_2} &= \sum_{j=1}^{k} \{(\nabla F_*)(e_j, e_j) + (\nabla F_*)(\sec\theta Qe_j, \sec\theta Qe_j)\} \\
&= \sum_{j=1}^{k} \{(\nabla F_*)(e_j, e_j) + \sec^2\theta(\nabla F_*)(Qe_j, Qe_j)\} \\
&= \sum_{j=1}^{k} \{(\nabla F_*)(e_j, e_j) + \sec^2\theta(\nabla F_*)(e_j, Q^2 e_j)\} \\
&= \sum_{j=1}^{k} \{(\nabla F_*)(e_j, e_j) - (\nabla F_*)(e_j, e_j)\} = 0.
\end{aligned}
$$

Therefore, the result follows.                                                    □

**Remark 25.** Comparing Theorem 151 with Lemma 74, we see that the conditions for such maps to be harmonic are the same between slant Riemannian maps and semi-slant Riemannian maps.

We study the condition for a semi-slant Riemannian map $F$ to be totally geodesic.

**Theorem 152.** *[222] Let $F$ be a semi-slant Riemannian map from a Riemannian manifold $(N, g_N)$ to a Kähler manifold $(M, g_M, J)$ with the semi-slant angle $\theta \in (0, \frac{\pi}{2})$. Then the map $F$ is totally geodesic if and only if:*

**(a)** *all the fibers $F^{-1}(y)$ are totally geodesic for $y \in M$,*

**(b)** *the horizontal distribution $(\ker F_*)^\perp$ is a totally geodesic foliation,*

**(c)** *for $X \in \Gamma((\ker F_*)^\perp)$, $V \in \Gamma((range F_*)^\perp)$, and $Y \in \Gamma(\mathcal{D}_1)$ with $\phi F_* Y = F_* Y'$ and $Y' \in \Gamma(\mathcal{D}_1)$, we have*

$$g_M(\widehat{\nabla}^F_X F_* Y', BV) + g_M(\bar{Q}\nabla^F_X \phi F_* Y, CV) = 0, \quad and$$

**(d)** *for $X, Y \in \Gamma(\mathcal{D}_2)$ and $V \in \Gamma((range F_*)^\perp)$, we have*

$$g_M(S_{\omega F_* Y} F_* X, BV) = g_M(\nabla^{F\perp}_X \omega F_* Y, CV) - g_M(\nabla^{F\perp}_X \omega \phi F_* Y, V).$$

*Proof.* Given $U_1, U_2 \in \Gamma(\ker F_*)$ and $X \in \Gamma((\ker F_*)^\perp)$, we have

$$g_M((\nabla F_*)(U_1, U_2), F_* X) = -g_M(F_* \nabla_{U_1} U_2, F_* X) = -g_N(\nabla_{U_1} U_2, X),$$

so that $(\nabla F_*)(U_1, U_2) = 0$ for $U_1, U_2 \in \Gamma(\ker F_*)$ if and only if (a) holds.

For $U \in \Gamma(\ker F_*)$ and $X, Y \in \Gamma((\ker F_*)^\perp)$, we obtain

$$g_M((\nabla F_*)(X, U), F_* Y) = -g_M(F_* \nabla_X U, F_* Y)$$
$$= -g_N(\nabla_X U, Y) = g_N(U, \nabla_X Y).$$

Hence, $(\nabla F_*)(X, U) = 0$ for $U \in \Gamma(\ker F_*)$ and $X \in \Gamma((\ker F_*)^\perp)$ if and only if (b) is satisfied. If $X \in \Gamma((\ker F_*)^\perp)$, $Y \in \Gamma(\mathcal{D}_1)$, and $V \in \Gamma((range F_*)^\perp)$, then by using Lemma 45, we get

$$\begin{aligned} g_M((\nabla F_*)(X, Y), V) &= g_M(\nabla^F_X F_* Y, V) \\ &= g_M(\nabla^F_X \phi F_* Y, JV) \\ &= g_M(\widehat{\nabla}^F_X F_* Y', BV) + g_M(\bar{Q}\nabla^F_X \phi F_* Y, CV) \end{aligned}$$

for some $Y' \in \Gamma(\mathcal{D}_1)$ with $\phi F_* Y = F_* Y'$ so that $(\nabla F_*)(X, Y) = 0$ for $X \in \Gamma((\ker F_*)^\perp)$ and $Y \in \Gamma(\mathcal{D}_1)$ if and only if we have (c). Given $X, Y \in \Gamma(\mathcal{D}_2)$ and $V \in \Gamma((range F_*)^\perp)$, we obtain

$$\begin{aligned} g_M((\nabla F_*)(X, Y), V) &= g_M(\nabla^F_X F_* Y, V) \\ &= -g_M(\nabla^F_X J(\phi F_* Y + \omega F_* Y), V) \\ &= \cos^2 \theta \, g_M(\nabla^F_X F_* Y, V) - g_M(\nabla^F_X \omega \phi F_* Y, V) \\ &\quad + g_M(\nabla^F_X \omega F_* Y, BV + CV), \end{aligned}$$

so that with some elementary calculations, $(\nabla F_*)(X, Y) = 0$ for $X, Y \in \Gamma(\mathcal{D}_2)$ if and only if (d) is satisfied. Therefore, we have the result. □

In the rest of this section, we are going to explore the relation between semi-slant Riemannian maps and pseudo-horizontally homothetic maps.

**Proposition 69.** *[222] Let $F$ be a semi-slant Riemannian map from a Riemannian manifold $(M_1, g_1)$ to an almost Hermitian manifold $(M_2, g_2, J)$. Then $F$ is a PHWC map.*

*Proof.* For $X \in \Gamma((ker F_*)^{\perp})$, we define $\tilde{J} F_*(X) = J F_*(PX) + \sec \theta \phi F_*(QX)$, then it is easy to see that $\tilde{J}$ is a complex structure on $(range F_*)$ and $range F_*$ is invariant with respect to $\tilde{J}$. Then we now define $\hat{J}X = {}^*F_* \circ J \circ F_*(PX) + \sec \theta {}^*F_* \phi F_*(QX)$, Then it follows that $\hat{J}$ is a complex structure on $(ker F_*)^{\perp}$. Thus $((ker F_*)^{\perp}, \hat{J})$ is an almost complex distribution. Considering $\hat{g} = g_1 |_{(ker F_*)^{\perp}}$, then by direct computation we obtain

$$\hat{g}(\hat{J}X, \hat{J}Y) = \hat{g}(X, Y)$$

for $X, Y \in \Gamma((ker F_*)^{\perp})$. Thus $\hat{g}$ is $\hat{J}$–Hermitian and $((ker F_*)^{\perp}, \hat{g}, \hat{J})$ is an almost Hermitian distribution. Therefore, $F$ is a PHWC map.                                   □

We now give necessary and sufficient conditions for a semi-slant Riemannian map $F$ from a Riemannian manifold $(M_1, g_1)$ to a Kähler manifold $(M_2, g_2, J)$ to be pseudo-horizontally homothetic map. First observe that we denote ${}^*F_* J F_*$,

$$\nabla^F_X J F_*(PY) - J F_*(P\mathcal{H}\nabla^1_X Y)$$

and

$$\nabla^F_X \phi F_*(QY) - \phi F_*(Q\mathcal{H}\nabla^1_X Y)$$

by $\tilde{Q}$, $(\nabla_X \phi_1) F_*(PY)$, and $(\nabla_X \phi_2) F_*(QY)$, respectively, where $\nabla^1$ denotes the Levi-Civita connection on $M_1$.

**Theorem 153.** *[222] Let $F$ be a semi-slant Riemannian map from a Riemannian manifold $(M_1, g_1)$ to a Kähler manifold $(M_2, g_2, J)$. Then $F$ is a pseudo-horizontally homothetic map if and only if*

$$(\nabla F_*)(X, \tilde{Q}(PY)) + \sec \theta (\nabla F_*)(X, Q(QX)) = (\nabla_X \phi_1) F_*(PY) + \sec \theta (\nabla_X \phi_2) F_*(QY)$$

*and*

$$g_2(F_*(PY), J(\nabla F_*)(X, U)) = \sec \theta g_2(\phi F_*(QY), (\nabla F_*)(X, U))$$

*for $X \in \Gamma((ker F_*)^{\perp})$ and $U \in \Gamma(ker F_*)$.*

*Proof.* First of all, we have

$$(\nabla_X \hat{J})Y = \nabla^1_X \hat{J}Y - \hat{J}\mathcal{H}\nabla^1_X Y$$

for $X, Y \in \Gamma((ker F_*)^\perp)$. Hence we obtain

$$
\begin{aligned}
(\nabla_X \hat{J})Y &= \nabla^{1*}_X F_* JF_*(PY) + \sec \theta^* F_* \phi F_*(QY) \\
&\quad -^* F_* JF_*(P\mathcal{H}\nabla^1_X Y) - \sec \theta^* F_* \phi F_*(Q\mathcal{H}\nabla^1_X Y).
\end{aligned}
$$

Then by direct computations, we have

$$(\nabla_X \hat{J})Y = \nabla^1_X \tilde{Q}(PY) + \sec \theta \nabla^1_X Q(QY) - \tilde{Q}(P\mathcal{H}\nabla^1_X Y) - \sec \theta Q(Q\mathcal{H}\nabla^1_X Y)$$

for $X, Y \in \Gamma((ker F_*)^\perp)$. Thus, using (1.71), we get

$$
\begin{aligned}
F_*(\nabla_X \hat{J})Y &= -(\nabla F_*)(X, \tilde{Q}(PY)) - \sec \theta ((\nabla F_*)(X, Q(QY)) \\
&\quad + \nabla^F_X JF_*(PY) - JF_*(P\mathcal{H}\nabla^1_X Y) \\
&\quad + \sec \theta \nabla^F_X \phi F_*(QY) - \sec \theta \phi F_*(Q\mathcal{H}\nabla^1_X Y)).
\end{aligned}
$$

On the other hand, since $QY, \tilde{Q}Y \in \Gamma((ker F_*)^\perp)$, we have

$$g_1((\nabla^1_X \hat{J})Y, U) = -\sec \theta g_1(Q(QY)Y, \nabla^1_X U) - g_1(\tilde{Q}(PX), \nabla^1_X U)$$

for $U \in \Gamma(ker F_*)$. Then, using adjoint map $^* F_*$ and (1.71), we obtain

$$g_1((\nabla^1_X \hat{J})Y, U) = \sec \theta g_2(\phi F_*(QY), (\nabla F_*)(X, U)) - g_2(F_*(PY), J(\nabla F_*)(, X, U)).$$

This completes the proof.    $\square$

## 7. Hemi-slant Riemannian maps to Kähler manifolds

In this section, we introduce hemi-slant Riemannian maps from Riemannian manifolds to almost Hermitian manifolds as a generalization of hemi-slant immersions, invariant Riemannian maps, anti-invariant Riemannian maps, and slant Riemannian maps to almost Hermitian manifolds. We provide examples, obtain a decomposition theorem, and find necessary and sufficient conditions for such maps.

**Definition 65.** [259] Let $M$ be a Riemannian manifold with Riemannian metric $g_N$, and $N$ be a $2m$-dimensional almost Hermitian manifold with Hermitian metric $g$ and an almost complex structure $J$. A Riemannian map $F : (M, g, J) \to (N, g_N)$ is called a *hemi-slant Riemannian map* if $(rangeF_*)$ of $F$ admits two orthogonal complementary distributions $\mathfrak{D}^\psi$ and $\mathfrak{D}^\perp$ such that $\mathfrak{D}^\psi$ is slant and $\mathfrak{D}^\perp$ is anti-invariant, i.e., we have

$$rangeF_* = \mathfrak{D}^\psi \oplus \mathfrak{D}^\perp. \tag{6.56}$$

In this case, the angle $\psi$ is called the *hemi-slant angle* of the Riemannian map.

From the above definition, we have the following result.

**Proposition 70.** *Let $F : (M, g_M) \to (N, g_N, J_N)$ be a hemi-slant Riemannian map between Riemannian manifold M and a Hermitian manifold N with integrable distribution $(rangeF_*)$. Then the leaves of the distribution $(rangeF_*)$ of a hemi-slant Riemannian map are hemi-slant submanifolds of N.*

We give some examples of hemi-slant Riemannian maps.

**Example 93.** Every slant submanifold of an almost Hermitian manifold is a hemi-slant Riemannian map to an almost Hermitian manifold with $ker F_* = \{0\}$.

**Example 94.** Every anti-slant (hemi-slant) submanifold of an almost Hermitian manifold is a hemi-slant Riemannian map to an almost Hermitian manifold with $ker F_* = \{0\}$.

**Example 95.** Every slant Riemannian map from a Riemannian manifold to an almost Hermitian manifold is a hemi-slant Riemannian map with $\mathfrak{D}^\perp = \{0\}$.

**Example 96.** Every semi-invariant Riemannian map from a Riemannian manifold to an almost Hermitian manifold is a hemi-slant Riemannian map with $\psi = \frac{\pi}{2}$.

Since invariant and anti-invariant Riemannian maps to almost Hermitian manifolds are also particular cases of semi-invariant Riemannian maps to almost Hermitian manifolds, they are also examples of hemi-slant Riemannian maps.

We say that the hemi-slant Riemannian map $F : (M, g, J) \to (N, g_N)$ is *proper* if $\mathfrak{D}^\perp \neq \{0\}$ and $\psi \neq 0, \frac{\pi}{2}$.

For any $V \in \Gamma(rangeF_*)$, we put

$$V = \mathfrak{P}V + \mathfrak{Q}V, \tag{6.57}$$

where $\mathfrak{P}V \in \Gamma(\mathfrak{D}^\psi)$ and $\mathfrak{Q}V \in \Gamma(\mathfrak{D}^\perp)$ and put

$$JV = \varphi V + \varpi V, \tag{6.58}$$

where $\varphi V \in \Gamma(rangeF_*)$ and $\varpi V \in \Gamma((rangeF_*)^\perp)$. Also, for any $\xi \in \Gamma((rangeF_*)^\perp)$, we have

$$J\xi = \mathfrak{B}\xi + \mathfrak{C}\xi, \tag{6.59}$$

where $\mathfrak{B}\xi \in \Gamma(rangeF_*)$ and $\mathfrak{C}\xi \in \Gamma((rangeF_*)^\perp)$.

As for hemi-slant Riemannian maps from almost Hermitian manifolds, the proof of the following theorem is exactly the same as that for hemi-slant submanifolds; see Theorem 3.2 of [238]. we therefore omit it.

**Theorem 154.** *[259] Let F be a Riemannian submersion from an almost Hermitian manifold $(M, g, J)$ onto a Riemannian manifold. Then F is a hemi-slant submersion if and only if there exists a constant $\lambda \in \Gamma([-1, 0]$ and a distribution $\mathcal{D}$ on $rangeF_*$ such that*
**(a)** $\mathcal{D} = \{V \in \Gamma(rangeF_*) \mid \varphi^2 V = \lambda V\}$, and
**(b)** *for any $V \in \Gamma(rangeF_*)$ orthogonal to $\mathcal{D}$, we have $\varphi V = 0$.*
*Moreover, in this case $\lambda = -\cos^2 \psi$, where $\psi$ is the slant angle of F.*

We now present an example of a proper Riemannian map.

**Example 97.** [259] Define a map $F : \mathbb{R}^6 \to \mathbb{R}^6$    by

$$F(x_1, ..., x_6) = \left(x_1, \frac{x_2 + x_3}{\sqrt{2}}, \frac{x_2 + x_3}{\sqrt{3}}, 0, \frac{x_5 + x_6}{\sqrt{2}}, 0\right).$$

Then the map $F$ is a proper hemi-slant Riemannian map such that

$$\mathcal{D}^\psi = span\{\partial y_1, \partial y_2 + \sqrt{\frac{2}{3}} \partial y_3, \},$$

with the slant angle $\psi = \arccos(\frac{3}{5})$ and $\mathcal{D}^\perp = span\{\partial y_5\}$. Moreover, we have

$$(rangeF_*)^\perp = span\{-\sqrt{\frac{2}{3}} \partial y_2 + \partial y_3, \partial y_4, \partial y_6\}, \partial y_i = \frac{\partial}{\partial y_i},$$

where $y_1, ..., y_6$ are the local coordinates on $\mathbb{R}^6$.

We now investigate the geometry of the leaves of distributions.

**Theorem 155.** *[259] Let F be a hemi-slant Riemannian map from a Riemannian manifold $(M, g_M)$ to a Kähler manifold $(N, g_N, J)$. The following assertions are equivalent:*
**(i)** *distribution $\mathcal{D}^\perp$ defines a totally geodesic foliation on N,*
**(ii)** $g_N((\nabla F_*)(X, {}^* F_*((\varphi F_*(Z)))), JF_*(Y)) = g_N(\nabla_X^{F\perp} \varpi F_*(Z), JF_*(Y))$ and

$$g_N((\nabla F_*)(X, {}^* F_*(\mathcal{B}V)), JF_*(Y)) = g_N(\nabla_X^{F\perp} JF_*(Y), \mathcal{C}V), \quad and \quad (6.60)$$

**(iii)** *F satisfies (6.60) and*

$$g_N((\nabla F_*)(X, Y), \varpi \varphi F_*(Z)) = g_N(\mathcal{B}\nabla_X^{F\perp} \varpi F_*(Z), F_*(Y))$$

*for $F_*(X), F_*(Y) \in \Gamma(\mathcal{D}^\perp), F_*(Z) \in \Gamma(\mathcal{D}^\psi)$ and $V \in \Gamma((rangeF_*)^\perp)$.*

*Proof.* For $F * (X), F_*(Y) \in \Gamma(\mathfrak{D}^\perp)$, from (3.5), (6.58), and (4.16), we have

$$g_N(\nabla_X^F F_*(Y), F_*(Z)) = -g_N(S_{JF_*(Y)} F_*(X), \varphi F_*(Z)) + g_N(\nabla_X^{F\perp} JF_*(Y), \varpi F_*(Z)).$$

Then (4.17) implies that

$$
\begin{aligned}
g_N(\nabla_X^F F_*(Y), F_*(Z)) &= -g_N(\nabla F_*)(X,^* F_*(\varphi F_*(Z))), JF_*(Y) \\
&\quad + g_N(\nabla_X^{F\perp} JF_*(Y), \varpi F_*(Z)).
\end{aligned}
\tag{6.61}
$$

On the other hand, (3.5), (6.59), and (4.16), we get

$$g_N(\nabla_X^F F_*(Y), V) = -g_N(S_{JF_*(Y)} F_*(X), \mathfrak{B}V) + g_N(\nabla_X^{F\perp} JF_*(Y), \mathfrak{C}V).$$

for $V \in \Gamma((range F_*)^\perp)$. Then (4.17) gives

$$
\begin{aligned}
g_N(\nabla_X^F F_*(Y), V) &= -g_N(\nabla F_*)(X,^* F_*(\mathfrak{B}V), JF_*(Y) \\
&\quad + g_N(\nabla_X^{F\perp} JF_*(Y), \mathfrak{C}V).
\end{aligned}
\tag{6.62}
$$

Equations (6.61) and (6.62) gives (i) $\Leftrightarrow$ (ii). For $X, Y \in \Gamma(\mathfrak{D}^\perp)$, $Z \in \Gamma(\mathfrak{D}^\psi)$, using (3.5) and (6.58), we have

$$
\begin{aligned}
g_N(\nabla_X^F F_*(Y), F_*(Z)) &= g_N(\nabla_X^F \varphi^2 F_*(Z), F_*(Y)) \\
&\quad + g_N(\nabla_X^F \varpi \varphi F_*(Z), F_*(Y)) - g_N(\nabla_X^F \varpi F_*(Z), JF_*(Y)).
\end{aligned}
$$

Now from (4.16) and Theorem 154, we derive

$$
\begin{aligned}
\sin^2 \psi g_N(\nabla_X^F F_*(Y), F_*(Z)) &= -g_N(S_{\varpi \varphi F_*(Z)} F_*(X), F_*(Y)) \\
&\quad - g_N(\nabla_X^{F\perp} \varpi F_*(Z), JF_*(Y)).
\end{aligned}
$$

Thus from (4.17), we obtain

$$
\begin{aligned}
\sin^2 \psi g_N(\nabla_X^F F_*(Y), F_*(Z)) &= -g_N((\nabla F_*)(X, Y), \varpi \varphi F_*(Z)) \\
&\quad - g_N(\nabla_X^{F\perp} \varpi F_*(Z), JF_*(Y)).
\end{aligned}
\tag{6.63}
$$

Equations (6.62) and (6.63) give (i) $\Leftrightarrow$ (iii). $\qquad\qquad\square$

In a similar way, we have the following result for $\mathfrak{D}^\psi$.

**Theorem 156.** *[259] Let F be a hemi-slant Riemannian map from a Riemannian manifold $(M, g_M)$ to a Kähler manifold $(N, g_N, J)$. The following assertions are then equivalent:*

**(i)** *distribution $\mathfrak{D}^\psi$ defines a totally geodesic foliation on N,*

**(ii)** $g_N((\nabla F_*)(Z_1,^* F_*((\varphi F_*(Z_2)))), JF_*(X)) = g_N(\nabla_{Z_1}^{F\perp} JF_*(X), \varpi F_*(Z))$ *and*

$$
\begin{aligned}
g_N((\nabla F_*)(Z_1,^* F_*(\mathfrak{B}V)), \varpi F_*(Z_2)) &= g_N(\nabla_{Z_1}^{F\perp} \varpi F_*(Z_2), \mathfrak{C}V) \\
&\quad - g_N(\nabla_{Z_1}^{F\perp} \varpi \varphi F_*(Z_2), V), \text{ and}
\end{aligned}
\tag{6.64}
$$

**(iii)** *F satisfies (6.64) and*

$$g_N((\nabla F_*)(Z_1, X), \varpi\varphi F_*(Z_2)) = g_N(\mathcal{B}\nabla_{Z_1}^{F\perp} \varpi F_*(Z_2), F_*(X))$$

*for $F_*(X) \in \Gamma(\mathcal{D}^\perp)$, $F_*(Z_1), F_*(Z_2) \in \Gamma(\mathcal{D}^\psi)$ and $V \in \Gamma((rangeF_*)^\perp)$.*

From Theorem 155 and Theorem 156, we have the following decomposition theorem.

**Theorem 157.** *[259] Let F be a hemi-slant Riemannian map from a Riemannian manifold $(M, g_M)$ to a Kähler manifold $(N, g_N, J)$ with integrable distribution $(rangeF_*)$. Then the leaf of $(rangeF_*)$ is a locally product Riemannian manifold $\mathfrak{M}_\perp \times \mathfrak{M}_\psi$ if and only if*

$$g_N((\nabla F_*)(U, {}^* F_*(\mathcal{B}V)), \varpi F_*(W)) = g_N(\nabla_U^{F\perp} \varpi F_*(W), \mathfrak{C}V)$$
$$-g_N(\nabla_U^{F\perp} \varpi\varphi F_*(W), V)$$

*and*

$$g_N((\nabla F_*)(U, V_1), \varpi\varphi F_*(Z)) = g_N(\mathcal{B}\nabla_U^{F\perp} \varpi F_*(Z), F_*(V_1))$$

*for $U, W \in \Gamma((ker F_*)^\perp)$, $F_*(Z) \in \Gamma((\mathcal{D}^\psi)$ and $V \in \Gamma((rangeF_*)^\perp)$, where $\mathfrak{M}_\perp$ and $\mathfrak{M}_\psi$ denote the leaves of $\mathcal{D}^\perp$ and $\mathcal{D}^\psi$, respectively.*

Finally, we obtain necessary and sufficient conditions for a hemi-slant Riemannian map to be totally geodesic.

**Theorem 158.** *[259] Let F be a hemi-slant Riemannian map from a Riemannian manifold $(M, g_M)$ to a Kähler manifold $(N, g_N, J)$.. Then F is totally geodesic if and only if the following conditions are satisfied:*
**(i)** *for $U, W \in \Gamma((ker F_*)^\perp)$ and $V \in \Gamma((rangeF_*)^\perp)$,*

$$g_N((\nabla F_*)(U, {}^* F_*(\mathcal{B}V)), \varpi F_*(W)) = g_N(\nabla_U^{F\perp} \varpi F_*(W), \mathfrak{C}V)$$
$$-g_N(\nabla_U^{F\perp} \varpi\varphi F_*(W), V),$$

**(ii)** *for $X, Y, Z \in \Gamma((ker F_*)^\perp)$,*

$$g_N(\nabla_X^{F\perp} \varpi F_*(Y), \varpi F_*(Z)) = g_N((\nabla F_*)(X, Z), \varpi\varphi F_*(Y))$$
$$+g_N((\nabla F_*)(X, {}^* F_*(\varphi Z)), \varpi F_*(Y)) + (1 + \cos^2 \psi)g_M(\mathcal{H}\nabla_X Y, Z),$$

**(iii)** *the distribution $(ker F_*)$ is totally geodesic, and*
**(iv)** *the distribution $(ker F_*)^\perp$ is integrable.*

*Proof.* For $X, Y, Z \in \Gamma((ker\, F_*)^\perp)$, using (3.5), (1.71), (4.15), and (6.58), we get

$$
\begin{aligned}
g_N((\nabla F_*)(X, Y), F_*(Z)) &= g_N(\nabla^F \varphi^2 F_*(Y), F_*(Z)) + g_N(\nabla_X^F \varpi\varphi F_*(Y), F_*(Z)) \\
&+ g_N(\nabla_X^F \varpi F_*(Y), \varphi F_*(Z)) + g_N(\nabla_X^F \varpi F_*(Y), \varpi F_*(Z)) \\
&- g_M(\mathcal{H}\nabla_X Y, Z).
\end{aligned}
$$

Then from (1.71), (4.16), and (4.17), we derive

$$
\begin{aligned}
(1 + \cos^2 \psi) g_N((\nabla F_*)(X, Y), F_*(Z)) &= -g_N((\nabla F_*)(X, Z), \varpi\varphi F_*(Y)) \\
-g_N((\nabla F_*)(X,{}^* F_*(\varphi Z)), \varpi F_*(Y)) &- (1 + \cos^2 \psi) g_M(\mathcal{H}\nabla_X Y, Z) \\
+g_N(\nabla_X^{F\perp} \varpi F_*(Y), \varpi F_*(Z)). &
\end{aligned} \tag{6.65}
$$

On the other hand, from (4.12) and (4.15), we have

$$
g_N((\nabla F_*)(U, V), F_*(X)) = -g(\mathcal{T}_U V, X) \tag{6.66}
$$

and

$$
g_N((\nabla F_*)(X, U), F_*(Y)) = g_M(U, \mathcal{A}_X Y) \tag{6.67}
$$

for $X, Y \in \Gamma((ker\, F_*)^\perp)$ and $U, V \in \Gamma(ker\, F_*)$. Then the proof is complete from (6.65), (6.66), (6.67) and Theorem 157.                                          □

**Remark 26.** By comparing our results of this section with the results of hemi-slant submersions, we can see that the results of this section are different from hemi-slant submersions as well as hemi-slant submanifolds.

**Remark 27.** We conclude this chapter by adding that the concept of Riemannian maps have also been onsidered to a few base spaces. This implies that there are many new problems to study in this research area.

(i)  Riemannian maps to almost quaternion Kaehler manifolds (see Park [220]).

(ii) Riemannian maps to trans-Sasakian manifolds (see Jaiswal-Pandey [149]).

# BIBLIOGRAPHY

1. R. Abraham, J.E. Marsden, and T. Ratiu. *Manifolds, Tensor Analysis, and Applications.* Springer, New York, 1988.
2. A. Adamów and R. Deszcz. On totally umbilical submanifolds of some class of Riemannian manifolds. *Demonstratio Math.*, 16:39–59, 1983.
3. M. A. Akyol and Y. Gündüzalp. Hemi-slant submersions from almost product Riemannian manifolds. *Gulf J. Math.*, To appear.
4. P. Alegre, B. Y. Chen, and M. I. Munteanu. Riemannian submersions, $\delta$-invariants, and optimal inequality. *Ann. Glob. Anal. Geom.*, 42:317–331, 2012.
5. S. Ali and T. Fatima. Integrability conditions for the distribution of anti-invariant Riemannian submersions. *J. Tensor Soc.*, 6:163–175, 2012.
6. S. Ali and T. Fatima. Anti-invariant Riemannian submersions from nearly Kaehler manifolds. *Filomat*, 27:1219–1235, 2013.
7. S. Ali and T. Fatima. Generic Riemannian submersions. *Tamkang J. Math.*, 44:395–409, 2013.
8. S. Ali and T. Fatima. Product theorems on anti-invariant Riemannian submersions. *Afrika Math.*, 26:471–483, 2015.
9. D. Allison. Geodesic completeness in static spacetimes. *Geom. Dedicata*, 26:85–97, 1988.
10. D. Allison. Lorentzian Clairaut submersions. *Geom. Dedicata*, 63:309–319, 1996.
11. C. Altafini. *Geometric Control Methods for Nonlinear Systems and Robotic Applications.* PhD thesis, The Royal Institute of Technology, Stockholm, 2001.
12. C. Altafini. Redundant robotic chains on Riemannian submersions. *IEEE Transactions on Robotics and Automation*, 20:335–340, 2004.
13. M. A. Aprodu. Phh harmonic submersions are stable. *Boll. Unione Mat.*, 8:1081–1088, 2007.
14. M. A. Aprodu and M Aprodu. Implicitly defined harmonic phh submersions. *Manuscripta Math.*, 100:103–121, 1999.
15. M. A. Aprodu, M Aprodu, and V. Br înzănescu. A class of harmonic submersions and minimal submanifolds. *Internat. J. Math.*, 11:1177–1191, 2000.
16. A. Arvanitoyeorgos. *An Introduction to Lie Groups and the Geometry of Homogeneous Spaces.* American Math. Soc., Rhode Island, 2003.
17. K. Aso and S. Yorozu. A generalization of Clairaut's theorem and umbilic foliations. *Nihonkai Math. J.*, 2:139–153, 1991.
18. A. C. Asperti, G. Lobos, and F. Mercuri. Pseudo-parallel submanifolds of a space form. *Adv. Geom.*, 2:57–71, 2002.
19. M. E. Aydın, A. Mihai, and I. Mihai. Some inqualities on submanifolds in statistical manifolds of constant curvature. *Filomat*, 29:465–477, 2015.
20. P. Baird and J. C. Wood. *Harmonic Morphisms Between Riemannian Manifolds.* Oxford Science Publications, Clarendon Press, Oxford, 2003.
21. A. Balmus. *Biharmonic Maps and Submanifolds.* Geometry Balkan Press, Bucharest, 2009.
22. J. K. Beem, Ehrlich P., and K. Easley. *Global Lorentzian Geometry.* Markel-Deccer Inc., New York, 1996.
23. R. G. Beil. Elektroweak symmetry on the tangent bundle. *Int. J. Theoretical Physics*, 40:591–601, 2001.
24. A. Bejancu. CR submanifolds of a Kaehler manifold. I. *Proc. Amer. Math. Soc.*, 69:135–142, 1978.
25. A. Bejancu. CR submanifolds of a Kaehler manifold. II. *Trans. Amer. Math. Soc.*, 250:333–3454, 1978.
26. A. Bejancu. *Geometry of CR-submaniflods.* Kluwer, Amsterdam, 1986.
27. A. Bejancu. Oblique warped products. *J. of Geometry and Physics*, 57:1055–1073, 2007.
28. A. Bejancu, K. Yano, and M. Kon. CR-submanifolds of complex space form. *J. Diff. Geom.*,

16:137–145, 1981.

29. C. Belta. *Geometric Methods for Mullti-robot Planning and Control.* PhD thesis, University of Pennsylvania, 2003.

30. A. Beri, İ. K. Erken, and C. Murathan. Anti-invariant Riemannian submersions from Kenmotsu manifolds onto Riemannian manifolds. *Turkish J. Math,* 40:540–552, 2016.

31. A. L. Besse. *Einstein Manifolds.* Springer, Berlin, 1987.

32. R. Bhattacharyaa and V. Patrangenarub. Nonparametic estimation of location and dispersion on Riemannian manifolds. *Journal of Statistical Planning and Inference,* 108:23–35, 2002.

33. A. Bhattacharyya, M. Tarafdar, and D. Debnath. On mixed super quasi-Einstein manifolds. *Differ. Geom. Dyn. Syst.,* 10:44–57, 2008.

34. R. L. Bishop and B. O'Neill. Manifolds of negative curvature. *Trans. Amer. Math. Soc.,* 145:149, 1969.

35. R.L. Bishop. Clairaut submersions. In *Differential geometry (in Honor of Kentaro Yano),* pages 21–31, 1972.

36. D. E. Blair. *Riemannian Geometry of Contact and Symplectic Manifolds.* Birkhäser, Boston, 2002.

37. D. E. Blair. D-hmothetic warping. *Publ. L'institut Math.,* 94:47–54, 2013.

38. D. E. Blair. D-homothetic warping and applications to geometric structures and cosmology. *Afr. Diaspora J. Math.,* 14:134–144, 2013.

39. D. E. Blair and B. Y. Chen. On CR-submanifolds of Hermitian manifolds. *Israel J. Math.,* 34:353–363, 1979.

40. J. P. Bourguignon. A mathematician's visit to kaluza-klein theory. *Rend. Sem. Mat. Univ. Poi. Torino,* Special Issue:143–163, 1988.

41. G. E. Bredon. *Topology and Geometry.* Springer, New York, 1993.

42. F. Brickell and R.S. Clark. *Differentiable manifolds.* Van Nostrand Reinhold Company, New York, 1970.

43. F. Bullo and A. D. Lewis. *Geometric Control of Mechanical Systems.* Springer, New York, 2004.

44. B.Ünal. Multiply warped products. *J. Geom. Phys,* 34:287–301, 2000.

45. B.Ünal. Doubly warped products. *Diff. Geom. Appl.,* 15:253–263, 2001.

46. D. Burns, F Burstall, P. De Bartolomeis, and J. Rawnsley. Stability of harmonic maps of Kähler manifolds. *J. Differential Geom.,* 30:579594, 1989.

47. J. L. Cabrerizo, A. Carriazo, L. M. Fernández, and M. Fernández. Semi-slant submanifolds of a Sasakian manifold. *Geom. Dedicata,* 78:183–199, 1999.

48. J. L. Cabrerizo, A. Carriazo, L. M. Fernández, and M. Fernández. Slant submanifolds in Sasakian manifolds. *Glasg. Math. J.,* 42:125–138, 2000.

49. R. Caddeo, S. Montaldo, and C. Oniciuc. Biharmonic submanifolds of $s^3$. *Internat. J. Math.,* 12:867–876, 2001.

50. A. V. Caldarella. On paraquaternionic submersions between paraquaternionic khler manifolds. *Acta Appl. Math.,* 112:1–14, 2010.

51. P. Candelas, G. Horowitz, A. Strominger, and E. Witten. Vacuum configurations for super strings. *Nucl. Phys,* 258:46–74, 1985.

52. H.D. Cao. Recent progress on ricci solitons. In *Recent Advances in Geometric Analysis,* pages 1–38. International Press, Somerville, Massachusetts, 2010.

53. A. Carriazo. Bi-slant immersions. In *Proc. ICRAMS,* pages 88–97, 2000.

54. J. Case, Y.J. Shu, and G Wei. Rigidity of quasi-Einstein metrics. *Differential Geometry and its Applications,* 29:93–100, 2011.

55. R. Castanõ-Bernard and D. Matessi. Lagrangian 3-torus fibrations. *J. Differential Geom.,* 81:483–573, 2009.

56. R. Castanõ-Bernard and D. Matessi. The fixed point set pf anti-symplectic involutions of Lagrangian fibrations. *Rend. Sem. Mat. Univ. Pol. Torino,* 68:235–250, 2010.

57. R. Castanõ-Bernard, D. Matessi, and J. P. Solomon. Symmetries of Lagrangian fibrations. *Advances in Math.,* 225:1341–1386, 2010.

58. G. Catino. Generalized quasi-Einstein manifolds with harmonic weyl tensor. *Math. Z.*, 271:751–756, 2012.

59. M. C. Chaki. On pseudo-symmetric manifolds. *Ştiint. Univ. Al. I. Cuza Iaşi Sect. I Mat.*, 33:53–58, 1987.

60. M. C. Chaki. On generalized quasi Einstein manifolds. *Publ. Math. Debrecen*, 58:683–691, 2001.

61. M. C. Chaki. On super quasi-Einstein manifold. *Publ. Math.*, 64:481–488, 2004.

62. M. C. Chaki and R. K. Maity. On quasi Einstein manifolds. *Publ. Math. Debrecen*, 57:297–306, 2000.

63. J. Cheeger and D.G. Ebin. *Comparison Theorems in Riemannian Geometry*. North-Holland Publishing Company, New York, 1975.

64. B. Y. Chen. *Geometry of submanifolds*. Marcel Dekker, New York, 1973.

65. B. Y. Chen. Extrinsic spheres in compact symmetric spaces are intrinsic spheres. *Michigan Math. J.*, 24:265–271, 1977.

66. B. Y. Chen. CR-submanifolds of a Kaehler manifold. i. *J. Differential Geom*, 16:305–322, 1981.

67. B. Y. Chen. CR-submanifolds of a Kaehler manifold ii. *J. Differential Geom*, 16:493–509, 1981.

68. B. Y. Chen. Differential geometry of real submanifolds in a khler manifold. *Monatsh. Math.*, 91:257–274, 1981.

69. B. Y. Chen. *Geometry of Submanifolds and Its Applications*. Science University of Tokyo, Tokyo, 1981.

70. B. Y. Chen. *Geometry of Slant Submanifolds*. Katholieke Universiteit, Leuven, 1990.

71. B. Y. Chen. Slant immersions. *Monatsh. Math.*, 41:135–147, 1990.

72. B. Y. Chen. Some pinching and classification theorems for minimal submanifolds. *Arch. Math.*, 60:568–578, 1993.

73. B. Y. Chen. A general inequality for submanifolds in complex space forms and its applications. *Arch. Math.*, 67:519–528, 1996.

74. B. Y. Chen. Riemannian submanifolds. In *Handbook of Differential Geometry, Vol. I*, pages 187–418. Elsevier, Amsterdam, 2000.

75. B. Y. Chen. Convolution of Riemannian manifolds and its applications. *Bull.Austral. math. Soc.*, 66:177–191, 2002.

76. B. Y. Chen. More on convolution of Riemannian manifolds. *Beiträage zur Algebra und Geometrie- Contributions to Algebra and Geometry*, 44:9–24, 2003.

77. B. Y. Chen. What can we do with nash's embedding theorem? *Soochow Journal of Math.*, 30:303–338, 2004.

78. B. Y. Chen. Examples and classification of Riemannian submersions satisfying a basic equation. *Bull. Aust. Math. Soc.*, 72:391–402, 2005.

79. B. Y. Chen. Riemannian submersions, minimal immersions and cohomology class. *Proc. Japan. Acad.*, 81:162–167, 2005.

80. B. Y. Chen. *Pseudo-Riemannian geometry, δ-invariants and applications*. World Scientific, Singapore, 2011.

81. B. Y. Chen and O. J. Garay. Pointwise slant submanifolds in almost Hermitian manifolds. *Turk. J. Math.*, 36:630–640, 2012.

82. B. Y. Chen and S. Ishikawa. Biharmonic pseudo-Riemannian submanifolds in pseudo-euclidean spaces. *Kyushu J.Math.*, 52:167–185, 1998.

83. B. Y. Chen and K. Ogiue. On totally real submanifolds. *Trans. Amer. Math. Soc.*, 193:257–266, 1974.

84. D. Chinea. Almost contact metric submersions. *Rend. Circ. Mat. Palermo*, 34:89–104, 1985.

85. M. Craioveanu and T. M. Rassias. *Old and New Aspects in Spectral Geometry*. Kluwer, Dortrecht, 2001.

86. M. Crampin. Tangent bundle geometry for Lagrangian dynamics. *J. Physics A*, 16:3755–3772, 1983.

87. M. Dajczer. *Submanifolds and Isometric Immersions*. Publish or Perish, Houston, 1990.

88. E. T. Davies. On the curvature of the tangent bundle. *Ann. Mat. Pura Appl.*, 81:193–204, 1969.

89. U. C. De. On weakly symmetric structures on a Riemannian manifold. *Facta Universitatis,Ser. Mechanics, Automatic Control and Robotics*, 3:805–819, 2003.

90. U. C. De and A. K. Gaji. On nearly quasi-Einstein manifolds. *Novi Sad J. Math.*, 38:115–121, 2008.

91. G. de Rham. Sur la reductibilite d'un espace de Riemann. *Comment. Math. Helv.*, 26:328–344, 1952.

92. S. Deng. *Homogeneous Finsler Spaces*. Springer, New York, 2012.

93. J. Deprez. Semi-parallel surfaces in euclidean space. *J. of Geometry*, 25:192–200, 1985.

94. S. Deshmukh, S. Ali, and S. I. Husain. Submersions of CR-submanifolds of a Kaehler manifold ii. *Indian J. Pure Appl. Math.*, 19:1185–1205, 1988.

95. S. Deshmukh, H. Ghazal, and H. Hashem. Submersions of CR-submanifolds on an almost Hermitian manifold i. *Yokohama Math. J.*, 40:45–57, 1992.

96. F. Dillen, J. Fastenakels, S. Haesen, J. V. Der Veken, and L. Verstraelen. Submanifold theory and parallel transport. *Kragujevac Journal of Math.*, 37:33–43, 2013.

97. M. Djoric and M. Okumura. *CR-submanifolds of Complex Projective Space*. Springer, New York, 2010.

98. P. Dombrowski. On the geometry of tangent bundle. *J. Reine Angew. Math.*, 210:73–88, 1962.

99. W. Drechsler and M. E. Mayer. *Fibre Bundle Techniques in Gauge Theories*. Springer, New York, 1977.

100. M. J. Duff, B.E.W Nilsson, and C. N. Pope. Kaluza-klein supergravity. *Physics Reports*, 130:1–142, 1986.

101. e. Boeckx, Kowalski O., and L. Vanhecke L. *Riemannian manifolds of conullity two*. World Scientific, Singapore, 1996.

102. J. Eells and L. Lemaire. *Selected topics in harmonic maps*. Amer. Math. Soc, Rhode Island, 1983.

103. J. Eells and H. J. Sampson. Harmonic mappings of Riemannian manifolds. *Amer.J.Math.*, 86:109–160, 1964.

104. C. Ehresmann. Sur les varieties presque complexes. *Proceedings International Congress of Math.*, 11:412–427, 1950.

105. İ. K. Erken and C. Murathan. On slant Riemannian submersions for cosymplectic manifolds. *Bull. Korean Math. Soc.*, 51:1749–1771, 2014.

106. İ. K. Erken and C. Murathan. Anti-invariant Riemannian submersions from cosymplectic manifolds onto Riemannian manifolds. *Filomat*, 29:1429–1444, 2015.

107. İ. K. Erken and C. Murathan. Slant Riemannian submersions from Sasakian manifolds. *Arab J. Math. Sci.*, 22:250–264, 2016.

108. G. Esposito. From spinor geometry to complex general relativity. *Int. J. Geom. Methods Mod. Phys.*, 2:675–731, 2005.

109. F. Etayo. On quasi-slant submanifolds of an almost Hermitian manifold. *Publ. Math. Debrecen*, 53:217–223, 1998.

110. M. Faghfouri and N. Ghaffarzadeh. Chen's inequality for invariant submanifolds in a generalized $(k, )$-space forms. *Glob. J. Adv. Res. Class. Mod. Geom.*, 4:86–101, 2015.

111. M. Falcitelli, S. Ianus, and A. M. Pastore. *Riemannian Submersions and Related Topics*. World Scientific, River Edge, NJ, 2004.

112. M. Falcitelli, S Ianus, A.M Pastore, and M. Visinescu. Some applications of Riemannian submersions in physics. *Revue Roumaine de Physique*, 48:627–639, 2003.

113. M. Fecko. *Differential Geometry and Lie Groups for Physicists*. Cambridge University Press, New York, 2006.

114. M. Fernández, M. Gotay, and A.Gray. Compact parallelizable four dimensional symplectic and and complex manifolds. *Proc.Amer.Math.Soc.*, 103:1209–1212, 1988.

115. M. Fernandez-Lopez, E. Garcia-Rio, D. Kupeli, and B. Ünal. A curvature condition for a twisted product to be a warped product. *Manusctria Math.*, 106:213–217, 2001.

116. D. Ferus. Immersions with parallel second fundamental form. *Math Z.*, 140:87–93, 1974.

117. A. E. Fischer. Riemannian maps between Riemannian manifolds. *Contemp. Math.*, pages 331–366, 1992.

118. A. Foussats, R. Laura, and O. Zandron. Tangent bundle approach for the factorization of gauge theories in a supergroup manifold. *Il Nuovo Cimento*, 92:13–22, 1986.

119. J. X. Fu. Harmonicity of Riemannian maps and gauss maps(chinese). *J. Hangzhou Univ. Natur. Sci. Ed.*, 23:15–22, 1996.

120. B. Fuglede. Harmonic morphisms between Riemannian manifolds. *Ann. Inst. Fourier (Grenoble)*, 28:107–144, 1978.

121. T. Fukami and S. Ishihara. Almost Hermitian structure on $s^6$. *Tohoku Math. J.*, 7:151–156, 1955.

122. E. Garca-Rio and D. N. Kupeli. *Semi-Riemannian Maps and Their Applications.* Kluwer, Dordrecht, 1999.

123. E. Garcia-Rio and D. N. Kupeli. On affine Riemannian maps. *Arch. Math.*, pages 71–79, 1998.

124. P. Gilkey, M. Itoh, and J. H. Park. Anti-invariant Riemannian submersions: A lie-theoretical approach. *Taiwanese J. Math.*, 20:787–800, 2016.

125. A. Gray. Pseudo-Riemannian almost product manifolds and submersions. *J. Math. Mech.*, 16:15–737, 1967.

126. W. Greub, S. Harperin, and R. Vanstone. *Connections, Curvature, and Cohomology*, volume 1. Academic Press, New York, 1972.

127. S. Gudmundsson and E. Kappos. On the geometry of tangent bundles. *Expo. Math.*, 20:1–41, 2002.

128. S. Gudmundsson and E. Kappos. On the geometry of the tangent bundle with the Cheeger-Gromoll metric. *Tokyo J. Math.*, 25:75–83, 2002.

129. S. Gudmundsson and J. C. Wood. Harmonic morphisms between almost Hermitian manifolds. *Boll. Un. Mat. Ital. B*, 7:185–197, 1997.

130. M. Gülbahar, E. Kılıç, S. Keleş, and M. M. Tripathi. Some basic inequalities for submanifolds of nearly quasi-constant curvature manifolds. *Differ. Geom. Dyn. Syst.*, 16:156–167, 2014.

131. Y. Gündüzalp. Anti-invariant semi-Riemannian submersions from almost para-Hermitian manifolds. *J. Funct. Spaces Appl.*, page 7, 2013.

132. Y. Gündüzalp. Anti-invariant Riemannian submersions from almost product manifolds. *Mathematical Sciences And Applications E-Notes (MSAEN)*, 1:58–66, 2013.

133. Y. Gündüzalp. Slant submersions from almost product Riemannian manifolds. *Turkish J. Math.*, 37:863–873, 2013.

134. Y. Gündüzalp. Slant submersions from lorentzian almost paracontact manifolds. *Gulf Journal of Mathematics*, 3:18–28, 2015.

135. Y. Gündüzalp. Slant submersions from almost paracontact Riemannian manifolds. *Kuwait J. Sci.*, 42:17–29, 2015.

136. Y. Gündüzalp. Semi-slant submersions from almost product Riemannian manifolds. *Demonstratio Math.*, To appear.

137. Y. Gündüzalp and B. Şahin. Paracontact semi-Riemannian submersions. *Turkish J. Math.*, 37:114–128, 2013.

138. R. S. Gupta. B. y. Chen's inequalities for bi-slant submanifolds in cosymplectic space forms. *Sarajevo J. Math.*, 21:117–128, 2013.

139. S. Haesen and L. Verstraelen. Properties of a scalar curvature invariant depending on two planes. *Manuscripta Math.*, 122:59–72, 2007.

140. S. M. K. Haider, M. Thakur, and Advin. Warped product skew CR-submanifolds of a cosymplectic manifold. *Lobachevskii J. Math.*, 33:262–273, 2012.

141. R. S. Hamilton. Three-manifolds with positive ricci curvature. *J. Diff. Geom.*, 17:255–306, 1982.

142. J. Hilgert and K. H. Neeb. *Structure and Geometry of Lie Groups.* Springer, Heidelberg, 2012.

143. P. A. Hogan. Kaluza-klein theory derived from a Riemannian submersion. *J. Math. Phys.*, 25:2301–2305, 1984.

144. Z.H. Hou and L. Sun. Geometry of tangent bundle with Cheeger-Gromoll type metric. *J. Math. Anal. Appl.*, 402:493–504, 2013.

145. D. Husemöller, M. Joachim, B. Jurco, and M. Schottenloher. *Basic Bundle Theory and K-Cohomology Invariants*. Springer, Heidelberg, 2008.

146. S. Ianus, R. Mazzocco, and G.E. Vilcu. Locally conformal Kähler submersions. *Acta Appl. Math.*, 104:83–89, 2008.

147. T. Ishihara. A mapping of Riemannian manifolds which preserves harmonic functions. *J. Math. Kyoto Univ.*, 19:215–229, 1979.

148. J. P. Jaiswal. Harmonic maps on Sasakian manifolds. *J. Geom.*, 104:309–315, 2013.

149. J. P. Jaiswal and B. A. Pandey. Non-existence harmonic maps on trans-Sasakian manifolds. *Lobachevskii J.Math.*, 37:185–192, 2016.

150. G. Y. Jiang. 2-harmonic isometric immersions between Riemannian manifolds. *Chinese Ann. Math. Ser. A*, 7:130–144, 1986.

151. G. Y. Jiang. 2-harmonic maps and their first and second variation formulas. *Chinese Ann. Math. Ser. A*, 7:389–402, 1986.

152. G. Y. Jiang. 2-harmonic maps and their first and second variation formulas, translated from the chinese by hajime urakawa. *Note Mat.*, suppl. n. 1:209–232, 2008.

153. R. H. Escobales Jr. and P. E. Parker. Geometric consequences of the normal curvature cohomology class in umbilic foliations. *Indiana Univ. Math. J.*, 37:389–408, 1988.

154. E. Kähler. Über eine bemerkenswerte Hermitesche metrik. *Abh. Math. Seminar Hamburg*, 9:173–186, 1933.

155. T. Kaluza. Zum unitätsproblem in der physik. *Sitzungsber. Preuss. Akad. Wiss*, pages 966–972, 1921.

156. U.H. Ki and S. Maeda. Notes on extrinsic spheres. *Bull. Korean Math. Soc.*, 35:433–439, 1998.

157. E. Kılıç, M. Gülbahar, and M. M. Tripathi. Chen-ricci inequalities for submanifolds of Riemannian and Kaehlerian product manifolds. *Annales Polonici Mathematici*, 116:37–56, 2016.

158. H. S. Kim, G. S. Lee, and Y. S. Pyo. Geodesics and circles on real hypersurfaces of type a and b in a complex space form. *Balkan J. Geom. Appl.*, pages 79–89, 1997.

159. O. Klein. Quantentheorie und fünfdimensionale relativitätstheorie. *Zeitschrift für Physik A*, 37(12):895–906, 1926.

160. S. Kobayashi. Submersions of CR submanifolds. *Tohoku Math. J.*, 39:95–100, 1987.

161. S. Kobayashi and K. Nomizu. *Foundations of Differential Geometry,I*. John Wiley- Sons, New York, 1963.

162. S. Kobayashi and K. Nomizu. *Foundations of Differential Geometry,II*. John Wiley- Sons, New York, 1969.

163. T. Koike, T. Oguro, and N. Watanabe. Remarks on some almost Hermitian structure on the tangent bundle. *Nihonkai Math. J.*, 20:25–32, 2009.

164. O. Kowalski. Curvature of the induced Riemannian metric on the tangent bundle of a Riemannian manifold. *J. Reine Angew. Math*, 250:124–129, 1971.

165. O. Kowalski and J. Szenthe. On the existence of homgeneous geodesics in homogeneous Riemannian manifolds. *Geometri Dedicata*, 81:209–214, 2000.

166. O. Kowalski and J. Szenthe. Erratum:on the existence of homgeneous geodesics in homogeneous Riemannian manifolds. *Geometri Dedicata*, 84:331–332, 2001.

167. V. H. S. Kumar and P. K. Suresh. Gravitons in kaluza-klein theory. *arXiv:gr-qc/0605016*, 2006.

168. C. W. Lee, J. W. Lee, and D. W. Yoon. Improved Chen inequality of Sasakian space forms with the tanaka-webster connection. *Filomat*, 29:1525–1533, 2015.

169. J. C. Lee, J. H. Park, B. Şahin, and D.Y. Song. Einstein conditions for the base space of anti-invariant Riemannian submersions and Clairaut submersions. *Taiwanese J. Math.Vol.*, 19:1145–1160, 2015.

170. J. M. Lee. *Riemannian Manifolds: An Introduction to Curvature*. Springer, New York, 1997.

171. J. W. Lee. Anti-invariant $\xi^{\perp}$-Riemannian submersions from almost contact manifolds. *Hacet. J. Math. Stat.*, 42:231–241, 2013.

172. J. W. Lee and B. Şahin. Pointwise slant submersions. *Bull. Korean Math. Soc.*, 51:1115–1126, 2014.

173. M. D. León and P. R. Rodrigues. *Methods of Differential geometry in Analytical Mechanics.* North-Holland Publishing, Amsterdam, 1989.

174. D. E. Lerner and P.D. Sommers. *Complex Manifold Techniques in Theoretical Physics.* Pitman Publishing, Boston, 1979.

175. H. Levy. Tensors determined by a hypersurface in Riemannian space. *Trans. Am. Math. Soc.,* 28:671–694, 1926.

176. X. Liu and X. Liang. Skew CR submanifolds of a Sasakian manifold. *Northeast. Math. J.,* 12:247–252, 1996.

177. E. Loubeau. Pseudo-harmonic morphisms. *Internat. J. Math.,* 8:943–957, 1997.

178. E. Loubeau and X. Mo. The geometry of pseudo harmonic morphisms. *Beiträge Algebra Geom.,* 45:87102, 2004.

179. Ü. Lumiste. *Semiparallel Submanifolds in Space Forms.* Springer, New York, 2009.

180. S. Maeda. Submanifold theory from the viewpoint of circles. *Mem. Fac. Sci. Eng. Shimane Univ. Ser. B Math. Sci.,* 40:15–32, 2007.

181. S. Maeta. k-harmonic maps into a Riemannian manifold withconstant sectional curvature. *Proc. Amer. Math. Soc.,* 140:1835–1847, 2012.

182. V. Mangione. Some submersions of CR-hypersurfaces of Kaehler-Einstein manifold. *Int. J. Math. Math. Sci.,* 18:1137–1144, 2003.

183. J. C. Marrero and J. Rocha. Locally conformal kähler submersions. *Geom. Dedicata,* 52:271–289, 1994.

184. D. Martin. *Manifold Theory.* Horwood Publishing, West Sussex, 2002.

185. T. Matos and J. A. Nieto. Topics on kaluza-klein theory. *Revista Mexicana de Fisica,* 39:81–131, 1993.

186. F. Memoli, G. Sapiro, and P. Thompson. Implicit brain imaging. *NeuroImage,* 23:179–188, 2004.

187. A. Mihai and I. N. Radulescu. Scalar and ricci curvatures of special contact slant submanifolds in Sasakian space forms. *Adv. Geom.,* 14:147–159, 2014.

188. I. Mihai. Special submanifolds in Hermitian manifolds. In *Topics in Modern Differential geometry, Vol. I,* pages 83–116. Simon Stevin Institute for Geometry, Tilburg, The Netherlands, 2010.

189. S. Montaldo and C. Oniciuc. A short survey on biharmonic maps between Riemannian manifolds. *Revista de la Union Matematica Argentina,* 47:1–22, 2006.

190. J. Morrow and K. Kodaira. *Complex manifolds.* Holt, Rinehart and Winston, New York, 1971.

191. M. I. Munteanu. Some aspects on the geometry of the tangent bundles and tangent sphere bundles of a Riemannian manifold. *Mediterr. J. math,* 5:43–59, 2008.

192. M. I. Munteanu. Old and new structure on tangent bundle. In *Eight International Conference on Geometry, Integrability and quantization,* pages 264–278, June 914, 2006.

193. R. M. Murray, Z Li, and S. S. Sastry. *A Mathematical Introduction to Robotic Manipulation.* CRC Press, New York, 1994.

194. S. B. Myers and N. Steenrod. The group of isometries of a Riemannian manifold. *Annals of Math.,* 40:400–416, 1939.

195. G. L. Naber. Gauge fields in physics and mathematis. *Journal of Dynamical Systems and Geometric Theories,* 1:19–34, 2002.

196. T. Nagano. Isometries on complex product spaces. *Tensor,* 9:47–61, 1959.

197. F. Narita. CR submanifolds of locally conformal kiihler manifolds and Riemannian submersions. *Colloquium Mathematicum,* LXX:165–179, 1996.

198. B. L. Neto. Generalized quasi-Einstein manifolds with harmonic anti-self dual weyl tensor. *Arch. Math.,* 106:489–499, 2016.

199. A. Newlander and L. Nirenberg. Complex analytic coordinates in almost complex manifolds. *Ann. Math.,* 65:391–404, 1957.

200. K. Nomizu and K. Yano. On circles and spheres in Riemannian geometry. *Math. Ann.,* 210:163–170, 1974.

201. T. Nore. Second fundamental form of a map. *Ann. Mat. Pura Appl.,* 146:281–310, 1987.

202. K. Ogiue. Differential geometry of kähler submanifold. *Advances in Math.*, 13:73–114, 1974.

203. Y. Ohnita. On pluriharmonicity of stable harmonic maps. *J. London Math. Soc.*, 35:563–568, 1987.

204. Z. Olszak. On almost complex structures with norden metrics on tangent bundles. *Periodica Mathematica Hungarica*, 51:59–74, 2005.

205. B. O'Neill. The fundamental equations of a submersion. *Mich. Math. J*, 13:458–469, 1966.

206. B. O'Neill. *Semi-Riemannian Geometry with Applications to Relativity*. Academic Press, New York, 1983.

207. C. Oniciuc. Biharmonic maps between Riemannian manifolds. *An. Stiint. Al.I.Cuza. Univ. Iasi*, XLVIII:237–248, 2002.

208. C. Oniciuc. *Biharmonic Submanifolds in Space Forms*. habilitation, Universitatea Alexandrau Ioan Cuza din Iaşi, 2012.

209. J. Oprea. *Differential Geometry and Its Applications*. Prentice Hall, New Jersey, 1997.

210. V. Oproiu. Some new geometric structures on the tangent bundles. *Publ. Math. Debrecen*, 55:261–281, 1999.

211. C. Özgür and A. De. Chen inequalities for submanifolds of a Riemannian manifold of nearly quasi-constant curvature. *Publ. Math. Debrecen*, 82:439–450, 2013.

212. C. Özgür and A. Mihai. Chen inequalities for submanifolds of real space forms with a semi-symmetric non-metric connection. *Canad. Math. Bull.*, 55:611–622, 2012.

213. B. Panday, J. P. Jaiswal, and R. H. Ojha. Necessary and sufficient conditions for the Riemannian map to be a harmonic map on cosymplectic manifolds. *Proc. Nat. Acad. Sci. India Sect. A*, 85:265–268, 2015.

214. N. Papaghiuc. Semi-slant submanifolds of a Kaehlerian manifold. *An. Stiint. Al.I.Cuza. Univ. Iasi*, 40:55–61, 1994.

215. K. S. Park. H-semi-invariant submersions. *Taiwanese J. Math.*, 16:1865–1878, 2012.

216. K. S. Park. H-slant submersions. *Bull. Korean Math. Soc.*, 49:329–338, 2012.

217. K. S. Park. Semi-slant Riemannian maps. *arXiv:1208.5362v2*, 2012.

218. K. S. Park. Almost h-semi-slant Riemannian maps. *Taiwanese J. Math.*, 17:937–956, 2013.

219. K. S. Park. H-semi-slant submersions from almost quaternionic Hermitian manifolds. *Taiwanese Journal of Mathematics*, 18:1909–1926, 2014.

220. K. S. Park. Almost h-semi-slant Riemannian maps to almost quaternionic Hermitian manifolds. *Communications in Contemporary Mathematics*, 17:23 pages, 2015.

221. K. S. Park and R. Prasad. Semi-slant submersions. *Bull. Korean Math*, 16:1865–1878, 2012.

222. K. S. Park and B. Şahin. Semi-slant Riemannian maps into almost Hermitian manifolds. *Czechoslovak Math. J.*, 64:1045–1061, 2014.

223. R. Penrose. The twistor geometry of light rays. *Classical Quantum Gravity*, 14:299–323, 1997.

224. S. Pigola, M. Rigoli, M. Rimoldi, and A. Setti. Ricci almost solitons. *Ann. Sc. Norm. Super. Pisa Cl. Sci*, 10:757–799, 2011.

225. R. Ponge and Reckziegel H. Twisted products in pseudo-Riemannian geometry. *Geom. Dedicata*, 48:15–25, 1993.

226. W. A. Poor. *Differential Geometric Structures*. McGraw-Hill, New York, 1981.

227. M. Pranović. On weakly symmetric Riemannian manifolds. *Pub. Math. Debrecen*, 46:19–25, 1995.

228. R. Prasad and S. Pandey. Slant Riemannian maps from an almost contact manifold. *Filomat*, To appear.

229. A. Pressley. *Elementary Differential Geometry*. Springer, London, 2010.

230. Deszcz R. On pseudosymmetric spaces. *Bull. Soc. Math. Belg*, 44:1–34, 1992.

231. Takagi R. An example of Riemannian manifold satisfying $r(x, y) \cdot r = 0$ but not $\nabla r = 0$. *Tôhoku Math. j.*, 24:105–108, 1972.

232. D. Rickles. Mirror symmetry and other miracles in superstring theory. *Found. Phys.*, 43:54–80, 2013.

233. D. Rickles. *A Brief History of String Theory*. Springer, Heidelberg, 2014.

234. G. S. Ronsse. Generic and skew CR-submanifolds of a Kaehler manifold. *Bulletin Inst. Math.*

*Acad. Sinica*, 18:127–141, 1990.

235. V. I. Rovenskii. *Foliations on Riemannian manifolds and submanifolds.* Birkhäuser, Boston, 1998.

236. S. K. Saha. Nearly Einstein manifolds. *Novi Sad J. Math.*, 45:17–26, 2015.

237. B. Şahin. Harmonic Riemannian maps on locally conformal Kaehler manifolds. *Proc. Indian Acad. Sci. Math. Sci.*, 118:573–581, 2008.

238. B. Şahin. Warped product submanifolds of Kaehler manifolds with a slant factor. *Ann. Polon. Math.*, 95:207–226, 2009.

239. B. Şahin. Anti-invariant Riemannian submersions from almost Hermitian manifolds. *Central European J.Math*, 8:437–447, 2010.

240. B. Şahin. Conformal Riemannian maps between Riemannian manifolds, their harmonicity and decomposition theorems. *Acta Appl. Math.*, 109:829–847, 2010.

241. B. Şahin. Invariant and anti-invariant Riemannian maps to Kähler manifolds. *Int. J. Geom. Methods Mod. Phys.*, 7:337–355, 2010.

242. B. Şahin. Skew CR-warped products of Kaehler manifolds. *Math. Commun.*, 15:189–204, 2010.

243. B. Şahin. Biharmonic Riemannian maps. *Ann. Polon. Math.*, 102:39–49, 2011.

244. B. Şahin. Semi-invariant Riemannian maps to Kähler manifolds. *Int. J. Geom. Methods Mod. Phys.*, 7:1439–1454, 2011.

245. B. Şahin. Slant submersions from almost Hermitian manifolds. *Bull. Math. Soc. Sci. Math. Roumanie*, 54:93–105, 2011.

246. B. Şahin. Anti-invariant Riemannian maps from almost Hermitian manifolds. *arXiv:1210.0401*, 2012.

247. B. Şahin. Semi-invariant Riemannian maps from almost Hermitian manifolds. *Indag. Math. (N.S.)*, 23:80–94, 2012.

248. B. Şahin. Riemannian submersions from almost Hermitian manifolds. *Taiwanese J. Math.*, 17:629–659, 2013.

249. B. Şahin. Semi-invariant Riemannian submersions from almost Hermitian manifolds. *Canadian Mathematical Bulletin*, 56:173–183, 2013.

250. B. Şahin. Slant Riemannian maps from almost Hermitian manifolds. *Quaest. Math.*, 36:449–461, 2013.

251. B. Şahin. Slant Riemannian maps to Kähler manifolds. *Int. J. Geom. Methods Mod. Phys.*, 10:12 pp, 2013.

252. B. Şahin. Warped product pointwise semi-slant submanifolds of Kähler manifolds. *Portugaliae Math.*, 70:251–268, 2013.

253. B. Şahin. Clairaut Riemannian maps. Presented in New Trends in Differential Geometry, September, Villasimius, Italy, 2014.

254. B. Şahin. Holomorphic Riemannian maps. *Zh. Mat. Fiz. Anal. Geom.*, 10:422–429, 2014.

255. B. Şahin. Generic Riemannian maps. *Preprint*, 2016.

256. B. Şahin. Notes on Riemannian maps. *U. P. B. Sci. Bull. Series A*, To appear.

257. B. Şahin. Chen first inequality for Riemannian maps. *Annales Polonici Math.*, 117 (2016), 249-258.

258. B. Şahin. Circles along a Riemannian map and Clairaut Riemannian maps. *Bull. Korean Math.Soc.*, To appear.

259. B. Şahin. Hemi-slant Riemannian maps. *Mediterranean J.Math.*, To appear.

260. B. Şahin. A survey on differential geometry of Riemannian maps between Riemannian manifolds. *An. Stiint. Al.I.Cuza. Univ. Iasi*, To appear.

261. B. Şahin and H. M. Taştan. Clairaut submersions from almost Hermitian manifolds. *Preprint*.

262. A. Salimov and K. Akbulut. A note on a paraholomorphic CheegerGromoll metric. *Proc. Indian Acad. Sci. (Math. Sci.)*, 119:187–195, 2009.

263. S. Sasaki. On the differential geometry of tangent bundles of Riemannian manifolds. *Tohōku M. J.*, 10:338–354, 1958.

264. J. M. Selig. *Geometric Fundamentals of Robotics.* Springer, New York, 2005.

265. S. A. Sepet and M. Ergüt. Pointwise slant submersions from cosymplectic manifolds. *Turkish*

*J. Math*, 40:582–593, 2016.

266. M. H. Shahid, F. Solamy, J.B. Jun, and M. Ahmad. Submersion of semi-invariant submanifolds of trans-Sasakian manifold. *Bull. Malays. Math. Sci. Soc.*, 36:63–71, 2013.

267. A. A. Shaikh. On pseudo quasi-Einstein manifold. *Period. Math. Hungar.*, 59:119–146, 2009.

268. A. A. Shaikh, R. Deszcz, M. Hotlos, J. Jelowicki, and H. Kundu. On pseudo-symmetric manifolds. *Publicationes Math. Debrecen*, 86:433–456, 2015.

269. S. Shenawy. A note on sequential warped product manifolds. *arXiv:1506.06056 [math.DG]*, 2015.

270. A. C. Silva. *Lectures on Symplectic Geometry*. Springer, Heidelberg, 2001.

271. P. Solórzano. *Group norms and their degeneration in the study of parallelism*. Phd thesis, Stony Brook University, 2011.

272. M. W. Spong, S. Hutchinson, and M. Vidyasagar. *Robot Modeling and Control*. Wiley, 2005.

273. S. E. Stepanov. On the global theory of some classes of mapping. *Ann. Global Anal. Geom.*, 13:239–249, 1995.

274. H. Stephani, D. Kramer, M. MacCallum, C. Hoenselaers, and E. Herlt. *Exact Solutions of Einstein's Field Equations*. Cambridge University Press, Cambridge, 2003.

275. M. Svensson. On holomorphic harmonic morphisms. *Manuscripta Math.*, 107:1–13, 2002.

276. R. Szabó. Structure theorems on Riemannian spaces satisfying $r(x, y) \cdot r = 0$ i, the local version. *J.Diff. Geom.*, 17:531–582, 1982.

277. R. Szabó. Structure theorems on Riemannian spaces satisfying $r(x, y) \cdot r = 0$ ii, the global version. *Geom Dedicate*, 19:65–108, 1985.

278. R. Szöke. Complex structures on tangent bundles of Riemannian manifolds. *Mathematische Annalen*, 291:409–428, 1991.

279. S. I. Tachibana and M. Okumura. On the almost-complex structure of tangent bundles of Riemannian spaces. *Tohoku Math. J.*, 14:156–161, 1962.

280. M. Tahara, L. Vanhecke, and Y. Watanabe. New structures on tangent bundles. *Note di Matematica*, 18:131–141, 1998.

281. M. Tahara and Y. Watanabe. Natural almost Hermitian, Hermitian and Kähler metrics on the tangent bundles. *Math. J. Toyama Univ.*, 20:149–160, 1997.

282. L. Tamassy and T. Q. Binh. On weakly symmetric and weakly projective symmetric Riemannian manifolds. *Coll. Math. Soc. Janos Bolyai*, 56:663–670, 1989.

283. H. M. Taştan. Anti-holomorphic semi-invariant submersions. *arXiv:1404.2385v1*, 2014.

284. H. M. Taştan. On langrangian submersions. *Hacettepe J. Math. Stat.*, 43:993–1000, 2014.

285. H. M. Taştan. Lagrangian submersions from normal almost contact manifolds. *Filomat*, To appear.

286. H. M. Taştan, B. Şahin, and Ş. Yanan. Hemi-slant submersions. *Mediterranean J. Math.*, 13:2171–2184, 2016.

287. M. M. Tripathi. Generic submanifolds of generalized complex space forms. *Publ. Math. Debrecen*, 50:373–392, 1997.

288. A. J. Tromba. *Teichmuller Theory in Riemannian Geometry*. Birkhauser-Verlag, Boston, 1992.

289. B. Ünal. *Doubly Warped Products*. PhD thesis, University of M issouri-Columbia, 2000.

290. H. Urakawa. *Calculus of Variations and Harmonic Maps*. American Math. Soc., Rhode Island, 1993.

291. H. Urakawa. Harmonic maps and biharmonic maps. *Symmetry*, 7:651–674, 2015.

292. V.I.Arnol'd. *Mathematical Methods of Classical Mechanics*. Springer., Heidelberg, 1996.

293. A. D. Vilcu and G. E. Vilcu. Statistical manifolds with almost quaternionic structures and quaternionic kähler-like statistical submersions. *Entropy*, 17:6213–6228, 2015.

294. G. E. Vilcu. 3-submersions from qr-hypersurfaces of quaternionic Kähler manifolds. *Ann. Polon. Math.*, 98:301–309, 2010.

295. G. E. Vilcu. Para-hyperhermitian structures on tangent bundles. *Proceedings of the Estonian Academy of Sciences*, 60:165–173, 2011.

296. G. E. Vilcu. On Chen invariants and inequalities in quaternionic geometry. *J. Inequal. Appl.*, 66:14pp, 2013.

297. J. Vilms. Totally geodesic maps. *J. Differential Geometry*, 4:73–79, 1970.

298. S. B. Wang. , the first variation formula for k-harmonic mapping. *Journal of Nanchang University*, 13, 1989.

299. Y. Wang. Multiply warped products with a semisymmetric metric connection. *Abst. and App. Analysis*, Article ID 742371:12 pages, 2014.

300. B. Watson. Almost Hermitian submersions. *J. Differential Geometry*, 11:147–165, 1976.

301. B. Watson. Riemannian submersions and instantons. *Mathematical Modelling*, 1:381–393, 1980.

302. P. West. *Introduction to Strings and Branes*. Cambridge University Press, Cambridge, 2012.

303. C. Von Westenholz. *Differential Forms in Mathematical Physic*. North-Holland Publishing Company, Amsterdam, 1978.

304. Y. Xin. *Geometry of Harmonic Maps*. Birkhäuser, Boston, 1996.

305. C. N. Yang and R. L. Mills. Conservation of isotopic spin and isotopic gauge invariance. *The Physical Review*, 96:191, 1954.

306. K. Yano. *Differential Geometry on Complex and Almost Complex Spaces*. Pergamon, New York, 1965.

307. K. Yano and S. Ishihara. Harmonic and relatively affine mappings. *J. Diff. Geometry*, 10:501–509, 1975.

308. K. Yano and M. Kon. *Anti-invariant submanifolds*. Marcel Dekker, New York, 1976.

309. K. Yano and M Kon. Contact CR-submanifolds. *Kodai Math. J.*, 5:238–252, 1982.

310. K. Yano and M. Kon. *CR-submanifolds of Kaehlerian and Sasakian Manifolds*. Birkhäuser, Boston, 1983.

311. K. Yano and M. Kon. *Structure on Manifolds*. World Scientific, Singapore, 1984.

312. S. T. Yau and S. Nadis. *The Shape of Inner Space: String Theory and the Geometry of the Universe's Hidden Dimensions*. Basic Books, New York, 2012.

313. G. B. Yosef and O. B. Shahar. A tangent bundle theory for visual curve completion. *IEE Trans. Patt. Analys. and Mach. Int.*, 34:1263–1280, 2012.

314. L. Zhang and P. Zhang. Notes on Chen's inequalities for submanifolds of real space forms with a semi-symmetric non-metric connection. *J. East China Norm. Univ. Natur. Sci. Ed.*, 1:6–15, 2015.

315. P. Zhang. Remarks on Chen's inequalities for submanifolds of a Riemannian manifold of nearly quasi-constant curvature. *Vietnam J. Math.*, 43:557–569, 2015.

316. P. Zhang, L. Zhang, and W. Song. Chen's inequalities for submanifolds of a Riemannian manifold of quasi-constant curvature with a semi-symmetric metric connection. *Taiwanese J. Math.*, 18:1841–1862, 2014.

317. H. Zhao, A. R. Kelly, J. Zhou, J. Lu, and Y. Y. Yang. Graph attribute embedding via Riemannian submersion learning. *Computer Vision and Image Understanding*, 115:962–975, 2011.

318. B. Zwiebach. *A First Course in String Theory*. Cambridge University Press, New York, 2009.

# Index

Printed in the United States
By Bookmasters